SCIENTIFIC AMERICAN
MOLECULAR
NEUROLOGY

SCIENTIFIC AMERICAN Introduction to Molecular Medicine
Edward D. Rubenstein, M.D., Series Editor

SCIENTIFIC AMERICAN Molecular Neurology
Edited by Joseph B. Martin, M.D., Ph.D.

Previously Published

SCIENTIFIC AMERICAN Molecular Oncology
Edited by J. Michael Bishop, M.D., and Robert A. Weinberg. Ph.D.

SCIENTIFIC AMERICAN Molecular Cardiovascular Medicine
Edited by Edgar Haber, M.D.

SCIENTIFIC AMERICAN Introduction to Molecular Medicine
*Edited by Philip Leder, M.D., David A. Clayton, Ph.D., and
Edward Rubenstein, M.D.*

SCIENTIFIC AMERICAN

MOLECULAR NEUROLOGY

Edited by

Joseph B. Martin, M.D., Ph.D.
Dean of the Faculty of Medicine
Harvard Medical School

Scientific American, Inc., New York

Library of Congress Cataloging-in-Publication Data
Scientific American molecular neurology/edited by Joseph B. Martin
p. cm.--(Scientific American introduction to molecular medicine)
Includes index.
ISBN 0-89454-030-0
1. Brain--Diseases--Molecular aspects. 2. Brain--Diseases--Genetic aspects. 3. Molecular
neurobiology. 4. Neurogenetics.
I. Martin, Joseph B., 1938– . II. Scientific American, inc. III. Series.
[DNLM: 1. Nervous System Diseases--genetics. 2. Nervous System Diseases--
physiopathology. WL 140 S415 1998]
RC347.S37 1998
616.8'047--dc21
DNLM/DLC
for Library of Congress   98-5123
                                                                                              CIP

Editorial Director	Aileen M. McHugh
Publishing Director	Linnéa C. Elliott
Project Editor	Ozzievelt Owens
Development Editor	Leslie LaPiana
	Tom Reynolds
Director of Art and Design	Elizabeth Klarfeld
Vice President, Associate Publisher/Production	Richard Sasso
Production Director	Christina Hippeli
Production Manager	Silvia Di Placido
Production Coordinator	Kelly Mercado
Electronic Composition	Jennifer Smith
	Carol Hansen

ISBN: 0-89454-030-0
Scientific American, Inc., 415 Madison Avenue, New York, NY 10017

Contributors

Robert H. Brown, Jr., D. Phil., M.D.   Associate Professor of Neurology, Harvard Medical School; Associate in Neurology, Massachusetts General Hospital

Manuel Buttini, Ph.D.   Postdoctoral Fellow, Molecular Neurobiology Program, the J. David Gladstone Institutes and Department of Neurology, University of California, San Francisco

Stephen C. Cannon, M.D., Ph.D.   Assistant Professor of Neurobiology, Harvard Medical School; Assistant Neurologist, Massachusetts General Hospital

Robert H. Edwards, M.D.   Professor, Departments of Neurology and Physiology, School of Medicine, University of California, San Francisco

Ricardo N. Fadic, M.D.   Assistant Professor, Department of Neurology, Catholic University of Chile

James F. Gusella, Ph.D.   Bullard Professor of Neurogenetics, Harvard Medical School; Director, Neurogenetics Unit, Massachusetts General Hospital

Stephen L. Hauser, M.D.   Chair and Betty Anker Fife Professor, Department of Neurology, School of Medicine, University of California, San Francisco

Jari Honkaniemi, M.D., Ph.D.   Assistant Professor, Department of Neurology, University of Tampere, Tampere, Finland

Mark A. Israel, M.D.   Professor, Department of Neurosurgery, and Brain Tumor Research Center, School of Medicine, University of California, San Francisco

Donald R. Johns, M.D.   Associate Professor of Neurology and Ophthalmology, Harvard Medical School; Director, Division of Neuromuscular Disease, Beth Israel Deaconess Medical Center

James R. Lupski, M.D., Ph.D.   Cullen Professor of Molecular and Human Genetics, Professor of Pediatrics, and Director, Medical Scientist Training Program, Baylor College of Medicine; Attending Physician, Texas Children's Hospital and Ben Taub General Hospital

Joseph B. Martin, M.D., Ph.D.   Dean of the Faculty of Medicine, and Caroline Shields Walker Professor of Neurobiology and Clinical Neuroscience, Harvard Medical School

Stephen M. Massa, M.D., Ph.D.   Assistant Professor of Neurology, University of California San Francisco; VA Medical Center, San Francisco

James O. McNamara, M.D.   Carl R. Deane Professor of Neuroscience, Departments of Medicine (Neurology), Neurobiology, and Pharmacology, Duke University Medical Center; Staff Physician, Durham VA Medical Center

Lennart Mucke, M.D.   Associate Professor, Department of Neurology and Neuroscience Program, School of Medicine, University of California, San Francisco; Head, Molecular Biology Program, the J. David Gladstone Institutes, San Francisco

David L. Nelson, Ph.D.   Associate Professor, Department of Molecular and Human Genetics, Baylor College of Medicine

Jorge Oksenberg, Ph.D.   Assistant Professor, Department of Neurology, School of Medicine, University of California, San Francisco

Stanley B. Prusiner, M.D.   Professor of Neurology and Professor of Biochemistry and Biophysics, School of Medicine, University of California, San Francisco

Frank R. Sharp, M.D.   Professor of Neurology, School of Medicine, University of California, San Francisco; VA Medical Center, San Francisco

Rudolph E. Tanzi, Ph.D.   Associate Professor of Neurology and Neuroscience, Harvard Medical School; Director, Genetics and Aging Unit, Massachusetts General Hospital

Anne B. Young, M.D., Ph.D.   Julieanne Dorn Professor of Neurology, Harvard Medical School; Chief, Neurology Service, Massachusetts General Hospital

Foreword

This book grew from my continuing fascination with the remarkable progress taking place in defining the genes that cause neurologic disease. My attention was first drawn to this field in 1979 when, as chairman of the Department of Neurology at Massachusetts General Hospital (MGH), I received from the National Institute of Neurological and Communicative Disorders and Stroke a request for proposals to establish a Huntington's disease (HD) center. After consultation with Alexander Rich, a renowned molecular geneticist from Massachusetts Institute of Technology, I was put in touch with David Housman, also from MIT, who first introduced me to the concept of using restriction fragment length polymorphisms (RFLPs) as DNA markers but, more important, arranged for me to meet with James Gusella, a Canadian student who was just completing his Ph.D. degree.

The Boston Huntington's disease center was funded, as was a center at Johns Hopkins University. Gusella joined the neurology department at MGH in July 1980 to begin developing a panel of RFLPs to test in families with HD. The rest of the story is well known and is discussed succinctly by Anne Young at the beginning of Chapter 3.

The goal of this monograph is to survey the remarkable progress being made in elucidating the pathogenesis and molecular pathology of the major neurologic disorders. Several classes of disorders are considered, including (1) the Mendelian inherited diseases represented by HD, Charcot-Marie-Tooth disease, and periodic paralysis, in which discovery of genes by positional cloning was critical in defining the genetic nature of the disorder; (2) the major neurologic syndrome complexes—stroke, epilepsy, multiple sclerosis, and brain tumors—in which pathogenesis has been enlightened by examination of molecular mechanisms; (3) disorders in which genetic and infectious etiologies overlap, as in human immunodeficiency virus (HIV) encephalopathy and prion disorders; (4) sporadic diseases that occasionally present as rare or uncommon inherited forms—Alzheimer's disease, amyotrophic lateral sclerosis (ALS), and Parkinson's disease; and (5) disorders of development caused by fragile X or genetic mitochondrial disorders.

Chapter 1 describes the principles of linkage analysis, positional cloning, and search for candidate genes that inform us of the enormous power and potential of the convergence of neurologic nosology, careful clinical studies, and molecular genetics. So far, these techniques have elucidated the genetic bases of more than 100 diseases, many of which can now be screened for in presymptomatic individuals. An appendix lists these disorders, current to October 1997.

Fragile X syndrome, an X-linked disease that is the most common form of mental retardation in boys after Down's syndrome, is the topic of Chapter 2. The first of the trinucleotide repeat disorders to be identified, fragile X syndrome results from a decrease or absence of the protein encoded by the *FMR1* gene, whose transcription is disrupted by the massive expansion of CGG repeats. The disorder represents a loss of function, as proved by the rare occurrence of point mutations in the *FMR1* gene. These mutations can cause an identical phenotype characterized by mental retardation, macro-orchidism, and peculiar facies. The phenomenon of dynamic mutations, in which thousands of expansions of the CGG repeat may occur, is also illustrated by fragile X syndrome. The disorder also demonstrates, in contrast to most other so-called triplet repeat disorders, a high incidence of new mutations.

Chapter 3 describes other trinucleotide repeat disorders, focusing in particular on those characterized by CAG repeats. The expanded glutamine tracts in the protein, encoded by the CAG repeats in the gene, appear to cause a toxic gain of function, in which selective neuronal populations become vulnerable in midlife to apoptotic cell death. These disorders are characterized by genetic anticipation, in which successive generations tend toward an earlier onset and a more severe clinical phenotype. The increased severity is associated with further expansion of the repeats, which occurs most commonly in DNA replication during spermatogenesis.

Chapters 4 through 9 describe advances in the genetics of common, severe neurologic diseases in which no single molecular pathogenesis has yet been proven. In the case of Alzheimer's disease (Chapter 4), much has been learned from the study of the uncommon forms of early onset dementia transmitted by autosomal dominant inheritance. Mutations in several genes have now been found, each resulting in the distinctive neuropathologic features originally described by Alois Alzheimer in 1906 and 1907: neurofibrillary tangles and senile (neuritic) plaques. The landmark discoveries in each of these subtypes demonstrate forcefully the fact that in brain, multiple metabolic/functional defects can lead to a common pathological phenotype; that is, different genotypes do not exclude a common phenotype. This is one of the most clearly defined examples of genetic heterogeneity.

Epilepsy (Chapter 5) is an extremely common disorder that can be caused by hundreds of different central nervous system assaults, ranging from developmental anomalies to trauma, stroke, metabolic disorder, and tumor. In this chapter, the focus is on genetic epilepsies of unknown cause (idiopathic) in which the clinical phenotype is a bilateral synchronous neuronal discharge. Genetic analysis has yielded some remarkable clues to the diversity of molecular mechanisms that may lead to epilepsy.

Chapter 6, on brain tumors, is largely devoted to the elucidation of the roles of genes in tumor formation and progression. Much of this work remains at the descriptive level; for example, linking genetic mutations found in brain tumors to those found in tumors of other tissues or exploring roles of neurotrophic factors and their receptors in brain tumor growth. Nevertheless, some important clues about tumor progression are found in the transition, for example, of astrocytoma to glioblastoma multiforme.

In many ways, the next two chapters were the most difficult. Stroke, the subject of Chapter 7, is in one sense a rather straightforward problem: a lack of blood supply depletes oxygen, leading rapidly to brain dysfunction and, if prolonged for only a few minutes, to brain cell death and tissue necrosis. Yet each infarct so produced has a striking irregularity of dimension and severity, with some zones of increasing vulnerability and other zones in which brain tissue is relatively spared. Interventions may dramatically change this "dance of death," but in exploring treatment options it is important to define and clarify the molecular and metabolic responses to energy deprivation. This chapter does so highly effectively.

Similarly, HIV infection (Chapter 8) seems a relatively straightforward matter. Yet controversy persists regarding how CNS invasion by the virus causes neurologic disorder. The virus does not enter neurons to any easily discernible degree, yet neurons die. How does this come about? The chapter is a comprehensible effort to delineate the hypotheses based on current scientific evidence at the cellular and molecular levels.

Parkinson's disease (Chapter 9) was the first neurodegenerative syndrome to yield to neurochemical and neuropathologic correlation. Discovery of the role of dopamine deficiency, caused by death of nerve cells in the zona compacta of the substantia nigra, led to the first effective treatment for a neurodegenerative disorder: Levodopa dramatically reverses many of the symptoms, at least early in the course of symptomatic neural cell loss. But what are the mechanisms of such selective cell loss? Although more than 95 percent of all cases of Parkinson's are considered sporadic, one gene for a familial form was discovered earlier this year—a point mutation in a synaptic protein called α-synuclein. The chapter explores current evidence for disregulation of the dopaminergic system and speculates on possible etiologic clues.

Nothing has so radically challenged basic biologic mechanisms as the work on prions (Chapter 10). These agents are so named because they are hypothesized to be entirely proteinaceous, and are infectious. Diseases caused by prions may be sporadic, inherited, or infectious; in many instances they are both infectious and inherited. The chapter is a detailed summary of current evidence supporting the protein-only etiologic hypothesis.

Why include a chapter on multiple sclerosis in a book on molecular neurology? At least two reasons can be cited. First, great efforts are being made to link the disease to genetic susceptibility, as reviewed in Chapter 11. Second, the molecular pathogenesis of MS provides instructive lessons on the complexities of autoimmune disorders.

The analytic power of neurogenetics has been particularly informative in the elucidation of motor neuron system degeneration (Chapter 12). The discovery that a mutation in the gene for SOD1, an enzyme known and studied for decades, can cause amyotrophic lateral sclerosis was not so surprising as the subsequent confounding of the explanations of how it does so. The lesson here is that one cannot always predict pathogenesis, even when the function of a candidate gene/protein has been thoroughly investigated. Equally puzzling has been the finding of several candidate genes on chromosome 5q to account for infantile and juvenile muscular atrophy. Again, the ultimate answer to how gene mutations cause disease will rely on sophisticated cell biology studies. These observations point to the fact that moving from gene mutations to disordered protein function to therapy will be a long and intellectually challenging adventure.

Discovery of the molecular defects in many of the inherited polyneuropathies (Chapter 13) has also yielded rich rewards. Frequently, careful clinical attention to the phenotypes in affected families has led correctly to discrimination among pedigrees, paving the way for gene identification. In other instances, clinical surprises have emerged, as in the molecular definition of HNPP (hereditary neuropathy caused by pressure palsy).

Mutations that result in ion channel defects in muscle are the subject of Chapter 14. The symptoms of muscle myotonia and periodic paralysis can result from many different channel defects, involving sodium, calcium, or chloride channels.

Perhaps no aspect of neurology has grown so quickly as the understanding of diseases caused by anomalies in mitochondria (Chapter 15). Mitochondrial disorders follow maternal inheritance patterns, and the mother may be only minimally or not at all affected. However, the complete cloning of the 16,000-plus base pairs of mitochondrial DNA made such screening relatively simple. Characteristically, these disorders affect CNS and other tissues.

A single monograph of moderate length cannot adequately cover all of the recent advances in molecular neurology. Aside from the discussion of fragile X syndrome, presented as the first disorder attributed to the expansion of trinucleotide repeats, I did not attempt to review many other defects caused by aberrations in developmental organization and neuronal migration in the CNS. As listed in the appendix, genes for several dozen of these have been cloned, including lissencephaly (a defect in cerebral cortical formation) and tuberous sclerosis (types 1 and 2). These entities, so interesting in their own right, could fill an entire volume equal in size and scope to this one.

Nor did I review the genetics of defects in the special senses, particularly vision and hearing. Again, most of these, where mutated genes are known, are included in the appendix. Particularly informative has been the extraordinary subdivision of genetic defects that cause the phenotype of retinitis pigmentosa. A mutation causing the most common form of macular degeneration has also been found. These vision disorders offer an excellent example of how neurogenetics has completely modified the classification of sensory disorders. Another fascinating development has been the delineation of gene mutations that cause hearing loss, some of which produce defects in proteins that link the hair cells of the cochlea, transforming the motion of sound waves into electrical impulses.

I excluded from the monograph the large group of infectious neurologic diseases caused by herpes simplex and zoster, cytomegalovirus, malaria, and many other agents. Considered even in part, this group of diseases would fill an entire volume.

I also excluded a consideration of the muscular dystrophies. Although most patients with muscular disorders traditionally consult neurologists, the field is somewhat disconnected from classical neurology because nervous tissue is not directly involved. Again, the advances have been so astonishing that adequate coverage was not possible here. An entire monograph might easily be devoted to Duchenne's muscular dystrophy alone.

Major advances are also being made in the research of disorders affecting the neuromuscular junction. These disorders may be either genetic or acquired, the former being most commonly manifested by development aberrations. They have not been considered here except for those in which muscle depolarization from ion channel defects causes failure of neuromuscular transmission (Chapter 14).

I must apologize for failing to deal in any substantive way with biomedical ethical concerns, presymptomatic genetic testing, confidentiality, regarding genetic predisposition, and the potential impact on insurability. These are exceedingly important matters about which much has been written but which could not, in my opinion, be adequately addressed within the space allotted this volume.

In conclusion, I must stress the importance of my associations with Massachusetts General Hospital; Harvard Medical School; and the University of California, San Francisco. My colleagues at these institutions expended most of the effort that made this work possible. I also express my appreciation to the other authors. Each was patient with me, often going through three or more revisions to make the text as clear as possible. I want to give special thanks to Leslie LaPiana and Tom Reynolds, who skillfully guided all of us through the experience.

Joseph B. Martin, M.D., Ph.D.

Contents

CHAPTER 1

Principles of Neurogenetics

James F. Gusella, Ph.D., Joseph B. Martin, M.D., Ph.D.

For most of its history, neurology has focused on description of the clinical features of disorders that affect nervous system function and of their biochemical or pathological correlates. The past two decades, however, have witnessed rapid, accelerating advances in our understanding of the genetic basis of many neurologic disorders and in our ability to decipher the biochemical and cellular mechanisms that underlie neurologic disease. This progress has already improved the capacity to diagnose and to classify neurogenetic disorders and, one hopes, presages great strides toward effective treatments. The foundations of this burgeoning knowledge in molecular neurology can be found in the biologic revolution brought about by the advent of recombinant DNA technology, which introduced the capacity to isolate, sequence, and manipulate human genes and the tools to analyze their functions.

The human genome consists of some 3×10^9 base pairs (bp) of DNA, packaged within chromosomes as long, continuous strings of DNA bundled with an equal mass of covering proteins. Each person has pairs of 22 autosomes (non–sex chromosomes), having received one of each type from each parent, and two sex chromosomes, either XX (female) or XY (male). Spread throughout the chromosomes are an estimated 60,000 to 100,000 genes whose DNA sequences carry the information to code for all the proteins that define the development, structure, activities, and interactions of cells and tissues in the human body. The coding sequence of each gene is present as small pieces of DNA called *exons*, separated by noncoding segments called *introns*. Depending on the amount of coding sequence and the size of the introns, a gene may span from less than 5,000 bp to more than 2,500,000 bp of DNA.

To make a protein, the gene is transcribed into a complementary RNA in the nucleus, and the exon sequences within the primary RNA transcript are spliced together, discarding the intron sequences [*see Figure 1*]. The mature processed RNA (messenger RNA or mRNA) is then transported to the cytoplasm of the cell, where it directs the synthesis of the corresponding protein with the amino acid sequence determined by the sequence of bases in the mRNA. Less than five percent of human DNA consists of actual coding sequence; the remainder is made up of regulatory elements that may flank the coding sequences or be present in introns, repetitive sequences that may occur in long tandem arrays or may be interspersed throughout the genome and be involved in chromosomal structure and organization, and other DNA sequences of no known function.

In the mid-1970s, technologies were developed to isolate or clone individual genes, either as artificially made DNA copies of the mature processed RNA (cDNA) or as fragments of the genome (genomic DNA), by inserting them into bacterial plasmids, circular DNA molecules that can be introduced into bacterial cells, where they replicate separately from the bacterial chromosomal DNA. Methods were also devised to determine directly the sequence of bases in cloned DNA, making it possible to predict the amino acid sequence of the protein encoded by a gene without first identifying and purifying the protein. These powerful technologies have since evolved, creating the field of molecular biology. There are now many different vector systems that enable the introduction of very small to extremely large DNA segments into a variety of organisms, including mammalian cells. For example, it is now possible to clone individual exons from DNA using so-called *exon-trapping vectors* or to isolate DNA stretches of more than 1,000,000 bp as yeast artificial chromosomes (YACs). Similarly, the invention of the *polymerase chain reaction* (PCR) has made it possible to amplify particular sequences from a mixture of DNA

Figure 1   *Gene to protein, illustrated by schematically double-stranded DNA in the cell nucleus with exon sequences (purple), promotor sequences (brown), and flanking and intron sequences (orange). The primary RNA transcript is produced by RNA polymerase beginning at the first exon and extending to the end of the last exon, where a sequence signal causes the addition of a number of consecutive adenosine ribonucleotides (A)n. This transcript undergoes splicing to attach consecutive exons directly to each other, removing intron sequences, and the consequent messenger RNA is transported to the cytoplasm, where it is translated on ribosomes into the protein that it encodes.*

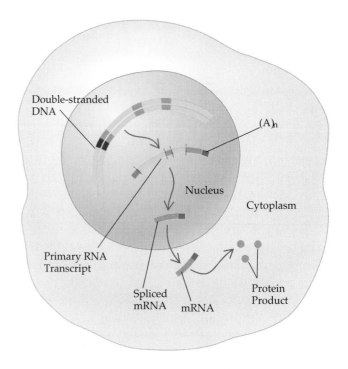

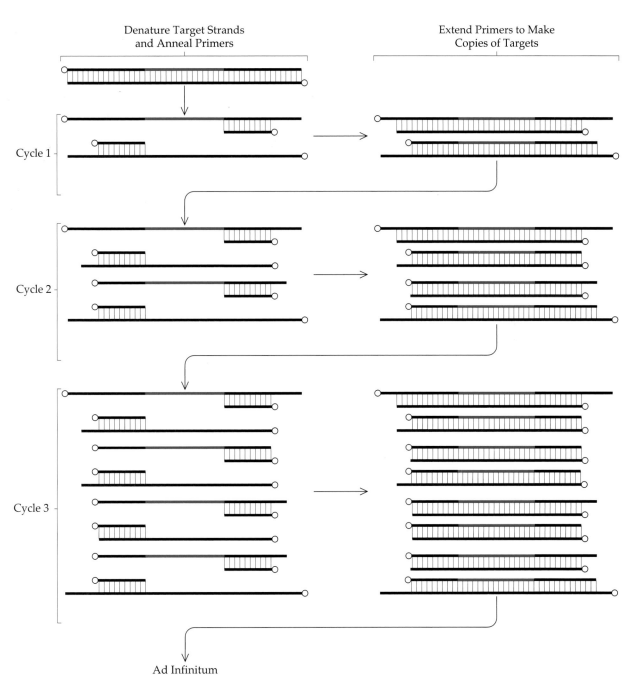

Denature Target Strands
and Anneal Primers

Extend Primers to Make
Copies of Targets

Cycle 1

Cycle 2

Cycle 3

Ad Infinitum

Figure 2 *The polymerase chain reaction. The polymerase chain reaction (PCR) is a common method of creating copies of specific fragments of DNA. PCR rapidly amplifies a single DNA molecule into many billions of molecules. In one application of the technology, small samples of DNA, such as those found in a strand of hair at a crime scene, can produce enough copies to perform forensic tests.*

fragments, permitting the chemical isolation of individual DNAs without the need for a biologic host [*see Figure 2*]. There are also specialized vector systems designed to provide regulated expression of the gene being introduced, integration of the gene into the host genome, or spread of the gene to adjacent cells. These vector systems offer the promise of genetic therapies for inherited disorders.

Molecular Biology of Neurologic Disease

Several thousand individual human genes have now been cloned and sequenced. More genes are expressed in the brain than in any other tissue, attesting to the great complexity of the nervous system. Individual human genes have been isolated on the basis of their function (e.g., glutamate receptors), their pattern of expression (e.g., brain-specific anonymous transcripts), their relationship to previously identified members of a family of related genes (e.g., ion channel genes), or their similarity to homologous genes in other organisms (e.g., cell death genes). Of greatest interest for molecular neurology specialists has been the isolation of genes with mutations that cause neurologic disease [*see Appendix*]. The identification of defects in neurogenetic diseases has occurred in two ways, which can be conveniently categorized as the *direct*, or candidate gene approach, and the *indirect* approach, in which a search is made of the entire genome.

Direct Approach: Candidate Genes in Neurologic Disease

Occasionally, neuropathologic or biochemical investigations implicate a particular protein as the culprit in a disorder. For example, Lesch-Nyhan syndrome was known to be associated with a deficiency in activity of the enzyme hypoxanthine-guanine phosphoribosyltransferase. Similarly, identification of the protein component of the amyloid plaques in Alzheimer's disease permitted isolation of the β-amyloid precursor protein gene and predicted it as a likely disease candidate. In such instances, if the gene has already been identified, it can be scanned directly for alterations in DNA sequence in patients with the disorder to discover pathogenic mutations. If the gene has not yet been described, the knowledge of the protein involved makes isolation of the gene straightforward, using schemes that employ either DNA probes predicted from the protein sequence or antibodies against the protein. The gene may then be sequenced in normal and diseased individuals to assess the possibility that it causes a disorder.

Indirect Approach: Positional Cloning

Unfortunately, the complexity of the nervous system, our incomplete knowledge of its normal operation, and the plethora of secondary effects that can result from the disease state have made it impossible to define with confidence a viable candidate gene in most neurologic disorders. In these cases, an indirect approach has been employed that merges the basic tenets of mendelian genetics with the power of molecular biology. This strategy, which has become known as *positional* or *location cloning*, involves first determining the chromosomal location of the genetic defect by genetic linkage analysis and then isolating and characterizing the disease gene on the basis of its map position, without any prior knowledge of the nature of the protein product.

STEP 1: PHENOTYPE AND MODE OF INHERITANCE

The first step in the location cloning approach is to identify families in which the disorder is present and to carefully examine their members for expression of disease symptoms. The diagnostic assessment of each individual is referred to as the individual's *phenotype*, which in simple disorders is

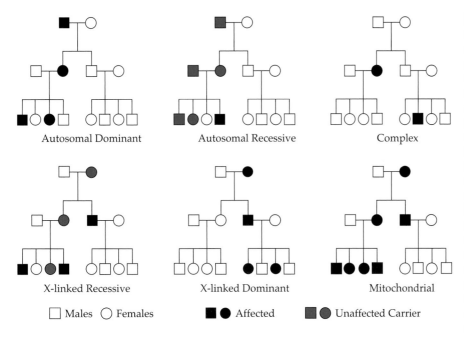

Figure 3 Modes of inheritance. Six different modes of inheritance of a disorder are depicted for the same family structure. The patterns in this case are based on the assumption of a disorder with complete penetrance and no decreased reproductive fitness. For autosomal dominant inheritance, males and females are affected equally, and any child of an affected individual has a 50 percent chance of inheriting the disease gene. In autosomal recessive disorders, males and females are affected equally, both parents must be carriers of the mutant gene, and each child of such parents has a 25 percent chance of being affected (two defective gene copies), a 50 percent chance of being an unaffected carrier (one defective gene copy from either parent), and a 25 percent chance of having no copy of the defective gene. X-linked recessive disorders are typically displayed in males who inherit the defective gene from an asymptomatic mother who carries a copy of the defective gene; thus, the defective gene cannot be transmitted by males to their sons. X-linked dominant disorders may appear similar to autosomal dominant disorders except that an affected male transmits the defect to all of his daughters and none of his sons. For mitochondrial inheritance, the disease gene is passed on by a female to all progeny but is never transmitted by a male. In complex disorders, multiple members of a pedigree may be affected with no evident simple mode of inheritance.

either normal or affected. In humans, a disease phenotype is said to be dominant if the presence of one copy of the underlying genetic defect on a single autosome is sufficient to cause clinical symptoms despite the presence of a normal version of the gene on the paired autosome. Autosomal dominant disorders are transmitted by either sex [*see Figure 3*]. Every child of an affected individual has a 50 percent chance of inheriting the disorder, which may therefore occur in every generation of a family. Consequently, genes displaying autosomal dominant inheritance may often be traced through large pedigrees with multiple affected individuals.

By contrast, if both copies of a gene on the paired autosomes must be mutant to cause disease, the phenotype is said to be recessive [*see Figure 3*]. Affected individuals must receive a mutant copy of the gene from each of their parents, both of whom are asymptomatic carriers. Recessive disorders are typically seen in one or more siblings of a single generation, with additional relatives displaying symptoms only in pedigrees with inbreeding or with a rare mating that involves the chance infusion of a defective gene from an unrelated carrier.

If a defective gene is present on the X chromosome, it will display sex-linked inheritance [*see Figure 3*]. In males, genes on the sole X chromosome are invariably expressed, whereas in females, one of the two X chromosomes is randomly inactivated in each cell and the alleles on it are not expressed. Consequently, if the defect is recessive, the abnormal phenotype is usually confined to males, whereas a dominant defect can display a phenotype in either sex. Although X-linked recessive disorders can theoretically be transmitted by either sex, the severity of the phenotype in most cases limits transmissions to those from female carriers. Unlike autosomal recessive disorders, X-linked recessive disorders are often exhibited in multiple generations of large pedigrees. Like autosomal dominant disorders, X-linked dominant disorders can be transmitted from either sex, but an affected male cannot transmit the defect to a male child.

In some disorders the disease phenotype is variable, with some individuals having different clinical signs or being more severely affected than others. An example of such a genetic defect, which is said to have *variable expressivity*, is von Recklinghausen's disease, or neurofibromatosis type 1 (NF1). Some individuals afflicted with this dominant disorder may have hundreds of cutaneous neurofibromas; others may display only an increased number of café au lait spots on the skin; still others may have malignant tumors, mental retardation, or scoliosis[1,2] [*see Chapter 6*]. In some disorders, not all carriers of the defect display an abnormal phenotype. The likelihood that a defect will cause a detectable abnormal phenotype is referred to as its *penetrance*. For example, late-onset dominant disorders, such as Huntington's disease (HD) and familial Alzheimer's disease, may show high penetrance only with advancing age. Other dominant disorders may show a high degree of nonpenetrance throughout life, such as early onset idiopathic torsion dystonia; only 30 to 40 percent of disease gene carriers ever display symptoms.[3-5]

A notable exception to the above modes of inheritance is seen in a subset of disorders in which the genetic defect does not occur in nuclear chromosomal DNA but in the 16559 bp closed circular DNA molecule found within mitochondria [*see Chapter 15*]. The mitochondria are transmitted through the cytoplasm of the female ovum independently of the nuclear chromosomes. Consequently, mitochondrial disorders may affect either sex but are transmitted exclusively through the female line [*see Figure 3*]. Because energy metabolism is crucial to neuronal function, a number of mitochondrial disorders, such as Leber's optic atrophy, the MELAS syndrome (mitochrondrial myopathy, encephalopathy, lactic acidosis, and strokelike episodes), and the MERRF syndrome (myoclonus with epilepsy and ragged red fibers), may have profound effects on nervous system function.[6-8]

Finally, there are many disorders for which risk is increased in the relatives of an affected individual but for which no clear mode of inheritance can be established [*see Figure 3*]. These disorders are referred to as complex because they are assumed to result from the interaction of multiple genetic variations, potentially with involvement of one or more environmental factors.

STEP 2: GENETIC LINKAGE ANALYSIS

Genetic linkage analysis involves tracking the inheritance of different chromosomal regions through a disease pedigree until a region is found whose pattern of inheritance correlates with that of the genetic defect.[9] By inference, the genetic defect must then lie in that chromosomal region. The application of this strategy depends on the observation that although each chromosome possesses a single contiguous double-stranded DNA molecule, these are not passed intact from generation to generation. Rather, they undergo exchange events known as *recombinations* or *crossovers* that effectively maintain the same gene order on the chromosome but shuffle the contributions from each generation so that the chromosomes present in an individual represent linear mosaics of the chromosomes present in the individual's parents [*see Figure 4*]. A crossover event may occur at any point along the chromosome, with the chances that it will occur between any two points being determined by how far apart they are on the DNA. On average, a one percent chance of recombination between two locations (or loci) corresponds to a separation of 10^6 bp of DNA. Genetic linkage can usually be detected for loci located within 2×10^7 bp of each other (20 percent recombination or recombination fraction = 0.2). By contrast, two loci separated by more than 5×10^7 bp will show 50 percent recombination or random segregation, making them indistinguishable from genes on separate chromosomes that are physically unlinked.

The broad application of the genetic linkage strategy also depends on the use of genetic markers as tags to track chromosomes as they are transmitted to successive generations. This capacity is provided by normal variation in the sequence of human DNA. Just as there is tremendous variation in the visible characteristics of individual members of the human race, there are normal differences in the sequence of DNA. On average, more than one base in 1,000 differs between any two human chromosomes, and any such difference can be used to track the inheritance of the chromosome segment in which it resides. The sequence variations or polymorphisms that are currently used most frequently as genetic markers involve tandem arrays of dinucleotide, trinucleotide, or tetranucleotide repeats that are interspersed frequently in the genome and often vary in number of repeat units, sometimes having more than a dozen different forms (or alleles). Each individual polymorphic repeat can be assayed by using oligonucleotides complementary to the genomic sequences flanking the repeat to prime its amplification by PCR, permitting measurement of its size [*see Figure 2*]. These repeats are so highly polymorphic that it is often possible to distinguish the corresponding chromosomal region from all four autosomes (two paternal, two maternal) involved in a given mating [*see Figure 4*].

Assessment of genetic linkage of a marker locus to a recessive or dominant disease gene is usually performed using the logarithm of the odds (LOD) score method.[10] This parametric method involves first specifying a number of parameters, including, among others, the mode of inheritance and estimated frequency of the disease allele and of the various alleles at the marker locus in the population. The investigator then calculates the likelihood of observing the specific pattern of inheritance of marker alleles in the pedigree if the marker and the locus are linked (at various assumed recombination frequencies from zero percent to 40 percent, such as $\theta = 0$, 0.01, 0.05, 0.1, 0.2, 0.3, and 0.4) relative to the likelihood of observing the same pattern by chance (no linkage, $\theta = 0.5$). The log of this likelihood ratio is called the LOD score (z). A score of $z > 3$ is considered significant evidence of linkage (> 1,000:1 odds relative to chance coinheritance of the marker and disease locus); a LOD score of $z < -2$ is considered to exclude the disease gene from the vicinity of the marker locus. Once linkage has been observed, the recombination fraction (θ) at which the maximum LOD score occurs is the maximum likelihood estimate of the true frequency of recombination between the marker and the disease and provides a rough guide to the physical distance separating them.

The first disease gene mapped to a chromosomal location using only genetic linkage to DNA markers was the defect causing HD in 1983[11,12] [*see Chapter 3*]. The dramatic success in a mystifying disorder that had proved unyielding to other approaches had a tremendous impact. It prompted many similar studies in other disorders and provided an early impetus to the proposal for the Human Genome Project, a large-scale project intended

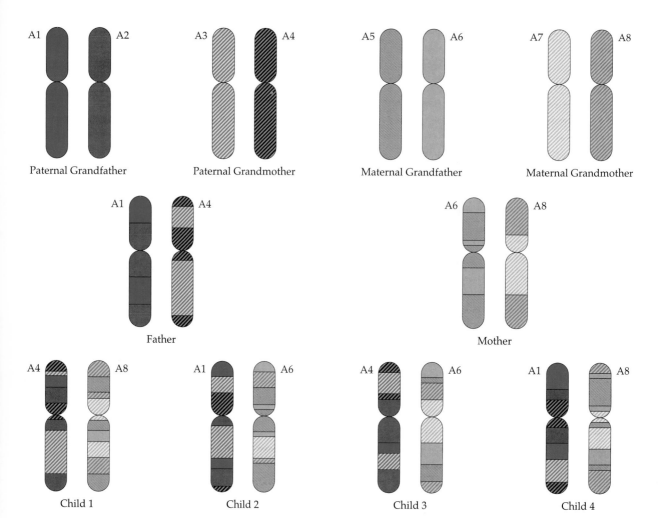

Figure 4   *Intergenerational transmission with meiotic recombination. An arbitrary chromosome pair (e.g., chromosome 7) as it is transmitted from grandparents to parents to children. In the grandparental generation, each chromosome is identified by a color. In meiotic recombination events, material is exchanged between the paired chromosomes, producing a different mosaic chromosome transmitted to each member of the successive generation. The use of genetic markers to track chromosomal regions through families and to test for genetic linkage to a disease gene is analogous to using the imaginary colors. Consequently, only markers with many alleles offer the potential to distinguish all eight chromosomes. In this depiction, a marker near the short arm telomere displays alleles A1, A2, A3, A4, A5, A6, A7, and A8. If a dominant disease gene were near the A4 allele (dark gray stripes) in the affected paternal grandmother, one would observe that the father, child 1, and child 3 were affected, whereas all others were unaffected. Similarly, if a recessive mutation were present near the A1 allele (light blue) in the paternal grandfather and the A6 allele (light gray) in the maternal grandfather, both father and mother would be carriers. Child 2 would exhibit the recessive disorder, children 3 and 4 would be unaffected carriers, and child 1 would be normal.*

to sequence the entire genome and identify all human genes. The project is slated for completion early in the twenty-first century. The subsequent combination of disease-specific studies and general human gene mapping has identified defects in numerous disorders [*see Appendix*] and has provided an ever-improving infrastructure of genetic markers, physical maps of chromosomes, and overlapping clone sets of genomic DNA that greatly facilitate similar studies in any additional inherited disorder.

It is now possible to approach any neurogenetic disorder using this strategy. The approach is critically dependent on careful phenotypic assessment to classify affected individuals and distinguish them from unaffected subjects. This can be complicated by nongenetic phenocopies, in which symptoms similar to a genetic disorder may have a nongenetic cause, or by nonallelic heterogeneity, in which similar symptoms are caused by mutations in different genes. However, with reasonable numbers of families, the power of the detailed human genetic map is now such that these obstacles can be resolved readily in simple mendelian disorders. Indeed, there are now so many markers on the human genetic map that many can be expected to fall in the vicinity of any given disease locus. Consequently, the calculation of linkage scores can be made more powerful by using the results from several marker loci simultaneously. In this multipoint method of linkage analysis, the calculation of location scores involves comparing the likelihood of observing a pattern of disease gene and marker allele inheritance, assuming various positions of the disease gene within a fixed marker map relative to nonlinkage. This generates a curve that is interpreted in the same manner as the two-point LOD score, with the peak designating the most likely location of the disease gene.

Genetic studies to identify disease genes are now being extended to complex disorders, such as psychiatric disorders, multiple sclerosis, Parkinson's disease, and late-onset Alzheimer's disease, in which large numbers of affected sibling pairs or other relatives are compared. In these investigations, in which mode of inheritance cannot be defined, various nonparametric methods of assessing linkage are used. These strategies test for significant deviation in the degree of sharing of particular alleles by affected members of the same family over the degree of sharing expected by chance. An alternative approach to complex multigenic disorders is to use association studies in which the inheritance of a genetic defect is compared in an entire population rather than within individual families. A powerful technique, the transmission disequilibrium test, compares the alleles present in affected individuals and those in their parents and tests for a significant difference in the pattern of alleles transmitted by the parents versus those not transmitted.[13] This strategy can therefore be used to detect evidence of linkage by examining many affected individuals even though none of these individuals has affected relatives available for comparison.

STEP 3: NARROWING THE SEARCH AND IDENTIFYING CANDIDATE GENES

Once a disease gene has been mapped to a chromosomal region, the task remains to identify the gene itself and thereby to predict the defective protein responsible for pathogenesis. In some cases, it is possible to use physical landmarks on the DNA map of the chromosome, such as deletions or translocations in individual patients that can mark the site of the disease gene. For example, two translocations on chromosome 17 in individual NF1 patients were found to inactivate a gene encoding a signal transduction modulator protein, which was named neurofibromin in reference to the role of its absence in this disorder.[1,2]

Usually, an investigator has no physical landmark to aim for and must find the disease gene on the DNA map of the chromosome by examining all genes in the region of interest. Consequently, it is imperative to reduce the candidate region to a manageable size of a few hundred thousand bp. Initially, this is done by using rare recombination events between the disease gene and linked markers as genetic landmarks. This places a further premium on accurate phenotyping because misdiagnoses can lead the gene quest to an incorrect location.

In many affected individuals who have a common ancestor, it is possible to use so-called historical recombination events as borders of the candidate region. Comparison of apparently unrelated affected individuals in a given population may reveal a common *haplotype*, or set of alleles, at consecutive markers across the region of interest that is infrequent on normal chromosomes. This shared haplotype represents the pattern of alleles present on the original chromosome where the disease mutation occurred and indicates that those who possess it are not unrelated but actually share a common ancestor. Deviations from the expected haplotype in affected individuals may then represent historical crossover events in the same manner that such crossovers are detected in families in which the parentage is known. The region of the haplotype that remains invariant in all affected individuals marks the position of the disease gene. This strategy is most easily applied in ethnically or geographically isolated populations. For example, haplotype analysis of familial dysautonomia and early onset idiopathic torsion dystonia in Ashkenazi Jews has pinpointed the locations of the corresponding disease genes.[5,14] However, the same strategy has also been successful in more complex populations. In the most obvious example, haplotype analysis in ethnically diverse patients revealed that approximately 40 percent of HD families share a common ancestor and led directly to a 150 kilobase (kb) segment of chromosome 4 containing the defect.[15] This approach will likely be essential for complex disorders in which individual recombination events cannot be defined precisely because multiple interacting genes contribute to the phenotype.

When a disease gene candidate region has been minimized, it must be isolated as cloned genomic DNA and the genes within it identified and scanned for evidence of defects in patient DNA [*see Figure 5*]. Numerous strategies are available for isolating the genomic DNA and building a physical map of the DNA in the candidate region, including using DNA probes or PCR to isolate overlapping YAC or bacterial artificial chromosome (BAC) clones or smaller but more manageable clones called *cosmids*. Similarly, a variety of approaches have been used to identify gene transcripts using this cloned genomic DNA, including direct screening of cDNA libraries, exon trapping, computer prediction of exons from DNA sequence, and evidence for evolutionary conservation of DNA sequence. As the Human Genome Project advances, it is gradually providing an infrastructure that makes these steps easier and has moved a significant portion of the disease cloning effort from the laboratory bench to the computer terminal, where on-line databases can be scanned for useful information. An early product of the Human Genome Project was the proliferation of genetic markers and maps to help in initial localization of disease genes and definition of a candidate region. However, the ultimate goal of determining the human DNA sequence required a method of building physical maps and providing ordered sets of genomic clones. This was approached by randomly generating thousands of

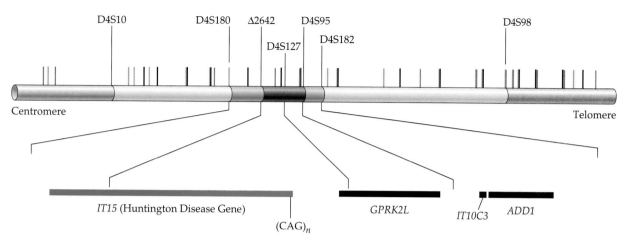

Figure 5 *An example of defining and narrowing a candidate region for a disease gene: the Huntington's disease (HD) gene region of chromosome band 4p16.3. Sequence recognition sites for cutting with restriction enzymes Not I (light blue lines), Mlu I (black lines), and Nru I (dark blue lines) are shown above the map, along with the locations of genetic markers used to localize the HD gene. The 2 x 10⁶ bp region telomeric to marker D4S10 and centromeric to marker D4S98 was defined by observation of recombination events in HD families. The ~500 kb segment between D4S180 and D4S182 was defined as a higher likelihood location by identifying a haplotype of shared marker alleles in ~33 percent of HD families. The ~150 kb segment between Δ2642 and D4S95 was identified as the highest likelihood candidate interval by observation of shared marker alleles in > 40 percent of HD families. Below the map, the intermediate candidate interval is expanded to show four of the genes that were scanned for mutation to eventually identify an unstable, expanded CAG trinucleotide repeat near the 5' end of IT15 as the cause of HD.*

sequence tagged sites (STSs), each a single site on the genome whose presence or absence in DNA could be assayed by the same type of PCR assay used for genetic markers. However, the STSs need not be polymorphic as they have been used to test the presence or absence of each site in large genomic clones, such as YACs and BACs, thereby permitting overlapping clones to be lined up to form a contig or overlapping set. The availability of predetermined sets of contigs for most chromosomal regions now obviates the need to isolate them anew once an investigator has defined a candidate region for a disease gene.

Another product of the Human Genome Project has been a massive number of random partial cDNAs, so-called *expressed sequence tags* (ESTs) that have been sequenced from various tissues. Although most are not yet mapped to specific chromosomal segments, those that are provide an increasingly dense set of genes for initial consideration as candidates within a disease region. Similarly, as the candidate region is sequenced, simple searching for matches in the EST database can be used to identify candidate genes among those ESTs that have not yet been mapped to a particular location. Eventually, the generation of the entire human DNA sequence should eliminate the need for any of the DNA cloning steps now necessary to prepare for evaluation of candidate genes.

Once the structure of a gene in the candidate region has been defined, it must be examined for evidence of a defect in the disease. Deletions or failure to produce a transcript are two obvious defects that can be readily detected, but many mutations are more subtle. A number of specialized techniques and gel electrophoresis systems have been elaborated to facilitate the comparison of DNA segments from normal and diseased individuals, but these all act simply as prescreens for direct comparison by DNA sequencing. Ultimately, it is necessary to identify changes in DNA sequence (i.e., mutations) that identify a candidate gene as the disease gene.

Mutant Genes Identified in Neurogenetic Disorders

When the disease gene is identified by the indirect approach of positional cloning, there is no way to predict in advance the nature of its protein product. For example, a number of disease genes [see Appendix] have been found to encode entirely novel proteins of unknown function. By contrast, a genetic defect in familial amyotrophic lateral sclerosis was found to involve mutations altering superoxide dismutase 1, an enzyme known and studied for decades without yielding any hint of its disease involvement[16] [see Chapter 12]. However, while not revealing the nature of the protein involved, often the mode of inheritance does provide a clue to the type of mutation expected in the disorder.

Recessive disorders typically involve *loss of function* of the mutated gene. Thus, affected individuals with two copies of the mutant gene possess no active protein product and suffer a consequent phenotype. Many recessive disorders are associated with loss of an enzyme activity, such as phenylketonuria (phenylalanine hydroxylase), Gaucher's disease (glucocerebrosidase), Tay-Sachs disease (hexosaminidase A), and gyrate atrophy of the choroid and retina (ornithine aminotransferase). The clinical features of such a recessive disorder may result either from the inability to produce a required metabolite downstream from the defect in the enzymatic pathway or from the failure to remove a compound produced upstream in the enzymatic pathway. In unaffected carriers of a recessive genetic defect, the one remaining normal allele produces sufficient activity to prevent the development of clinical symptoms. The actual molecular lesion that causes gene inactivation in a recessive disorder can be any of several types of alteration, including deletion of DNA from the gene; insertion of exogenous DNA into the gene; inversion of DNA within the gene; or mutation of single base pairs that alter splicing signals to prevent normal RNA processing, that introduce a premature stop codon, which causes truncation of the protein product, or that produce a missense substitution, which destroys the activity of the protein product. Most of these inactivating changes result in failure to express any stable protein product.

Dominant disorders may also involve loss of function of one allele if the remaining normal allele cannot fully mask the effects of the deficiency. For example, in DOPA-responsive dystonia, inactivation of one copy of the GTP cyclohydrolase 1 gene results in deficiency of biopterin, leading to reduced activity of tyrosine hydroxylase and decreased dopamine levels[17] [see Chapter 9]. Three different types of nervous system tumors illustrate definitive loss of function effects: retinoblastoma, NF1, and neurofibromatosis type 2 (NF2) [see Chapter 6]. In each, the family pedigrees conform to autosomal dominant inheritance. But the genetic mechanism in each tumor is loss of function in both alleles. This may occur as a result of two mutations—one in the germline, followed by a second mutation during somatic cell division. This produces a so-called double hit inactivating both genes, as first proposed by

Knudson. The tumors result from loss of function of a normal growth suppressor protein. In retinoblastoma, the Rb protein exerts a powerful growth inhibitory effect in many cells; however, the tumor that arises from its absence most commonly affects the retina. Tumors of bone (osteosarcoma) and other nonneuronal tissues have been described but are rare, whereas tumors of the retina, usually bilateral, are almost invariant. Absence of Rb protein in dividing retinoblasts results in their failure to differentiate.

Dominant mutations more commonly involve expression of a protein product with altered function or *gain of function*. Alteration of normal function is typified by a number of channel disorders, such as hyperkalemic periodic paralysis and paramyotonia congenita, which are both attributed to missense substitution mutations of the muscle sodium channel α-subunit that alter the kinetics of channel function[18] [*see Chapter 14*]. In these disorders, both normal and mutant alleles are expressed, but the mutant protein allows leakage of sodium ions that causes repetitive muscle firing (myotonia) and depolarization, which in turn can inactivate all of the sodium channels, both mutant and normal. Consequently, the muscles are weak as they become electrically inexcitable. As only one mutant allele is required to make a portion of the sodium channels abnormal, the disease phenotype is dominant.

Another type of altered function that can be seen in dominant disorders is a so-called *dominant negative mutation*. In this case, both normal and abnormal protein products are expressed, but the mutant protein interacts directly or indirectly with the normal protein and prevents it from performing its normal function. Such mutations are often associated with structural proteins in which a single abnormal subunit can inactivate an entire structure. For example, dominantly inherited retinitis pigmentosa can be caused by a variety of amino acid substitution mutations in the rhodopsin gene that disrupt the structure of the photoreceptors. Since loss of function is the ultimate basis of the phenotype, a mutation that causes inactivation of rhodopsin without a dominant negative effect can cause recessively inherited retinitis pigmentosa.[19]

Gain-of-function mutations, sometimes referred to as *toxic gain of function*, are exemplified in a series of neurodegenerative disorders, including Huntington's disease, dentatorubropallidoluysian atrophy, Kennedy's syndrome, and spinocerebellar ataxias 1, 2, 3, and 6[20] [*see Chapter 3*]. In each case the mutation involves expansion of a normally polymorphic stretch of repeating CAG trinucleotides that results in lengthening of a stretch of consecutive glutamine residues in the corresponding protein. Investigations in HD of its gene product, huntingtin, provide strong evidence that this mutation confers a new deleterious property (toxic gain of function) without interfering with the gene's normal function.

Complex disorders involve multiple gene-gene interactions or gene-environment effects, in which individual genes may be inactivated or otherwise altered by mutations that predispose to the disorder but are insufficient to cause symptoms on their own. In this circumstance, the genetic risk factor may simply be a normal polymorphic variant of a gene, such as the apolipoprotein E_4 locus (APOE) that is a risk factor in late-onset Alzheimer's disease[21,22] [*see Chapter 4*].

Implications of Disease Gene Identification

The discovery of a new disease gene can have profound implications for research, diagnosis, and clinical care. It provides the critical triggering event that begins the cascade of pathogenesis leading to clinical symptoms and thereby places researchers at the starting point of a path in which every additional step may provide a target for developing a rational therapy. With the disease gene identified, the molecular neurologist can attempt to understand the nature of the genetic defect and the mechanism whereby it leads to disease.

Research

One of the first steps is to use the analysis of mutations in the disease gene to aid in disease categorization. Similar disorders may be caused by different genes, such as Charcot-Marie-Tooth disease, because of separate loci on chromosomes 1, 3, 8, 11, 17, and X [*see Chapter 13*], and familial Alzheimer's disease, attributable to loci on chromosomes 1, 14, and 21 [*see Chapter 4*]. Thus, it is sometimes impossible to recognize the distinctions until at least one of the genes has been identified. Once similar clinical entities have been recognized as separate genetic disorders, it is possible to draw distinctions in the potential management of the disorders and concentrate research on each pathogenic mechanism individually.

On the other hand, different mutations in the same gene can sometimes produce different clinical pictures, such as the mutations in hyperkalemic periodic paralysis and paramyotonia congenita[23] [*see Chapter 14*]. In such cases, delineation of individual mutations in comparison with clinical symptoms may reveal the structure-function relationships of the disease protein's domains and yield clues to the protein's physiologic role. A striking example of genotype-phenotype correlation is the relationship between the expanded CAG trinucleotide repeat in HD and the patient's age at neurologic onset of the disorder [*see Figure 6; see Chapter 3*]. This change appears to be the universal cause of HD as no other mutations in this gene have been found in any HD patients. The strong inverse correlation of CAG repeat length with age of onset implicates the expanded polyglutamine stretch in huntingtin as the primary determinant of pathogenesis.

The isolation of a disease gene also usually sets off a variety of biochemical and cell biologic approaches to determining the protein product's normal function and the mode of action of disease mutations. Comparisons are made with previously reported genes and proteins in public databases to discern whether the gene is a member of a family or whether the protein has recognizable motifs that would suggest a function. The generation of polyclonal or monoclonal antibodies is pursued to study the structure and localization of the protein in normal and diseased human tissues and in cell culture models. Studies may also involve manipulating the expression of the normal and mutant protein in cell culture by the use of expression vectors introduced into cells to produce protein tagged with an epitope that permits easy monitoring of the product. Ultimately, delineation and comparison of the protein's structure, description of its pattern of tissue and developmental expression, definition of its subcellular localization, and iden-

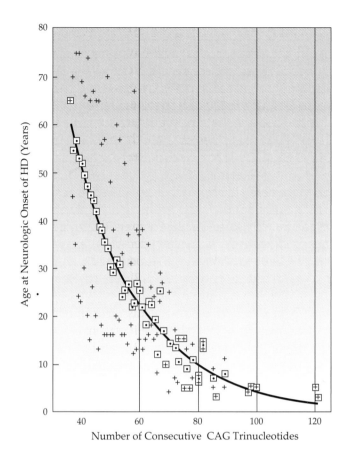

Figure 6   *The graph shows the mean age at neurologic onset (squares) associated with different CAG repeat lengths in 1,228 Huntington's disease (HD) patients reported in the literature.[20] These data show a good fit to an exponential decay model (line; $r^2 = 0.97$). Also shown are the maximum and minimum ages at onset reported for each repeat length (+). For CAG repeat numbers with a single observation, the + symbol coincides with the center of the square.*

tification of interacting proteins can provide the basis for specific hypotheses concerning its physiologic role.

A complement of the biochemical and cell biologic approach is the study of the homologous gene in other organisms that are amenable to genetic manipulation and construction of animal models. Occasionally, existing mutants are found to represent naturally occurring models of human disease, such as the mutation in the tuberous sclerosis gene in the Eker rat or the loss of the aniridia locus in the Dickie small-eye mutant mouse.[24-26] The primary organism for animal model construction has been the mouse (*Mus musculus*), in which it is possible to inactivate or specifically alter genes in embryonic stem cells in culture and to reimplant these cells for development into a designed mutant animal.[27] However, the fruit fly (*Drosophila melanogaster*) and the roundworm (*Caenorhabditis elegans*) also represent useful genetic systems in which a gene's functions and interactions can be explored.

The tools of molecular biology have vastly improved our capacity to identify genes and to evaluate and manipulate their expression in both cultured cells and whole organisms. However, determining the function of a novel disease protein remains the most challenging task in the investigation of human genetic disorders.

Diagnosis

Advances in molecular neurology also provide the ability to deliver diagnostic information not previously available. Even before a disease gene is isolated, linked genetic markers can be used in specific circumstances to infer its presence. For example, linkage of HD to chromosome 4 in 1983 allowed predictive testing of presymptomatic individuals in extended families.[28] The same test permitted prenatal diagnosis in these families. Similarly, linked markers for early onset dystonia have been used to infer the presence of this disease gene in members of the Ashkenazi Jewish population, in which the defect displays a penetrance of less than 50 percent.[3-5] Once the disease gene has been cloned, it is possible to perform direct testing of individuals by examining the gene for mutations using any of a variety of techniques. However, the effectiveness of this approach depends partly on the nature of the genetic defect. In HD, all individuals have the same type of mutation—an expanded CAG trinucleotide—so measuring the length of this stretch provides an accurate, inexpensive diagnostic test.[29] However, in many disorders, each family possesses a mutation at a different site in the gene. Consequently, the entire gene must be scanned for changes, an approach that is costly and less than 100 percent accurate with current technology. Thus, for individuals with rare disorders, genetic testing may not be commercially available despite prior identification of the disease gene. Within the near future, technological advances in scanning of DNA sequence are expected to eliminate this difficulty and make genetic testing less costly, more efficient, and more widely applicable.

Although technological advances will make DNA diagnosis more effective, they will not overcome the major difficulties involved in the delivery of this sensitive information to patients. In late-onset disorders, such as HD, predictive testing can represent a mixed blessing because it can deliver to a currently healthy individual a virtual death sentence if there is no effective treatment.[28] The intensive genetic counseling required can be exceedingly difficult for both patient and health care giver. In disorders with psychiatric components, there may even be a danger that imparting a stress-inducing genetic diagnosis may trigger behavioral complications. Even in disorders in which knowledge of genetic status may be critical for clinical management and life planning, such a diagnosis may threaten the retention of insurance and employment. Genetic testing poses enormous ethical concerns, including issues of confidentiality, suicide risk, testing of minors, and prenatal testing. Adequate support services may require the skills of neuropsychologists, counselors, and social workers. These issues are at the forefront of a debate that extends far beyond neurology to all areas of modern medicine and that in some cases will be resolved only by effective legislation.

Treatment

A resolution of the quandary presented by presymptomatic DNA diagnosis is the development of treatments that would make predictive testing universally beneficial. Identification of disease genes and delineation of modes of pathogenesis can provide insights into the development of rational therapies. Moreover, in many cases, such therapies may take novel forms because the tools of molecular biology have introduced the possibility of using genes themselves as therapeutic agents. In recessive loss-of-function disorders, gene replacement is the goal, whereas in dominant gain-of-function disorders, prevention of the effects of the mutant protein must be achieved. For each case, a variety of engineered mammalian viruses designed as DNA vectors may prove to be effective delivery systems for nucleic acids targeted to effect specific alteration of cell metabolism.[30] The degree to which the search for a treatment should be based on traditional pharmaceutical approaches for identifying small molecule therapeutics, biotechnological approaches for protein therapy, or genetic therapy approaches will depend on the precise nature of the defect and its mode of action in each disorder. The goal of a cure for most inherited neurologic disorders is far from being realized. However, it is safe to predict that the molecular neurology revolution will continue to yield a rapid, accelerating expansion in our understanding of the genetic and biochemical alterations underlying neurologic disorders and in the development of more effective tools to alleviate their consequences.

References

1. Shen MH, Harper PS, Upadhyaya M: Molecular genetics of neurofibromatosis type 1 (NF1). *J Med Genet* 33:2, 1996

2. von Deimling A, Krone W, Menon AG: Neurofibromatosis type 1: Pathology, clinical features and molecular genetics. *Brain Pathol* 5:153, 1995

3. Risch NJ, Bressman SB, deLeon D, et al: Segregation analysis of idiopathic torsion dystonia in Ashkenazi Jews suggests autosomal dominant inheritance. *Am J Hum Genet* 46:533, 1990

4. Bressman SB, de Leon D, Brin MF, et al: Idiopathic dystonia among Ashkenazi Jews: evidence for autosomal dominant inheritance. *Ann Neurol* 26:612, 1989

5. Risch N, de Leon D, Ozelius L, et al: Genetic analysis of idiopathic torsion dystonia in Ashkenazi Jews and their recent descent from a small founder population. *Nat Genet* 9:152, 1995

6. Brown MD, Voljavec AS, Lott MT, et al: Leber's hereditary optic neuropathy: a model for mitochondrial neurodegenerative diseases. *FASEB J* 6:2791, 1992

7. Bindoff LA, Desnuelle C, Birch-Machin MA, et al: Multiple defects of the mitochondrial respiratory chain in a mitochondrial encephalopathy (MERRF): a clinical, biochemical and molecular study. *J Neurol Sci* 102:17, 1991

8. Ciafaloni E, Ricci E, Shanske S, et al: MELAS: clinical features, biochemistry, and molecular genetics. *Ann Neurol* 31:391, 1992

9. Gusella JF: DNA polymorphism and human disease. *Annu Rev Biochem* 55:831, 1986

10. Ott J: *Analysis of Human Genetic Linkage*, rev. Johns Hopkins University Press, Baltimore, 1991

11. Gusella JF, Wexler NS, Conneally PM, et al: A polymorphic DNA marker genetically linked to Huntington's disease. *Nature* 306:234, 1983

12. Gusella JF, Tanzi RE, Anderson MA, et al: DNA markers for nervous system diseases. *Science* 225:1320, 1984

13. Spielman RS, McGinnis RE, Ewens WJ: Transmission test for linkage disequilibrium: the insulin gene region and insulin-dependent diabetes mellitus (IDDM). *Am J Hum Genet* 52:506, 1993

14. Blumenfeld A, Slaugenhaupt SA, Axelrod FB, et al: Localization of the gene for familial dysautonomia on chromosome 9 and definition of DNA markers for genetic diagnosis. *Nat Genet* 4:160, 1993

15. Gusella JF, MacDonald ME: Hunting for Huntington's disease. *Mol Genet Med* 3:139, 1993

16. Rosen DR, Siddique T, Patterson D, et al: Mutations in Cu/Zn superoxide dismutase gene are associated with familial amyotrophic lateral sclerosis. *Nature* 362:59, 1993

17. Ichinose H, Ohye T, Takahashi E, et al: Hereditary progressive dystonia with marked diurnal fluctuation caused by mutations in the GTP cyclohydrolase I gene. *Nat Genet* 8:236, 1994

18. Cannon SC: From mutation to myotonia in sodium channel disorders. *Neuromuscul Disord* 7:241, 1997

19.   Rosenfeld PJ, Cowley GS, McGee TL, et al: A null mutation in the rhodopsin gene causes rod photoreceptor dysfunction and autosomal recessive retinitis pigmentosa. *Nat Genet* 1:209, 1992

20.   Gusella JF, Persichetti F, MacDonald ME: The genetic defect causing Huntington's disease: repeated in other contexts? *Mol Med* 3:238, 1997

21.   Mayeux R, Schupf N: Apolipoprotein E and Alzheimer's disease: the implications of progress in molecular medicine. *Am J Public Health* 85:1280, 1995

22.   Lendon CL, Ashall F, Goate AM: Exploring the etiology of Alzheimer disease using molecular genetics. *JAMA* 227:825, 1997

23.   Hudson AJ, Ebers GC, Bulman DE: The skeletal muscle sodium and chloride channel diseases. *Brain* 118:547, 1995

24.   Glaser T, Lane J, Housman D: A mouse model of the aniridia-Wilms tumor deletion syndrome. *Science* 250:823, 1990

25.   Yeung RS, Xiao GH, Jin F, et al: Predisposition to renal carcinoma in the Eker rat is determined by germ-line mutation of the tuberous sclerosis 2 (*TSC2*) gene. *Proc Natl Acad Sci USA* 91:11413, 1994

26.   Kobayashi T, Hirayama Y, Kobayashi E, et al: A germline insertion in the tuberous sclerosis (*Tsc2*) gene gives rise to the Eker rat model of dominantly inherited cancer. *Nat Genet* 9:70, 1995

27.   Soriano P: Gene targeting in ES cells. *Annu Rev Neurosci* 18:1, 1995

28.   Meissen GJ, Myers RH, Mastromauro CA, et al: Predictive testing for Huntington's disease with use of a linked DNA marker. *N Engl J Med* 318:535, 1988

29.   Huntington's Disease Collaborative Research Group: A novel gene containing a trinucleotide repeat that is expanded and unstable on Huntington's disease chromosomes. *Cell* 72:971, 1993

30.   Smith AE: Viral vectors in gene therapy. *Annu Rev Microbiol* 49:807, 1995

Acknowledgments

Figure 1   Tom Moore and Dana Burns-Pizer.

Figure 2   Tom Moore.

Figures 3 and 6   Marcia Kammerer.

Figures 4 and 5   Seward Hung.

CHAPTER 2

Molecular Basis of Mental Retardation: Fragile X Syndrome

David L. Nelson, Ph.D.

Mental retardation is defined by reduced general intellectual functioning (an IQ score two standard deviations below the mean, typically 70 or below) and by significant limitations in adaptive skills, apparent in individuals before they reach 18 years of age. It is estimated that some 2.5 to three percent of the population of the United States meet this definition.[1] Individuals with significant impairment, defined by IQ scores below 50, are found at a rate of three or four per thousand. There are many causes of mental retardation, and at least one third of cases are of unknown etiology. Most mental retardation results from environmental factors, such as poor maternal health, prematurity, and perinatal insult. There are many genetic causes of mental retardation, and among these, chromosome abnormalities, such as trisomy 21 found in Down syndrome, account for a significant fraction.

An excess of males among the retarded population has long been noted, and it has been proposed that this is attributable to single gene defects on the X chromosome.[2,3] Many families have been identified with X-linked patterns of inherited mental retardation, and numerous genetic linkages have been established that suggest separate gene defects. Numerous X-linked genes have been identified that, when defective, result in mental retardation. Often, these defects also produce a disease that appears to primarily affect organ systems other than the brain. Examples include the hypoxanthine phosphoribosyltransferase and Duchenne's muscular dystrophy genes. In most cases, the connection between the cellular or biochemical defect and mental retardation remains unknown. Fragile X syndrome accounts for a substantial portion of X-linked mental retardation, and its high prevalence appears related to the unusual nature of the genetic defect.

Fragile X syndrome (OMIM #309550) is among the most common of human single gene disorders and is the leading cause of inherited mental retardation. It is inherited as an X-linked dominant disorder with reduced penetrance. Males and females can be affected; however, the degree of mental retardation is usually more severe in males.

The first clinical report of the syndrome derived from identification of an English family in 1943.[4] This family exhibited an X-linked pattern of mental retardation with characteristic facial features. After other families were identified, the disorder became known as the Martin-Bell syndrome. The current name, fragile X syndrome, derives from the observation of a folate-sensitive chromosomal fragile site at band position Xq27.3 in chromosomes at metaphase prepared from afflicted individuals. This site was identified initially in 1969,[5] although more thorough characterization of the parameters of its appearance was not completed until the mid to late 1970s.[6] Fragile sites appear as nonstaining gaps in the chromosome and are occasionally observed to result in chromosome breakage during preparation of the material for study. Although some other fragile sites appear to have the potential for in vivo breakage, resulting in deletions and translocations, there is no evidence of chromosomal fragility in patients' cells in vivo. The fragile site has served as a diagnostic tool since the 1970s, although the availability now of direct DNA tests, which have higher reliability, has greatly reduced the utility of cytogenetics for diagnosis.

The defective gene in fragile X syndrome (*FMR1*) was identified by positional cloning in 1991[7] following efforts to locate the chromosomal anomaly.[8,9] Both the fragile site and the gene defect derive from expansion and methylation of a trinucleotide repeat (CGG) located in the 5′ untranslated region of the *FMR1* messenger RNA (mRNA). This was the first example of a large number of unstable trinucleotide repeats found to cause human genetic disorders[10] and remains among the best understood for the mutation's effect on gene function [*see Chapter 3*].

The prevalence of fragile X syndrome has been somewhat controversial. Estimates that relied on cytogenetic analyses (which can frequently result in false positive findings) ranged as high as 1:1,200 males.[11] However, more recent surveys using DNA-based tests have suggested a disease frequency closer to 1:4,000.[12]

Clinical Features

The association of the fragile X site with families exhibiting X-linked mental retardation allowed improved definition of the clinical syndrome. This description continues to be refined with the advent of DNA-based diagnosis; however, it remains a highly variable phenotype. The syndrome is difficult to diagnose in newborns. Physical features gradually accumulate with age. However, even in fully affected males, the facies are subtle and generally unremarkable to those unfamiliar with the syndrome, especially in younger patients.

Adult male patients generally exhibit a long, narrow face with moderately increased head circumference (> 50th percentile). Prominence of the jaw and forehead and particularly large, mildly dysmorphic ears are typical.

Some of the phenotype is reminiscent of a mild connective tissue disorder, exhibiting hyperextensible joints, high arched palate, pes planus, pectus excavatum, and mitral valve prolapse. Macro-orchidism (enlarged testicular volume) is a common finding in postpubescent affected males. Almost 90 percent of such males exhibit testicular volumes in excess of 25 ml, as measured by an orchidometer.

Mental retardation and developmental delay are the most prominent clinical features of fragile X syndrome. Prepubescent males have globally delayed developmental milestones, and some may display avoidance behavior similar to autism as well as hyperactivity and attention deficit. The latter two are frequent presenting complaints in boys with fragile X syndrome. Development of speech and language is almost always involved but to variable degrees. Whereas absence of speech is rare, milder communication difficulties are common, including a characteristic jocular, litany-like speech. Mental retardation ranges from profound to borderline, with an average IQ in the moderately retarded range. It has been estimated that fragile X syndrome accounts for as much as 20 percent of all boys with IQ scores between 30 and 55.[13]

Fragile X syndrome is much milder in females than in males. Somatic signs may be absent or mild, although the facies of older females may tend to resemble those of affected males. The mental retardation, in particular, is less severe in most female patients, falling in the mild to borderline range. A number of studies suggest increased emotional lability in both affected and carrier females. These are manifested as both behavioral and psychiatric abnormalities. Although these are highly variable, several studies have identified schizotypal features, depression, social avoidance, anxiety, and shyness among girls and women with IQs in both the normal and affected ranges.[14]

Studies of pathology in fragile X patients have not provided significant insight into the disease process. Testes show some edema but otherwise are normal in overall structure.[15] Brain structure in fragile X patients has been studied by both postmortem and neuroimaging techniques.[16] The single reproducible postmortem finding has been abnormalities (thinning and short synaptic lengths) of dendritic spines in cortical neurons, although this has been reported only for a handful of patients.[17] Neuronal densities appear unchanged. Neuroimaging studies have shown reductions of the posterior cerebellar vermis, enlargement of the caudate nucleus and lateral ventricle, and age-related increases in the size of the hippocampus. High levels of *FMR1* mRNA and its protein are expressed in the hippocampus and in the Purkinje cells of the cerebellum.[18-20] How functions of the protein relate to the clinical manifestations is unknown. A knockout mouse model of fragile X syndrome shows macro-orchidism but no discernible neuroanatomical abnormalities.[21,22] It does recapitulate some behavioral aspects of the human phenotype, particularly hyperactivity and defects in learning.

Diagnosis

Patients are typically referred for developmental delay, and because of the relatively mild phenotype in young patients, diagnosis is rarely achieved on clinical grounds alone. Any family with an X-linked pattern of mental retardation should be considered for fragile X testing. Cytogenetic testing for the presence of the fragile site was the only diagnostic test available until the discovery of the gene and mutation. Because the CGG repeat expansion mutation is found in virtually all fragile X patients, the use of DNA testing for diagnosis is highly reliable and has largely supplanted the use of cytogenetic testing for the fragile site. However, because of the prevalence of mental retardation resulting from other cytogenetic abnormalities, it is important to consider a routine karyotype in the evaluation of any case of developmental delay.

Molecular Genetics

Mutations

The vast majority (> 99.5 percent) of fragile X patients carry the same mutation, which is a massive expansion of a trinucleotide repeat (CGG) located in the 5′ untranslated region of the *FMR1* gene. This repeat is polymorphic in the human population and outside of fragile X families ranges in length from as few as five to as many as 50 triplets.[23] The most common alleles carry either 29 or 30 repeats. Approximately 70 percent of females are heterozygous at this locus, and normal-sized alleles are transmitted with high fidelity from parent to child.

Affected individuals are found to carry more than 200 repeats and typically exhibit more than 500 repeats[24]; there can be as many as 2,000. They are often mosaic in length; that is, many different lengths are observed in a single sample from an individual, resulting from the presence of different lengths in different cells within the sample (length mosaicism). These large numbers of repeats are termed *full mutations* and are usually found to be methylated at C residues within the repeat sequence as well as in the nearby CG island that marks the gene's promoter region. Methylation has been found to correlate with loss of mRNA production, presumably through diminished transcriptional initiation.[25,26] This results in loss of the product of the *FMR1* gene (FMRP) and the disease. Thus, the expansion mutation in fragile X syndrome leads to loss of function of the *FMR1* gene.

Loss-of-function mutations not associated with CGG expansions have also been identified in fragile X syndrome. These confer the same phenotype.[27] Other genes are unlikely to be affected by methylation in the region. Several patients with deletions of the gene and surrounding sequences show no added phenotypic features, suggesting that nearby genes are not involved.[28-34] To date, a single patient with a missense mutation has been identified.[35] This patient is much more severely affected, with an IQ estimated below 20, massive macro-orchidism, and other dysmorphic features. The mutation found in this patient alters an isoleucine residue found in one of the likely functional domains of the protein, which may account for the severity of the disease. The patient's cells express a mutant FMR1 protein, which may have a gain of function or dominant negative biochemical defect caused by the amino acid change.

Mosaicism in the pattern of methylation is seen in some patients, and methylation status most closely correlates with disease severity.[36] However, the extent of variation of the mutation among tissues in a single individual complicates such correlation studies, which typically rely on blood leukocytes for DNA testing. Evidence indicates that translation of mRNAs carrying long (≥ 200) CGG repeats is also diminished, suggesting that large, unmethylated alleles may also confer the disease.[37]

Premutations

Alleles of a size intermediate between those found in the general population and those in affected individuals have been termed *premutations*. These are found in unaffected carriers (both male and female) of the disorder in fragile X families. No instance of expansion to full mutation size has been identified to arise from a normal allele, and premutations appear to be maintained for many generations in fragile X lineages without selective disadvantage. Premutations can range in length from 44 to ~ 200 repeats and are typically found to be between 70 and 100 repeats. They are not methylated in males, although like normal alleles, they are subject to methylation associated with X inactivation in females.

Premutations are found to change in size in nearly all transmissions from parent to offspring, yet they are usually somatically stable. Thus, they exhibit an extremely high mutation frequency. Mutations can take the form of changes within the premutation size range, or they can involve massive expansions (transitions) to the disease-causing full mutation [*see Figure 1*]. However, these latter events are found to occur exclusively upon transmission from mothers to their children, never from fathers to their daughters. This sex specificity had been recognized in empirical pedigree studies of fragile X families, along with the tendency for the probability of this event increasing in subsequent generations of a fragile X family.[38,39] The molecular genetic basis for the female transition specificity is as yet unknown,[40,41] although the most recent studies point to selection for reduction of full mutations to premutations in the male germline as one factor. The increasing likelihood of the disease in a family can be accounted for by the observation of increasing risk of transition to the full mutation with escalating size of premutation in a female carrier, coupled with the tendency for the repeats to increase in size while transmitted in the premutation size range [*see Table 1*].[23] This unstable character of the CGG repeat explains most of the peculiar (nonmendelian) inheritance patterns found in fragile X syndrome. Similar phenomena in other human genetic disorders caused by unstable triplet repeats can also explain the deviations from mendelian inheritance in those diseases[42,43] [*see Chapter 3*].

Instability of the *FMR1* CGG repeat has been found to increase as the number of repeats is increased; however, interrupting AGG triplets have been found to play an important role in maintaining stability of the repeat[44,45] both in human pedigrees and in evolutionary time.[46] A threshold of ~ 35 uninterrupted CGG repeats seems to mark the transition from reasonably stable to completely unstable transmissions and explains the observation of stable alleles with longer repeat tracts than some unstable alleles. The *FMR1* gene has conserved the CGG repeats throughout mammalian evolution, suggesting that these have a specific function in the gene. A role of specific DNA binding proteins at the CGG site suggests one possibility for function of the gene.[47]

a

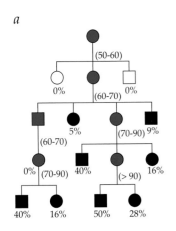

b

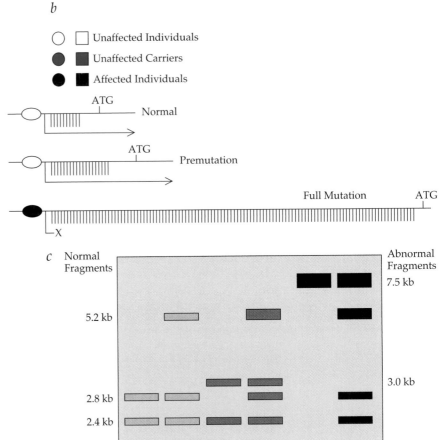

Figure 1 *(a) Numbers in parentheses represent increasing length of CGGs with passage between generations, and correlation with increased risk of mental retardation (percentages). Correlation with repeat lengths is from observed risks of expansion to full mutation (see Table 1). (b) The first exon of the FMR1 gene is depicted in normal, premutated, and fully mutated states. The regions show position and lengths of CGG repeats relative to the promoter (ovals), transcript (arrows), and start codon (ATG). The normal exon is unmethylated and has 30 repeats. The premutation exon is unmethylated and has 90 repeats. Transcription is abolished (X) in the fully mutated and methylated (black oval) exon, which typically has more than 200 repeats. (c) Shown are typical Southern blot assays of hybridization results in normal (light gray), premutation (dark gray), and full mutation (black) samples from males and females using the pE5.1 probe after digestion with ECO RI and Bss HII. Band sizes are indicated in kilobases (kb). Broader fragments represent mosaic, full-mutation bands.*

Prevalence of the Premutation

The high incidence of fragile X syndrome, coupled with the rather unusual inheritance pattern of premutation alleles, suggests a very high prevalence of premutations in the general population. Small-scale studies have suggested allele frequencies of 1:500 to 1:600. A large 1995 study found a carrier frequency for premutations (defined as > 55 repeats) of 1:259 women (~ 1:500 X chromosomes).[48] This finding underscores the impact of this common genetic disorder.

FMR1 Gene, Protein, and Related Genes

The sequence of the *FMR1* gene and substantial flanking DNA has been determined. The gene spans 38 kilobases (kb), is transcribed in a proximal to distal direction, and is divided into 17 exons [*see Figure 2*]. To date, no other genes have been described in the immediate vicinity, and the closest gene known (*FMR2*) is 600 kb distal (telomeric). A variety of alternatively spliced transcripts has been observed in humans and mice.[49,50] These lead to a number of potential protein isoforms, some of which have been identified as separate bands on Western blots ranging in size from 67 to 80 kilodaltons (kd).[51]

The protein product of the *FMR1* gene, FMRP, has been found in many adult and fetal tissues but is concentrated in central nervous system neurons and seems to be absent in glial populations. High levels of FMRP are

Table 1   Risk of Full Mutation of the FMR1 Gene

Repeat Number	Maternal Carrier	Paternal Carrier
< 60	< 1%	0%
61–70	17%	0%
71–80	71%	0%
81–90	86%	0%
> 91	> 99%	0%

also found in cells of the testis, primarily in spermatogonia. These two sites of expression are consistent with the major phenotypic aspects of the disorder. Widespread expression is observed in both human and mouse embryos.[18,19]

Within the cell, FMRP appears predominantly cytoplasmic; however, isoforms lacking exon 14 seem to be limited to the nucleus.[52,53] This finding is consistent with a nuclear export signal found in exon 14.[52,53] In the cytoplasm, the protein appears to interact with ribosomes[54-56] and may form complexes with two proteins of strikingly similar amino acid sequence, FXR1P and FXR2P.[55,57-59]

Subsequent to the original isolation of *FMR1*, a class of proteins with RNA-binding activity was defined by sequence similarity; the hnRNP K homology (KH) domain and RGG (Arg-Gly-Gly) box were found shared among a growing family of proteins characterized from organisms ranging from archeabacteria to humans.[60-62] The presence of two KH domains and RGG boxes led to the hypothesis that *FMR1* is an RNA-binding protein.[63]

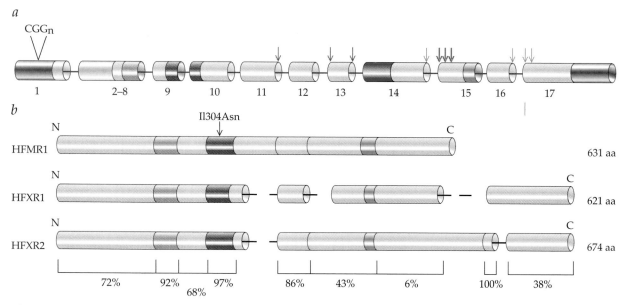

Figure 2   (a) Diagram of the FMR1 gene shows exons 1 to 17, functional domains KH1 (grayish blue) and KH2 (slate blue), RGG boxes (medium gray), and nuclear localization (NLS) (light gray) and export (NES) (blue) signals. Arrows indicate alternative splicing sites. Dark-gray areas indicate noncoding sequences. (b) Similarities among human FMR1, FXR1, and FXR2 genes are indicated by the positions of their respective KH1 and KH2 functional domains and RGG boxes. The position of the only known missense mutation in a case of fragile X syndrome (Ile304Asn) is indicated by an arrow. N is the N-terminus. C is the C-terminus. aa indicates amino acids.

Two studies demonstrate in vitro RNA-binding activity for *FMR1*.[64,65] Ashley and colleagues also found that FMRP appears to have a selectivity for a fraction of mRNAs expressed in brain,[64] including its own message. RNA binding has been disrupted by deletions of the C-terminus of the protein, which ablate the RGG sequences.[65]

The patient carrying the missense mutation in *FMR1* [35] has an Ile residue changed to Asn at position 304 of the most common isoform. The normal Ile in this position is conserved in proteins carrying a KH domain, suggesting a critical role for this residue in the function of the domain. This mutant protein shows reduced RNA-binding activity[56,66] while retaining its ability to form protein-protein interactions with FXR1P and FXR2P.[55] Evidence derived from structural analysis of another member of the KH family, human vigilin, suggests that this mutation prevents the protein from folding properly.[67]

FXR1P and FXR2P also contain KH domains and RGG boxes, and RNA binding has been demonstrated in vitro for each.[54,55,58] Specificity of binding has yet to be tested for FXR1P and FXR2P; however, their ability to interact with *FMR1* as partners in yeast two-hybrid analysis and in coimmunoprecipitation suggests the likelihood that functional multimers are present in the large (> 600 kd) complexes in which FMRP is found in vivo.[56] High-level expression of *FXR1* is found in heart, skeletal muscle, and testis, although it appears to be present in all tissues tested by Northern and PCR analyses.[57,58] FXR2P shows an affinity for FMRP,[59] suggesting it may have a more significant role in interacting with FMRP. *FXR1* and *FXR2* are located on autosomes at 3q28 and 17p13.1, respectively.

Homologs of all three members of this gene family have been identified from a variety of vertebrates, and these show much higher levels of amino acid conservation than is typical for orthologous proteins at these evolutionary distances.[68] Human, mouse, chicken, and *Xenopus laevis* complementary DNAs (cDNAs) have been characterized for *FMR1*. Amino acid identities between human and chicken coding sequences are 92 percent,[68] and similarly high conservation is seen between human and *Xenopus FMR1*.[58] Strong homologies are also noted between distant species' *FXR1* and *FXR2* genes.[59] Similarity among the three proteins is higher in the N-terminus, averaging ~ 68 percent, although there are regions of very high similarity (> 90 percent identity in KH domains). One curious finding is high levels of nucleic acid conservation in 5′ and 3′ untranslated regions of orthologs, suggesting a conserved regulatory function.[57,58,68]

FMR1 Function

The emerging evidence that the FMR1 protein plays a role in cytoplasmic mRNA metabolism, possibly in mRNA targeting or control of translation,[52,54-56] is somewhat at odds with the pathophysiology of fragile X syndrome. The disorder has profound effects on intellect, yet few structural or pathological abnormalities are found, suggesting that development is essentially normal. It might be expected that loss of a protein involved in a fundamental process such as RNA transport or translational control would lead to much more widespread anomalies. It may therefore be the case that for most tissues, sufficient redundancy of function is provided by other similar proteins (e.g., *FXR1* and *FXR2* products), and that only neurons find absence of

FMR1 protein to be problematic. A knockout mouse model of fragile X syndrome is similarly mildly affected, with subtle learning defects accompanied by enlarged testes.[21] Thus, the very highly conserved FMR1 protein can be regarded as dispensable for normal growth and structural development (if not normal intellectual development). It remains to be discovered how defects in this function might lead to mental retardation.

An intriguing lead is provided by work on the structural and functional aspects of neurons. Neurons exhibit considerable asymmetry in mRNA localization, and a number of mRNAs have been found localized to dendrites, including those for proteins that might be expected to respond to postsynaptic signals, such as the α-subunit of Ca^{2+}/calmodulin-dependent kinase II.[69,70] Ribosomes are found in postsynaptic regions of dendrites, and translation occurs in these regions,[71] possibly under the control of synaptic transmission.[72,73] Weiler and colleagues have demonstrated localization of the mRNA for *FMR1* in cDNA libraries prepared from dendrites,[74] and stimulation-responsive translation. The FMR1 protein is present in rat hippocampal dendrites at the same locations that neuronal polysomes are observed (spine necks, heads, dendritic branch points).[74,75] These observations may relate to the abnormalities in dendritic spine architecture in fragile X patients[17] and in *FMR1* knockout mice.[76] If the FMR1 protein is involved in delivering mRNAs to ribosomes, as is suggested by its nuclear/cytoplasmic shuttling and ribosome associations, could it be responsible for targeting RNAs to dendritic ribosomes? Could the FMR1 protein act to mask mRNAs in dendrites until a signal (e.g., a postsynaptic response) is received? Any of these functions individually or in combination might be expected to result in a defect in learning because synaptic plasticity involves intracellular signaling requiring both translation and transcription.[77,78] If the FMR1 protein is receiving signals, what are they?

One possible signal would involve protein dimer/oligomer formation, with homomeric multimers having different affinities for RNA or ribosomes than heteromeric multimers, suggesting a role for *FXR1* or *FXR2*. Protein phosphorylation is a common signal transducer; however, no evidence for phosphorylation of FMR1 has been found. An alternative secondary modification of FMR1 that could transduce a signal might be arginine methylation. The RGG boxes of RNA binding proteins, including FMR1, are good substrates for arginine methylase.[79] Because this is a conserved motif among RNA-binding proteins, it is possible that functional aspects of RNA-protein interactions might be mediated by this activity; however, no evidence exists at present. Although FMR1 protein might be directly involved in signal transduction associated with synaptic plasticity, the phenotypic consequences may be equally significant if absence of the protein led to mislocalization of a set of mRNAs normally located in dendrites [*see Figure 3*]. The finding of an *FMR1* missense mutation with a more severe phenotype might then be explained by gain of function of the protein such that it interferes with the localization or signal response processes through aberrant interactions with cofactors such as interacting proteins, ribosomes, and mRNAs.

FMR2 and FRAXF

Two additional rare, folate-sensitive fragile sites, FRAXE and FRAXF, are found distal to FRAXA, the site in the *FMR1* gene responsible for fragile X

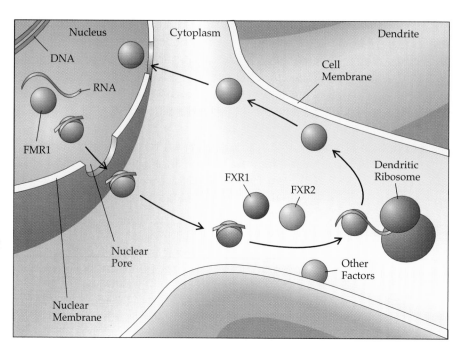

Figure 3   *FMR protein joins with RNA in the cell nucleus and transports the RNA through the nuclear membrane and through the cytoplasm to the dendritic ribosomes. Cofactors FXR1, FXR2, and possibly others may be involved at several steps. It is possible that absence of FMR protein results in mislocalization of RNAs or an inadequate translational response in dendritic ribosomes.*

syndrome [*see Figure 4*]. FRAXE and FRAXF were assumed to be FRAXA by cytogenetic testing and were distinguished only after the development of DNA probes at FRAXA.[80,81] Each of these sites also results from expansion and methylation of CGG repeats,[82,83] and FRAXE is associated with a rare form of mild mental handicap distinguishable from fragile X syndrome.[84] FRAXF appears to cause no abnormalities in individuals carrying expanded repeats.

A gene (*FMR2*) whose expression is reduced by FRAXE CGG repeat expansion has been identified.[85,86] This gene is quite large (~ 500 kb, with a transcript of 9.5 kb), and its transcription initiates at the FRAXE site. The repeat is likely to be carried in the 5′ untranslated sequence of the mRNA in a manner identical to that of *FMR1*; however, the repeat orientation is opposite, and in the reading frame of the mRNA would be a GCC repeat. Other, non-CGG expansion mutations in this gene have been found that support a loss of function as the primary mechanism for disease. The repeat expansion mutation is quite rare, probably less than 1:50,000[87,88]; however, the large target size of the gene suggests that other mutations might be found in families with apparent X-linked mental retardation. Scanning for mutations in such a large gene represents a significant technical challenge. Expression of *FMR2* is limited to brain in adult tissues, and within the brain, hippocampus shows highest expression.[89] The repeat expansion silences expression when methylated in fibroblasts.[90] The gene product bears no resemblance to the *FMR1* gene, and its function is not suggested by its sequence, although it shares similarity with transcriptional activators.

Other Trinucleotide Disorders

Since the identification of the fragile X CGG repeat, more than one dozen additional human genetic disorders have been found that are attributable to expanded trinucleotide repeat sequences. These include the CAG repeats found expressed as polyglutamine-encoding stretches in a number of neuro-degenerative disorders with late age of onset, such as Huntington's disease,[91] and the recent intronic GAA repeat expansion found in Friedreich's ataxia[92] [*see Chapters 1 and 3*]. A CTG repeat sequence in the 3′ untranslated region of the myotonin protein kinase gene has been found to confer myotonic muscular dystrophy when expanded to hundreds or thousands of triplets.[93] Of relevance to this discussion is the finding of diminished mentation in some individuals with myotonic dystrophy, and that this aspect of the phenotype correlates with repeat length.[94] Clearly, trinucleotide repeat mutations have a significant role in neurologic and neuromuscular disorders. It remains to be seen whether this class of mutations will prove significant in additional genetic forms of intellectual deficit (beyond the three known to date) or possibly in the neuropsychiatric diseases of adolescent or adult onset.[95]

Management and Treatment

Currently, a variety of educational interventions are employed to treat children who have fragile X. These can have significant benefits.[96] Pharmacological treatment has yet to be standardized, and trials of a variety of psychoactive substances, particularly serotonin reuptake inhibitors, are ongoing. Additional understanding of the normal function of the FMR1 protein may provide suggestions for effective drug intervention. The potential for treating the disease through gene therapy is tempered by the observation of affected females, which strongly suggests that the gene product acts in a cell-autonomous fashion. In females, the random pattern of X inactivation found in the critical tissue (presumably neurons) affects the intellectual potential of women carrying full mutations. This has significant bearing on the potential for treatment with gene therapy because this modality would necessitate targeting of neurons and would require a rather high efficiency (50 percent or higher) to exceed the average number of expressing neurons found in female patients.

Future Directions

Of the large number of human genetic disorders known to be caused by expanding trinucleotide repeats, fragile X syndrome is the best characterized with respect to the effect of the mutation on the gene. The challenges that remain, however, are large. They include development of a better understanding of the natural history of the fragile X mutation and improved methods for its detection. With improved tests, the potential for popula-

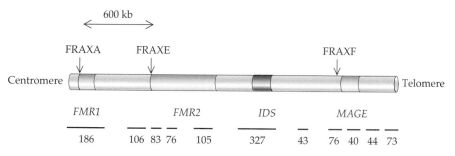

Figure 4 *Three fragile sites, FRAXA, FRAXE, and FRAXF, are shown in the Xq27.3-q28 region of the X chromosome. The two genes associated with FRAXA and FRAXE are FMR1 and FMR2, respectively. IDS indicates iduronate sulfatase, the gene defective in Hunter's syndrome. MAGE indicates one of several members of a family of genes in proximal Xq28 found expressed on the surface of melanomas and other tumors. The normal functions of these genes are unknown. The numbered (kilobase pairs) segments indicate sequenced regions. Distances are approximate.*

tion-based screening for this common mutation will be a reality; however, along with such a scheme comes a significant challenge in providing meaningful information to those at risk. The very dynamic nature of this mutation is confusing to professionals, and this can be quite difficult to convey to individuals at risk. Given the high frequency of the disease and its profound cost to families and society, the demand for population-based screening will be high. It will be vital to develop inexpensive and accurate tests in support of such efforts, along with effective vehicles to convey the information to the public.

Clearly, a large focus on the function of the *FMR1* gene product is required for developing rational treatment strategies. What is the role of *FMR1* in the development of neuronal circuits? What are the consequences of its absence, and how can these be modulated? Despite six years of attention to this interesting protein, these investigations remain in their infancy. The added benefit of such studies may be an improved understanding of fundamental cell biologic processes, possibly ones integral to learning and memory.

References

1. Boyle C, Yeargin-Allsopp M, Doernberg N, et al: Prevalence of selected developmental disabilities in children 3-10 years of age: the Metropolitan Atlanta Developmental Disabilities Surveillance Program, 1991. *MMWR CDC Surveill Summ* 45:1, 1996

2. Lehrke R: Theory of X-linkage of major intellectual traits. *Am J Ment Defic* 76:611, 1972

3. Turner G: Intelligence and the X chromosome. *Lancet* 347:1814, 1996

4. Martin JP, Bell J: A pedigree of mental defect showing sex-linkage. *J Neurol Psychiatry* 6:154,1943

5. Lubs HA: A marker X chromosome. *Am J Hum Genet* 21:231, 1969

6. Sutherland GR: Fragile sites on human chromosomes: demonstration of their dependence on the type of tissue culture medium. *Science* 197:265, 1977

7. Verkerk AJMH, Pieretti M, Sutcliffe JS, et al: Identification of a gene (*FMR-1*) containing a CGG repeat coincident with a breakpoint cluster region exhibiting length variation in fragile X syndrome. *Cell* 65:905, 1991

8. Yu S, Pritchard M, Kremer E, et al: Fragile X genotype characterized by an unstable region of DNA. *Science* 252:1179, 1991

9. Oberlé I, Rousseau F, Heitz D, et al: Instability of a 550-base pair DNA segment and abnormal methylation in fragile X syndrome. *Science* 252:1097, 1991

10. Warren ST: The expanding world of trinucleotide repeats. *Science* 271:1374, 1996

11. Webb TP, Bundey SE, Thake AI, et al: Population incidence and segregation ratios in the Martin-Bell syndrome. *Am J Hum Genet* 23:573, 1986

12. Morton J, Bundey S, Webb T, et al: Fragile X syndrome is less common than previously estimated. *J Med Genet* 34:1, 1997

13. Turner G, Daniel A, Frost M: X-linked mental retardation, macro-orchidism, and the Xq27 fragile site. *J Pediatr* 96:837, 1980

14. Lachiewicz AM: Females with fragile X syndrome: a review of the effects of an abnormal FMR1 gene. *Ment Retard Dev Disabil Res Rev* 1:292, 1995

15. Johannisson R, Rehder H, Wendt V, et al: Spermatogenesis in two patients with the fragile X syndrome I. Histology: light and electron microscopy. *Hum Genet* 76:141, 1987

16. Abrams MT, Reiss AL: The neurobiology of fragile X syndrome. *Ment Retard Dev Disabil Res Rev* 1:269, 1995

17. Rudelli RD, Brown WT, Wisniewski HM: Adult fragile X syndrome cliniconeuropathologic findings. *Acta Neuropathol (Berl)* 67:289, 1985

18. Hinds HL, Ashley CT, Sutcliffe JS, et al: Tissue specific expression of FMR-1 provides evidence for a functional role in fragile X syndrome. *Nat Genet* 3:36, 1993

19. Abitbol M, Menini C, Delezoide A-L, et al: Nucleus basalis magnocellularis and hippocampus are the major sites of FMR-1 expression in the human fetal brain. *Nat Genet* 4:147, 1993

20. Devys D, Lutz Y, Rouyer N, et al: The FMR-1 protein is cytoplasmic, most abundant in neurons and appears normal in carriers of a fragile X premutation. *Nat Genet* 4:335, 1993

21. Bakker CE, Verheij C, Willemsen R, et al: Fmr1 knockout mice: a model to study fragile X mental retardation. *Cell* 78:23, 1994

22. Willems PJ, Reyniers E, Oostra BA: An animal model for fragile X syndrome. *Ment Retard Dev Disabil Res Rev* 1:298, 1995

23. Fu YH, Kuhl DPA, Pizzuti A, et al: Variation of the CGG repeat at the fragile X site results in genetic instability: resolution of the Sherman paradox. *Cell* 67:1047, 1991

24. Rousseau F, Heitz D, Biancalana V, et al: Direct diagnosis by DNA analysis of the fragile X syndrome of mental retardation. *N Engl J Med* 325:1673,1991

25. Pieretti M, Zhang F, Fu YH, et al: Absence of expression of the *FMR-1* gene in fragile X syndrome. *Cell* 66:817, 1991

26. Sutcliffe JS, Nelson DL, Zhang F, et al: DNA methylation represses *FMR-1* transcription in fragile X syndrome. *Hum Mol Genet* 1:397, 1992

27. Lugenbeel KA, Carson NL, Chudley AE, et al: Absence of FMR1 protein in two mentally retarded fragile X males without CGG repeat expansion. *Am J Hum Genet* 55:A230, 1994

28. Wöhrle D, Kotzot D, Hirst MC, et al: A microdeletion of less than 250 kb, including the proximal part of the FMR-1 gene and the fragile-X site, in a male with the clinical phenotype of fragile-X syndrome. *Am J Hum Genet* 51:299, 1992

29. Gu Y, Lugenbeel K, Vockley J, et al: A de novo deletion in FMR1 in a patient with developmental delay. *Hum Mol Genet* 3:1705, 1994

30. Hirst M, Grewal P, Flannery A, et al: Two new cases of FMR1 deletion associated with mental impairment. *Am J Hum Genet* 56:67, 1995

31. Quan F, Zonana J, Gunter K, et al: An atypical case of fragile X syndrome caused by a deletion that includes the FMR1 gene. *Am J Hum Genet* 56:1042, 1995

32. Meijer H, de Graaff E, Merckx D, et al: A deletion of 1.6 kb proximal to the CGG repeat of the FMR1 gene causes the clinical phenotype of the fragile X syndrome. *Hum Mol Genet* 3:615, 1994

33. Gedeon A, Baker E, Robinson H, et al: Fragile X syndrome without CGG amplification has an FMR1 deletion. *Nat Genet* 1:341, 1992

34. Tarleton J, Richie R, Schwartz C, et al: An extensive de novo deletion removing FMR1 in a patient with mental retardation and the fragile X syndrome phenotype. *Hum Mol Genet* 2:1973, 1993

35. De Boulle K, Verkerk AJMH, Reyniers E, et al: A point mutation in the FMR-1 gene associated with fragile X mental retardation. *Nat Genet* 3:31, 1993

36. McConkie-Rosell A, Lachiewicz AM, Spiridigliozzi GA, et al: Evidence that methylation of the FMR-1 locus is responsible for variable phenotypic expression of the fragile X syndrome. *Am J Hum Genet* 53:800, 1993

37. Feng Y, Zhang F, Lokey LK, et al: Translational suppression by trinucleotide repeat expansion at FMR1. *Science* 268:731, 1995

38. Sherman SL, Morton NE, Jacobs PA, et al: The marker (X) syndrome: a cytogenetic and genetic analysis. *Ann Hum Genet* 48:21, 1984

39. Sherman SL, Jacobs PA, Morton NE, et al: Further segregation analysis of the fragile X syndrome with special reference to transmitting males. *Hum Genet* 69:289, 1985

40. Nelson DL, Warren ST: Trinucleotide repeat instability: when and where? *Nat Genet* 4:107, 1993

41. Malter H, Iber J, Willemsen R, et al: Characterization of the full fragile X syndrome mutation in fetal gametes. *Nat Genet* 15:165, 1997

42. Richards RI, Sutherland GR: Dynamic mutations: a new class of mutations causing human disease. *Cell* 70:709, 1992

43. Nelson DL. Six human genetic disorders involving mutant trinucleotide repeats. *Genome Rearrangement and Stability*. Davies KE, Warren ST, Eds. Cold Spring Harbor Laboratory Press, Cold Spring Harbor, NY, 1993, p 1

44. Eichler EE, Holden JJA, Popovich BW, et al: Length of uninterrupted CGG repeats determines instability in the FMR1 gene. *Nat Genet* 8:88, 1994

45. Kunst CB, Warren ST: Cryptic and polar variation of the fragile X repeat could result in predisposing normal alleles. *Cell* 77:853, 1994

46. Eichler EE, Kunst CB, Lugenbeel KA, et al: Evolution of the cryptic FMR1 CGG repeat. *Nat Genet* 11:301, 1995

47. Richards RI, Holman K, Yu S, et al: Fragile X syndrome unstable element, p(CCG)n, and other simple tandem repeat sequences are binding sites for specific nuclear proteins. *Hum Mol Genet* 2:1429, 1993

48. Rousseau F, Rouillard P, Morel M-L, et al: Prevalence of carriers of premuation-size alleles of the FMR1 gene—and implications for the population genetics of the fragile X syndrome. *Am J Hum Genet* 57:1006, 1995

49. Verkerk AJMH, de Graaff E, De Boulle K, et al: Alternative splicing in the fragile X gene FMR1. *Hum Mol Genet* 2:399, 1993

50. Ashley CT, Sutcliffe JS, Kunst CB, et al: Human and murine FMR-1: alternative splicing and translational initiation downstream of the CGG-repeat. *Nat Genet* 4:244, 1993

51. Small K, Warren ST: Analysis of FMRP, the protein deficient in fragile X syndrome. *Ment Retard Dev Disabil Res Rev* 1:245, 1995

52. Eberhart DE, Malter HE, Feng Y, et al: The fragile X mental retardation protein is a ribonucleoprotein containing both nuclear localization and nuclear export signals. *Hum Mol Genet* 5:1083, 1996

53. Sittler A, Devys D, Weber C, et al: Alternative splicing of exon 14 determines nuclear or cytoplasmic localisation of FMR1 protein isoforms. *Hum Mol Genet* 5:95, 1996

54. Khandjian EW, Corbin F, Woerly S, et al: The fragile X mental retardation protein is associated with ribosomes. *Nat Genet* 12:91, 1996

55. Siomi MC, Zhang Y, Siomi H, et al: Specific sequences in the fragile X syndrome protein FMR1 and the FXR proteins mediate their binding to 60S ribosomal subunits and the interactions among them. *Mol Cell Biol* 16:3825, 1996

56. Tamanini F, Meijer N, Verheij C, et al: FMRP is associated to the ribosomes via RNA. *Hum Mol Genet* 5:809, 1996

57. Coy JF, Sedlacek Z, Bachner D, et al: Highly conserved 3' UTR and expression pattern of FXR1 points to a divergent gene regulation of FXR1 and FMR1. *Hum Mol Genet* 4:2209, 1995

58. Siomi MC, Siomi H, Sauer WH, et al: FXR1, an autosomal homolog of the fragile X mental retardation gene. *EMBO J* 14:2401, 1995

59. Zhang Y, O'Connor P, Siomi M, et al: The fragile X mental retardation syndrome protein interacts with novel homologs FXR1 and FXR2. *EMBO J* 14:5358, 1995

60. Burd CG, Dreyfuss G: Conserved structures and diversity of functions of RNA-binding proteins. *Science* 265:615, 1994

61. Mattaj IW: RNA recognition: a family matter? *Cell* 73:837, 1993

62. Siomi H, Matunis MJ, Michael WM, et al: The pre-mRNA binding K protein contains a novel evolutionarily conserved motif. *Nucleic Acids Res* 21:1193, 1993

63. Gibson TJ, Rice PM, Thompson JD, et al: KH domains within the FMR1 sequence suggest that fragile X syndrome stems from a defect in RNA metabolism. *TIBS* 18:331, 1993

64. Ashley CT Jr., Wilkinson KD, Reines D, et al: FMR1 protein contains conserved RNP-family domains and demonstrates selective RNA binding. *Science* 262:563, 1993

65. Siomi H, Siomi MC, Nussbaum RL, et al: The protein product of the fragile X gene, FMR1, has characteristics of an RNA-binding protein. *Cell* 74:291, 1993

66. Siomi H, Choi M, Siomi M, et al: Essential role for KH domains in RNA binding: impaired RNA binding by a mutation in the KH domains of FMR1 that causes fragile X syndrome. *Cell* 77:33, 1994

67. Musco G, Stier G, Joseph C, et al: Three-dimensional structure and stability of the KH domain: molecular insights into the fragile X syndrome. *Cell* 85:237, 1996

68. Price DK, Zhang F, Ashley JCT, et al: The chicken FMR1 gene is highly conserved with a CCT 5' untranslated repeat and encodes an RAN-binding protein. *Genomics* 31:3, 1996

69. St Johnston D: The intracellular localization of messenger RNAs. *Cell* 81:161, 1995

70. Steward O: mRNA localization in neurons: a multipurpose mechanism? *Neuron* 18:9, 1997

71. Steward O: Dendrites as compartments for macromolecular synthesis. *Proc Natl Acad Sci USA* 91:10766, 1994

72. Weiler IJ, Greenough WT: Metabotropic glutamate receptors trigger postsynaptic protein synthesis. *Proc Natl Acad Sci USA* 90:7168, 1993

73. Weiler IJ, Greenough WT: Potassium ion stimulation triggers protein translation in synaptoneurosomal polyribosomes. *Mol Cell Neurosci* 2:305, 1991

74. Weiler I, Irwin S, Klintsova A, et al: Fragile X mental retardation protein is translated near synapses in response to neurotransmitter activation. *Proc Natl Acad Sci USA* 94:5395, 1997

75. Feng Y, Gutekunst C-A, Eberhart DE, et al: Fragile X mental retardation protein: nucleo-cytoplasmic shuttling and association with somatodendritic ribosomes. *J Neurosci* 17:1539, 1997

76. Comery T, Harris J, Willems P, et al: Abnormal dendritic spines in fragile X knockout mice: maturation and pruning deficits. *Proc Natl Acad Sci USA* 94:5401, 1997

77. Link W, Konietzko U, Kauselmann G, et al: Somatodendritic expression of an immediate early gene is regulated by synaptic activity. *Proc Natl Acad Sci USA* 92:5734, 1995

78. Frey U, Morris RGM: Synaptic tagging and long-term potentiation. *Nature* 385:533, 1997

79. Liu Q, Dreyfuss G: In vivo and in vitro arginine methylation of RNA-binding proteins. *Mol Cell Biol* 15:2800, 1995

80. Sutherland GR, Baker E: Characterisation of a new rare fragile site easily confused with the fragile X. *Hum Mol Genet* 1:111, 1992

81. Hirst MC, Barnicoat A, Flynn G, et al: The identification of a third fragile site, FRAXF in Xq27-28 distal to both FRAXA and FRAXE. *Hum Mol Genet* 2:197, 1993

82. Knight SJL, Flannery AV, Hirst MC, et al: Trinucleotide repeat amplification and hyper-methylation of a CpG island in FRAXE mental retardation. *Cell* 74:127, 1993

83. Parrish JE, Oostra BA, Verkerk AJMH, et al: Isolation of a GCC repeat showing expansion in FRAXF, a fragile site distal to FRAXA and FRAXE. *Nat Genet* 8:229, 1994

84. Mulley JC, Yu S, Loesch DZ, et al: FRAXE and mental retardation. *J Med Genet* 32:162, 1995

85. Gu Y, Shen Y, Gibbs RA, et al: Identification of FMR2, a novel gene associated with the FRAXE CGG repeat and CpG island. *Nat Genet* 13:109, 1996

86. Gecz J, Gedeon AK, Sutherland GR, et al: Identification of the gene FMR2, associated with FRAXE mental retardation. *Nat Genet* 13:105, 1996

87. Allingham-Hawkins DJ, Ray PN: FRAXE expansion is not a common etiological factor among developmentally delayed males. *Am J Hum Genet* 56:72, 1995

88. Brown WT: The FRAXE syndrome: is it time for routine screening? *Am J Hum Genet* 58:903, 1996

89. Chakrabarti L, Knight SJL, Flannery AV, et al: A candidate gene for mild mental handicap at the FRAXE fragile site. *Hum Mol Genet* 5:275, 1996

90. Gecz J, Oostra B, Hockey A, et al: FMR2 expression in families with FRAXE mental retardation. *Hum Mol Genet* 6:435, 1997

91. The Huntington's Disease Collaborative Research Group: A novel gene containing a trinucleotide repeat that is expanded and unstable on Huntington's disease chromosomes. *Cell* 72:971, 1993

92. Campuzano V, Montermini L, Molto MD, et al: Friedrich's Ataxia: autosomal recessive disease caused by an intronic GAA triplet repeat expansion. *Science* 271:1423, 1996

93. Caskey CT, Pizzuti A, Fu YH, et al: Triplet repeat mutations in human disease. *Science* 256:784, 1992

94. Turnpenny P, Clark C, Kelly K: Intelligence quotient profile in myotonic dystrophy, inter-generational deficit, and correlation with CTG amplification. *J Med Genet* 31:300, 1994

95. Morris AG, Gaitonde E, McKenna PJ, et al: CAG repeat expansions and schizophrenia: association with disease in females and with early age-at-onset. *Hum Mol Genet* 4:1957, 1995

96. Hagerman RJ, Silverman AC: *Fragile X Syndrome: Diagnosis, Treatment and Research.* Johns Hopkins University Press, Baltimore, 1991

Acknowledgments

The author wishes to acknowledge his many friends and colleagues for advice and assistance in preparation of this manuscript, and to extend thanks to the many fragile X families whose samples and support have aided in the studies described. Work described in this manuscript was supported in part by grants from the U.S. National Institutes of Health (R01-HD29256 and P01-GM52982).

Figures 1, 2, and 4   Marcia Kammerer.

Figure 3   Seward Hung.

CHAPTER 3

Huntington's Disease and Other Trinucleotide Repeat Disorders

Anne B. Young, M.D., Ph.D.

The progress of the past 20 years in neurogenetics has brought a number of hereditary neurologic diseases to a new level of understanding. Much of this success can be traced to a 1978 Hereditary Disease Foundation workshop in Los Angeles at which Drs. David Botstein, David Housman, and Ray White, all renowned geneticists, suggested that the gene for Huntington's disease (HD) could be localized and identified using new genetic strategies.[1] They proposed to identify anonymous DNA markers located at reasonably spaced intervals throughout the human genome and then collect DNA samples from large HD families to link the HD gene to the location of one of these anonymous markers. To initiate the project, Dr. Joseph Martin, then chief of neurology at the Massachusetts General Hospital, hired Dr. James Gusella, who had just finished his postdoctoral fellowship with Dr. Housman. The key to success depended not only on the molecular genetics but also on obtaining samples from a large multigenerational family. Dr. Nancy Wexler, then working at the National Institute of Neurological Disease and Stroke, suggested that the study of a large Venezuelan family with HD could serve as the single most important source of samples for the linkage study. Over the next five years, the Venezuelan HD pedigree was confirmed as the largest multigenerational HD family in the world. Numerous blood and skin samples from branches of the pedigree were collected and sent to Dr. Gusella's laboratory.[1] In 1983, years earlier than anticipated, the work paid off when an anonymous marker on chromosome 4p was discovered to be closely linked to the HD gene.[2]

A long, laborious search for the actual gene followed this early linkage success. Investigators skilled in various aspects of yeast, fly, worm, and human genetics were recruited to the Huntington's Disease Collaborative Research Group. Over a 10-year period, a variety of approaches and strategies for homing in on specific genes were defined and refined. During the course of these studies in HD, many other molecular geneticists used these positional cloning strategies to identify other disease genes. In 1993, the actual mutation in the HD gene was finally identified as an unstable, expanded triplet repeat, a type of defect that had been discovered in the previous two years to cause three diseases: fragile X syndrome, myotonic dystrophy, and spinobulbar muscular atrophy (SBMA).[3]

Investigation of the genetic basis of a large number of inherited degenerative diseases of the nervous system is now well advanced. With the advent of modern genetic and molecular techniques, the underlying mutations responsible for many disorders have been identified, and transgenic and cell-based disease models are becoming available. In the next decade, we are likely to learn a great deal about the mechanisms by which these genetic mutations lead to selective neurodegeneration. Armed with knowledge of molecular pathophysiology, successful therapeutic approaches will be forthcoming. This chapter discusses the advances in the genetics and neurobiology of triplet repeat disorders and, in particular, Huntington's disease and related CAG repeat disorders [*see Table 1*].

Triplet Repeat Disorders

In 1991, the genetic defect in fragile X syndrome was identified as an unstable, expanded (CGG)n repeat in the 5'-untranslated (5'-UT) region of the fragile X gene on the X chromosome [*see Chapter 2*].[4-6] In the normal gene, a repeat of 5 to 54 CGG triplets was identified that resulted in no discernible phenotypic variation, but in certain individuals, the repeat was expanded many times to a range of 230 to 4,000. In these individuals, a variable phenotype of mental retardation, large ears, and large testicles was observed. The severity of the disorder was greater in those with larger expansions of the repeat. Individuals with a large expansion displayed fragile X chromosomes on karyotype analysis, hence the name *fragile X syndrome*.

Table 1 Neurologic Diseases Caused by Unstable,
Expanded Trinucleotide Repeats

Disorder	Gene	Repeat	Consequence
Fragile X	*FMR1*	CGG	5'UT-decreased protein
Myotonic dystrophy	*DM*	CTG	3'UT-decreased protein
Friedreich's ataxia	*FRDA*	GAA	Intron 1–decreased protein
Spinobulbar muscular atrophy	*AR*	CAG	Polyglutamine tract
Huntington's disease	*IT15*	CAG	Polyglutamine tract
Spinocerebellar atrophy types 1, 2, and 6	*SCA1, 2, and 6*	CAG	Polyglutamine tract
Machado-Joseph disease (spinocerebellar atrophy type 3)	*SCA3*	CAG	Polyglutamine tract
Dentatorubropallidoluysian atrophy	*DRPLA*	CAG	Polyglutamine tract

Table 2 Neurologic Diseases Caused by Unstable,
Expanded CAG Repeats

Disorder	Protein	Normal Repeats	Mutant Repeats	Transmission
SBMA	Androgen receptor	11–33	40–62	X-linked
HD	Huntingtin	11–34	37–120	4p dominant
SCA1	Ataxin 1	25–36	41–81	6p dominant
SCA2	Ataxin 2	15–24	35–59	12q dominant
MJD (SCA3)	Ataxin 3	13–36	62–82	14q dominant
SCA6	Ataxin 6*	4–16	21–27	19p dominant
DRPLA	Atrophin 1	7–25	49–85	12p dominant

*α_{1A}-subunit of a voltage-dependent calcium channel. Allelic disorder with hereditary paroxysmal cerebellar ataxia and episodic ataxia.
DRPLA—dentatorubropallidoluysian; HD—Huntington's disease; SBMA—spinobulbar muscular atrophy; SCA1—spinocerebellar atrophy type 1; SCA2—spinocerebellar atrophy type 2; SCA3— spinocerebellar atrophy type 3 (Machado-Joseph disease); SCA6—spinocerebellar atrophy type 6.

Shortly thereafter, the genetic defect in myotonic dystrophy was identified as a (CTG)n repeat in the gene for a protein kinase.[7] The size of the expansion again correlated with the severity of the disease. The expansion occurs in the 3′-untranslated (3′-UT) region of the gene and leads to a decreased expression of the protein.

In 1991, the first of the CAG repeat disorders was identified.[8] The gene responsible for the X-linked disorder Kennedy's syndrome, or SBMA, was discovered to be the androgen receptor that normally contains a sequence of (CAG)n in which n varies from 11 to 33 [*see Table 2*]. In SBMA, the n varied from 40 to 62. Subsequently, other adult-onset neurodegenerative disease genes have been found to be caused by unstable, expanded CAG repeats in different genes. These diseases include HD, several of the spinocerebellar ataxias (SCAs), and dentatorubropallidoluysian atrophy (DRPLA).[5,6]

One other triplet repeat disorder involving an intronic GAA repeat has been identified in Friedreich's ataxia. The expanded GAA repeat in intron 1 in a gene encoding the protein frataxin.[5] The expanded repeat leads to underexpression of the protein and subsequent disease. The frataxin protein is targeted to the mitochondria and, when mutated, may lead to an abnormality in energy metabolism or increased free radical formation.[9]

CAG Repeat Disorders

Shortly after the identification of the CAG repeat expansion in the gene for the androgen receptor in patients with Kennedy's syndrome, the mutation causing spinocerebellar ataxia type 1 (SCA1) was found to be an expanded CAG repeat [*see Table 2*]. Subsequently, similar CAG expansions were found in the *HD* gene, the spinocerebellar ataxia type 3 (*SCA3*), or Machado-Joseph disease (*MJD*), gene, and the *DRPLA* gene.[6] More recently, CAG expansions have been discovered in spinocerebellar ataxias types 2, 6, and 7 (SCA2, SCA6, and SCA7).[10-12] Clinicians had previously identified a common set of unusual clinical features in each of these diseases that distinguished them from other inherited disorders [*see Table 3*].

Table 3 Common Features of CAG Trinucleotide
Repeat Disorders

Autosomal dominant (except for spinal and bulbar muscular atrophy–X-linked)	Anticipation
Mid-life onset (25–45 years)	Expansions of repeats correlate with severity and earlier onset
Progression with fatal outcome	Genetic screening possible
Bilateral and symmetric "systems" neuronal loss	

Common Clinical Features Shared by CAG Repeat Disorders

Variable Age of Onset

In each of the above disorders, clinical analysis of the pedigrees demonstrated considerable variability in the age of disease onset [*see Figure 1*]. The mean age of onset for all the CAG repeat disorders occurs in the third and fourth decades of life. In each, however, a small percentage of individuals have onset in the first decade, and a similarly small percentage have onset as late as the eighth or ninth decade.

Anticipation

In many of the pedigrees, a phenomenon called "anticipation" was observed. In these cases, the age of onset within a family appeared to be younger with succeeding generations. Indeed, in certain instances, a child of an individual might have onset prior to that of the affected parent. This phenomenon was most clearly described for HD. Furthermore, in HD, 80 to 90 percent of individuals with the most marked anticipation and juvenile onset inherited the gene from their father.[13] Much less anticipation appeared to occur with maternal descent. Anticipation is also noted in SBMA, MJD, SCA1, SCA2, SCA6, and DRPLA, although anticipation with both maternal and paternal descent is seen in SCA2.[14]

Phenotypic Variation

Marked differences in phenotype are noted in HD and the other CAG repeat disorders, leading clinicians to classify subtypes of many of the diseases, including Westphal's variant of HD; juvenile HD; senile chorea; MJD types 1, 2, and 3; and DRPLA with progressive myoclonic epilepsy. Although it was tempting to speculate that such phenotypic variation is caused by different mutations in the same gene, a confounding factor was that for these disorders, the whole range of phenotypic variation can be seen in a single pedigree. Juvenile onset also appeared to be associated with more severe and widespread clinical symptomatology. For instance, in HD, late onset is associated with more pure chorea and less prominent dystonia, rigidity, and cognitive dysfunction, whereas juvenile onset often is associated with rigidity, dystonia, seizures, and ataxia.

New Mutations

Until the discovery of the HD gene, new mutations were considered to be either nonexistent or very rare. In most foci of HD observed around the world, it seemed possible to trace likely founders to ancestors of European descent. Rare individuals, however, were identified who had the symptoms and neuropathologic hallmarks of the disease but no apparent family history.[10] The genetically confirmed parents of such individuals lived without symptoms of HD into their 70s or 80s. In these families there was no neuropathologic confirmation of lack of disease. With the discovery of the mutation in HD, it was possible to analyze these sporadic cases and their families in detail.[15] Similar phenomena have been observed in the other CAG repeat disorders.

Genetic Characteristics

CAG Repeat Number and Age of Onset

All the CAG repeat disorders show a strong inverse correlation between the age of onset and the size of the CAG repeat.[10] The correlation is highest for large repeat numbers and early age of onset; the correlations are weaker as the repeat number approaches the lower limits associated with clinical disease [*see Figure 1*]. In large data sets, the age-of-onset curve versus CAG repeat size is best fit by an exponential function.[10] Although the correlation is statistically strong, accounting for over 50 percent of the overall variability in age of onset, repeat numbers cannot be used reliably to predict age of onset in a given individual. In HD, for instance, individuals with a repeat number of 40 have been identified with ages of onset from 22 to 75 years.[3] Additional genetic or environmental factors likely influence the age of onset. These specific factors, some of which may powerfully attenuate disease onset and progression, have yet to be identified. Individuals homozygous for the CAG repeat mutation have been identified in HD, MJD, SCA1, SCA2, and DRPLA.[10] In HD, SCA1, and SCA2, the length of the highest repeat versus age of onset in homozygotes falls on the age-of-onset curve generated for heterozygotes [*see Figure 1*]. In contrast, homozygotes for MJD and DRPLA have earlier ages of onset than predicted by the length of the largest CAG repeat.[10] Thus, in HD, SCA1, and SCA2, the clinical onset appears to be dependent only on the contribution of the longest repeat. This suggests a saturation of the pathogenic mechanism; in MJD and DRPLA, the repeats of both alleles appear to contribute to the development of pathology.

Expansion Via Paternal Descent

In HD, comparisons of the repeat number in the offspring of affected males and females demonstrate that instability of the CAG repeat occurs most often in transmission from affected fathers.[16] The affected progeny of females with HD have repeat numbers within ±4 of the maternal repeat number, and the regression of the correlation is close to 1. Among the progeny of affected males, many have repeat numbers close to the paternal number, but many others inherit an expanded CAG repeat number. Within individuals, there is little somatic variation; the repeat number varies by zero to four repeats between tissues and in different brain areas. The remarkable exception is the repeat number in sperm. Sperm samples from affected males sometimes show repeat numbers similar to those seen in their

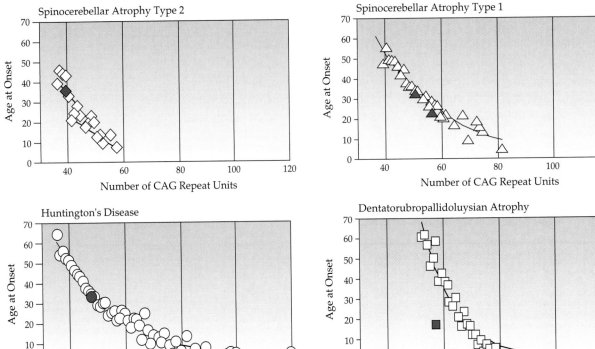

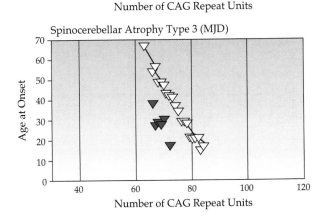

Figure 1   *The age of onset versus the number of CAG repeats is depicted for different CAG repeat disorders. Heterozygotes are represented by the open symbols and homozygotes by the closed symbols. The age of onset of the homozygotes is plotted against the number of CAG repeats in the larger of the two disease alleles. Each panel also shows the curve fit. MJD is Machado-Joseph disease.*

lymphocytes or other tissues. In certain individuals, however, large expansions of the CAG repeat can be found in sperm. Single sperm analyses indicate a broad spectrum of expanded repeat numbers in sperm samples.[17] It also appears that the older the male, the more likely it is to find sperm with an enlarged expansion. The cause of such expansion is currently under intense investigation to ascertain whether environmental factors influence the expansion or perhaps even reduce it.

Similar CAG instability occurs in SCA1, MJD, and DRPLA but not in SBMA, SCA2, or SCA6. The extent of expansion in the ova has not been directly investigated.

New Mutations

New mutations for all the triplet repeat disorders have been identified. The most extensive analyses have been carried out in HD. It was long believed that new HD mutations were very rare or nonexistent. Numerous instances have now been documented, however, in which one unaffected elderly parent of the genetically and clinically identified HD proband has a repeat number slightly below that associated with symptomatic disease but considerably above the mean observed in unaffected individuals.[15,18-20] Siblings of the proband may also have this intermediate repeat but remain asymptomatic. Presumably, the CAG repeat had expanded into this intermediate range in the forebears of the presumed new mutation case but did not cause symptomatic disease. Then, upon passage of the gene to the offspring, a further expansion into the symptomatic range of repeat number occurred. Interestingly, in most such instances, the parent with the intermediate expansion and affected offspring has been the father. Detailed studies of large numbers of HD cases now show that CAG repeats under 28 are stably transmitted to the next generation and are not sufficient to cause symptomatic HD.[21] Repeat numbers in the range of 29 to 35 are not stably transmitted and may be expanded in the next generation but do not lead to symptomatic HD. Repeat numbers of 36 to 39 are also not stably transmitted but may lead to symptomatic HD. (The word *may* is used here because individuals with 36 repeats have been identified with symptomatic HD, and others with 39 repeats have been identified in their 90s without symptoms. Thus, in the range of 36 to 39 repeats, the gene mutation is not fully penetrant.) Repeats equal to or greater than 40 are not stably transmitted and are associated with the development of clinical HD.

Extent of Pathology

Each of the CAG repeat disorders displays a characteristic selective neuronal degeneration.[6,10] In HD, the medium spiny projection neurons of the striatum are the most severely affected, and the pathology progresses in a posteroanterior, dorsoventral, mediolateral manner across the caudate-putamen.[22] The deep layers of cortex are also affected, as are the thalamus and other regions of the basal ganglia.[23,24] In SBMA, the anterior horn cells, brain stem motor nuclei, and to a lesser extent, the dorsal root ganglia are affected. In MJD, the dentate nucleus, anterior horn cells, and dorsal root ganglia are affected, as are the globus pallidus, substantia nigra, and certain brain stem motor nuclei, but the inferior olive is characteristically spared. In SCA1, the cerebellar Purkinje cells, the inferior olive, and certain brain stem motor nuclei degenerate as they do in SCA2. In DRPLA, the dentate nucleus, globus pallidus, red nucleus, subthalamic nucleus, and substantia nigra degenerate.[6,10]

Of all the CAG repeat disorders, only SBMA retains a relatively selective neuronal degeneration even at the higher repeat numbers, presumably because of the low-level expression of the androgen receptor in neurons other than anterior horn cells and brain stem nuclei. For the other CAG repeat disorders, the affected genes are expressed more broadly in the brain.[6] More widespread pathology occurs in individuals with larger repeat numbers [*see Figure 2*]. Thus, juvenile-onset HD is associated not only with more severe pathology in the characteristic structures but also with extensive involvement of the cerebellum, globus pallidus, and cerebral cortex. Juvenile patients are more likely to manifest distinct clinical signs with ataxia, seizures,

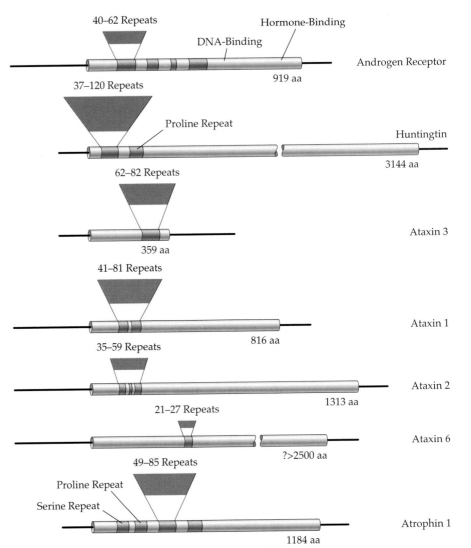

Figure 2 *Selected features of the genes for the CAG repeat disorders are shown, including the relative location of the CAG repeat (blue), the location of other amino acid repeats, and the location of hormone-binding sites. For SCA1 (ataxin 1) and SCA2 (ataxin 2), the CAG repeat in the wild-type allele is shown as interrupted because in each of these disorders, the repeat is not continuous but contains one or more intervening amino acids. In the mutant genes for SCA1 and SCA2, the repeat becomes continuous. The ataxin 3 gene is also known as the MJD-1 gene. The ataxin 6 gene codes for an α_{1A}-subunit of a voltage-gated calcium in the brain.*

rigidity, and dystonia.[25] In DRPLA, juvenile onset with large CAG expansions leads to striatal, cortical, and white-matter pathology; myoclonic epilepsy; and dystonia.[26] In MJD, late adult onset is characterized by amyotrophy, sensory neuropathy, and ataxia, but juvenile-onset patients display prominent choreoathetosis and dystonia superimposed on the characteristic adult-onset symptoms.[27]

The phenomenon of more extensive, widespread, and even overlapping pathology in those individuals with large CAG repeat expansions suggests that common mechanisms likely underlie the neuronal degeneration. The protein context in which the repeat is expressed defines the selective regional pathology at smaller CAG repeat expansions. At larger expansions, however, the protein context in which the repeat is expressed becomes less critical, and additional neuronal populations are affected.

Neuropathology

The gross pathology of HD involves atrophy of the caudate-putamen as well as cell loss in the deep layers of the cerebral cortex, thalamus, globus pallidus, and occasionally the cerebellum.[22-24] At the cellular level, the brunt of the striatal pathology is borne by the medium spiny neurons that project to medial and lateral globus pallidus and pars reticulata and pars compacta of substantia nigra.[28] The medium spiny neurons make up 90 percent of the cells in the striatum, and they are the primary recipients of excitatory glutamatergic input from cortex and thalamus and dopamine input from substantia nigra. Two types of medium spiny neurons can be distinguished by the neuropeptides and receptors they contain.[28,29] All use γ-aminobutyric acid (GABA) as a neurotransmitter, but one group contains predominantly enkephalin and D_2 dopamine receptors, and the other group contains predominantly substance P, dynorphin, and D_1 dopamine receptors. These two groups are differentially affected in HD. The GABA-enkephalin group projecting to lateral globus pallidus and the GABA-substance P group projecting to substantia nigra are affected before the GABA-substance P group projecting to medial globus pallidus.[28,29] Early pathology in the medium spiny neurons is observed as swollen and recurved dendrites.[30] Increased numbers of striatal cells show apoptotic changes compared with control striatum.[31] Striatal cells are organized in biochemically and anatomically confined compartments called *striosomes* and *matrix*. Data from Hedreen and Folstein suggest that the wave of cell death and gliosis in the HD striatum is preceded by clusters of glial fibrillary acidic protein–positive cells and neuronal loss in striosomes.[32]

Ten percent of striatal neurons are large aspiny cholinergic interneurons, medium aspiny somatostatin-neuropeptide Y-nitric oxide synthase–positive interneurons, and medium aspiny GABA-ergic interneurons containing either calretinin or parvalbumin.[33-35] Interestingly, these interneurons are either relatively or absolutely spared in the disease. There are prominently increased numbers of oligodendrocytes in HD striatum in addition to astrocytic gliosis.[22,23] In cerebral cortex, neurons in layers 5 and 6 are depleted in midstage disease, but neurons in superficial layers are not clearly affected.[23,24]

Distribution studies of huntingtin (the protein encoded by the *HD* gene) show apparently ubiquitous staining of central nervous system neurons with less staining in glia.[36-39] Detailed double-label studies in humans and rats show prominent labeling of the medium spiny projection neurons that die in the disease but little or no huntingtin immunoreactivity in the striatal interneurons spared in the disease.[40,41] How the regional expression of the huntingtin protein relates to disease in other areas is currently unknown.

Potential Mechanisms of Pathology

Each of the CAG disease genes encodes a novel protein. The SBMA gene encodes the androgen receptor, and the SCA6 gene encodes the α_{1A}-subunit of a brain voltage-gated calcium channel.[11] The functions of the proteins encoded by the *HD* (huntingtin), *SCA1* (ataxin 1), *MJD* (ataxin 3), *SCA2* (ataxin 2), and *DRPLA* (atrophin 1) genes are unknown.[6,10] The structures of the proteins differ, and the location of the polyglutamine repeat within the protein differs [*see Figure 3*]. In SCA1 and SCA2, the polyglutamine stretch is interrupted by one or more CAA sequences, and only in the mutant forms does the repeat become one long, continuous tract of glutamines. This ex-

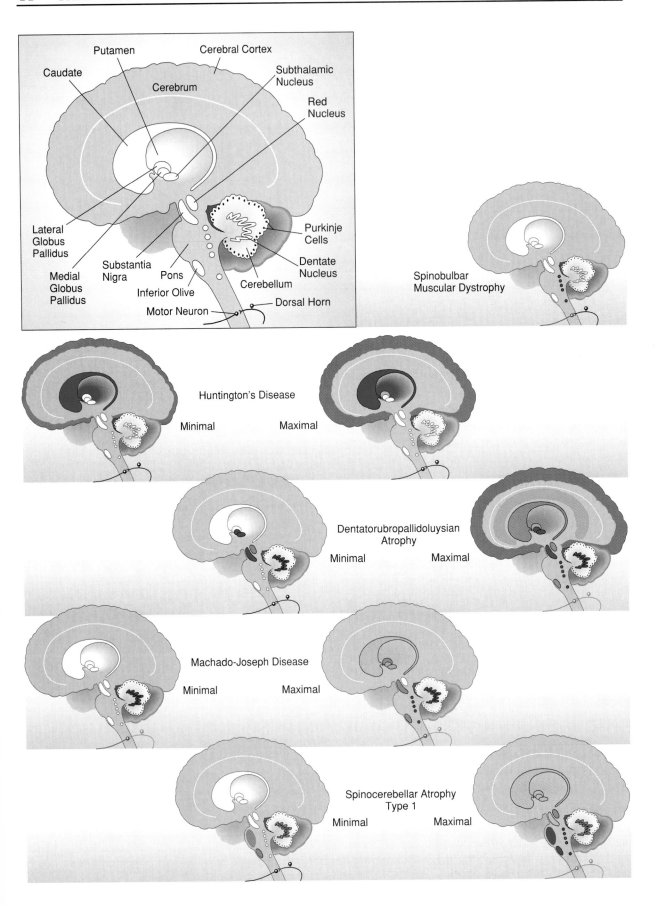

Figure 3 *Brain pathology in the various CAG repeat disorders is illustrated. Both minimal pathology, as seen with lower numbers of CAG repeats, and maximal pathology, as seen with larger CAG repeats, are shown. Dark blue areas depict severe pathology, and various lighter shades of blue indicate different degrees of intermediate pathology.*

panded stretch of glutamines appears to confer a structure that significantly alters its migration on gel electrophoresis. The wild-type and mutant bands are more widely separated than expected based simply on the molecular weight of the additional glutamines [*see Figure 4*]. There is no evidence of differences in the glycosylation, ubiquitination, or phosphorylation of mutant versus wild-type protein measured in preparations from synaptic terminals.[42] In initial studies, there was also no evidence of intracellular accumulations of abnormal protein products in any of the CAG repeat disorders [*see Table 4*]. Recent studies, however, have identified ubiquitinated, intranuclear inclusions of mutant protein in transgenic animal models of HD and in HD and MJD brain.[43-46]

PROTEIN-PROTEIN INTERACTIONS

β-Sheets/Polar Zippers Structural analysis of polyglutamine peptides indicates that they may form so-called β-*sheets* or *polar zippers*.[47] When two stretches of polyglutamine are oriented opposite each other, or if one long stretch forms a loop, the oxygen of the -CONH2 group at the end of the glutamine can form a hydrogen bond with the hydrogen attached to the carbon of the main peptide backbone or with the hydrogen of the NH2 group

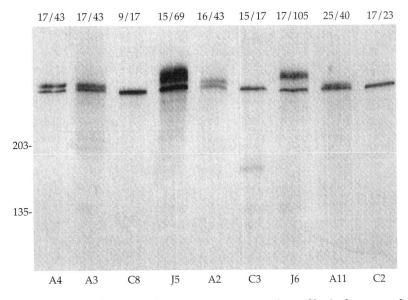

Figure 4 *Western blots of crude synaptosome preparations of brain from several control subjects (C) and individuals with adult-onset (A) and juvenile-onset (J) Huntington's disease. The blots are stained with an antibody to the huntingtin protein. The single band in the control subject lanes and the lower band in the HD case lanes represent the wild-type huntingtin protein. The upper band in the HD case lanes represents the mutant huntingtin protein. The CAG repeat numbers for the two alleles are given above each lane. The separation of the two bands correlates roughly with the size of the expanded repeat.*

Table 4   Proposed Mechanisms
of Pathology

Altered transcriptional regulation/neuronal intranuclear
   inclusions
Transglutamidation of polyglutamine with other proteins
Protein-protein interactions
   β-sheets or polar zippers
   Huntingtin-associated protein 1 (HAP1)
   Huntingtin-interacting proteins 1, 2, and 3 (HIP1, 2,
      and 3)
   Glyceraldehyde-3-phosphate dehydrogenase
   Apopain
Abnormal energy metabolism

of glutamine on the opposite strand, causing a strong noncovalent link between the two peptide strands. If such polar zippers are formed between mutant huntingtin molecules or between mutant huntingtin and other proteins, the functions of the proteins might be impaired or altered. At present, this hypothesis warrants in-depth consideration in light of the findings of nuclear inclusions seen in transgenic HD models and human HD brain.[43-46] Perutz has shown that the longer the polyglutamine stretch, the more stable the polar zipper.[48] Interestingly, polar zippers are structures that commonly bind together subunits of transcription factor complexes. Androgen receptors are known transcription factors. Whether the other CAG repeat proteins function as such with the expanded polyglutamine stretch is under investigation.

Transamidation   Glutamine and polyglutamine are good substrates for transaminase enzymes that covalently link glutamine residues from one peptide with a lysine residue on another peptide, thereby covalently linking peptides.[49] Synthetic peptides composed of increasing lengths of polyglutamine flanked by polar residues to enhance solubility are good substrates for transglutaminase enzymes from liver and brain.[50] The longer the polyglutamine stretch, the better the substrate. Whether such reactions take place between huntingtin and other proteins in brain is unknown. Such a process is of interest, however, because many transglutaminases are calcium dependent and activated during cell death processes. Transglutaminases are active in brain, but their role in brain function is not known.

HAP1, Ubiquitin, Apopain, and GAPDH   Intense efforts in numerous laboratories have focused on determining the function of the novel huntingtin, atrophin, and ataxin proteins by searching for proteins with which they interact. Using the yeast two-hybrid and other systems, a number of proteins that bind to huntingtin have been identified. One protein of known function is the enzyme glyceraldehyde-3-phosphate dehydrogenase (GAPDH), a component of the glycolytic pathway.[51] It was hypothesized that interaction with GAPDH might lead to abnormal energy metabolism and thereby selective neuronal pathology. Unfortunately, abnormal GAPDH activity has not been found in HD tissues. Huntingtin also interacts with several other proteins, termed *huntingtin associated protein 1* (HAP1) and *huntingtin interacting proteins 1, 2, and 3* (HIP-1, HIP-2, and HIP-3). HIP-2 is a ubiquitin-conjugating enzyme that does not distinguish mutant from wild-type protein.[52] HAP1 is a novel protein that binds to huntingtin and binds more avidly to mutant than to wild-type protein.[53, 54] HAP1 has a distribution similar to that of nitric oxide synthase.[54] Huntingtin also binds indirectly to calmodulin in a calcium-dependent manner and could thereby affect a variety of calcium-dependent cell functions.[55] Apopain, a human enzyme related to the nematode cysteine protease cell death gene *ced-3*, has been found to cleave the huntingtin protein in cell extracts, and the rate of cleavage increases with the length of the polyglutamine repeat.[56] If such cleavage is shown to occur in HD, it would suggest that the pathophysiological mechanism of neuronal degeneration in the disease may be attributed to enhanced apoptosis.

Excitotoxins

One of the most popular and long-standing hypotheses about the pathophysiology of HD has been the notion of cell death based on glutamate-induced excitotoxic cell death. In 1977, Coyle and Schwarcz[57] and the McGeers[58] showed that infusion of the glutamate agonist kainate into the striatum could cause axon-sparing lesions reminiscent of those in HD. Later studies showed that selective agonists at the *N*-methyl-D-aspartate (NMDA) subtype of glutamate receptor could provide an even better model because such agonists killed the medium spiny striatal projection neurons but spared the interneurons for HD.[59] NMDA receptors are decreased in HD striatum to a greater extent than other neurotransmitter receptors, supporting the notion that the cells that die in HD have a high density of NMDA receptors.[60, 61] To date, however, no abnormalities in glutamate receptors per se have been identified in HD. Furthermore, NMDA receptors are present in high concentrations in other parts of the brain not affected by the disease process.

Energy Metabolism

Another long-standing hypothesis concerning the pathogenesis of HD proposes that there is an abnormality in cellular energy metabolism. The first suggestions in support of this hypothesis arose from the observation that persons with HD lose weight even though they consume a large number of calories per day. Obese persons with HD seem to fare better than leaner persons.[62] Until recently, however, little data have been reported regarding energy metabolism in HD.[63] Magnetic resonance imaging spectroscopy studies have now shown that lactic acid levels are increased in HD.[64, 65] Postmortem studies also show abnormalities in complex II, III, and IV activity in the brain.[66, 67] Coenzyme Q10, a free-radical scavenger that shuttles electrons between mitochondrial oxidases, decreases brain lactate in HD patients.[65]

Reversible and irreversible inhibitors of complex II of the mitochondrial electron transport chain cause prominent striatal damage that spares the interneurons just as in HD.[68] In fact, the complex II inhibitor 3-nitropropionic acid can cause selective striatal damage when given systemically to primates, humans, and rodents. Long-term administration of small amounts of the agent to monkeys over two years resulted in selective striatal degeneration and a movement disorder.[69] Abnormalities in energy metabolism are completely compatible with the excitotoxic hypothesis because oxidative stress renders neurons more susceptible to excitotoxic neuronal injury.[60, 63]

Abnormalities in energy metabolism have also been observed in MJD, but the other CAG repeat disorders have not been studied in this respect.[70] This subject has been somewhat ignored, but several toxic substances that act at various parts of the electron transport chain produce differing patterns of selective neuronal degeneration.[71] Cyanide, a cytochrome oxidase inhibitor, produces prominent pallidal, subthalamic, and cerebellar pathology.[71,72] The malate-aspartate shunt inhibitor 3-acetylpyridine produces inferior olive, substantia nigra, and brain stem nuclei lesions.[71,73] It has also been clear for years that different mitochondrial mutations cause different but regionally specific pathologies.

How might an increased stretch of polyglutamines lead to abnormalities in energy metabolism? Huntingtin or an N-terminal fragment of huntingtin might be mistargeted to the nucleus and there alter the transcription of nuclear encoded proteins destined to be involved in energy metabolism. This hypothesis is particularly intriguing based on the recent observation on nuclear inclusions of the huntingtin protein in transgenic HD mice and human HD brain[43-46] [*see Epilogue*]. Alternatively, the glutamines might bind to proteins involved in the trafficking of other proteins to specific cellular targets or may affect trafficking and transport of vesicles or mitochondria.[74] If, for instance, the polyglutamine tract in atrophin interacts with nuclearly encoded cytochrome oxidase subunits destined for the mitochondria, such interactions might retard trafficking, reduce the complex IV activity, and lead to selective pallidal degeneration. Similar interactions might occur between huntingtin and subunits of succinate dehydrogenase (complex II). As the number of glutamines in either huntingtin or atrophin increases, huntingtin might affect cytochrome oxidase subunits and atrophin might affect succinate dehydrogenase subunits, leading to more widespread pathology. Such ideas are currently only speculation but are being tested experimentally.

Transgenics

The most exciting recent development since the discovery of the genetic defects responsible for the different CAG repeat disorders has been the creation of transgenic animals for several of the diseases. A transgenic line of mice expressing the human *SCA1* gene with variable numbers of CAG repeats behind a promoter that is Purkinje cell specific develops progressive ataxia and Purkinje cell loss. The disorder begins at variable times after birth, dependent on the number of CAG repeats.[75] Although the model has been criticized because it results in expression of the mutant gene only in a restricted neuronal class and thus does not mimic the human disease, it promises to be extremely useful in defining the pathophysiology of cell death caused by increased CAG repeats and in screening potential therapeutic agents.

A transgenic mouse expressing the first exon of the human huntingtin gene with variable numbers of repeats driven by the human promoter has been created and manifests a movement disorder in early adult life followed by a progressive decline and death within weeks.[76] The first symptom is weight loss followed by irregular involuntary movements and excessive grooming and stereotyped behavior. The animals also appear to have seizures. They eat voraciously but do not gain weight, and the brain is small. Brain atrophy precedes body atrophy. Studies of various markers of specific neurons and neuronal features in the brains of these animals indicate that nuclear expression of the protein occurs prior to the weight loss and the onset of clinical signs.[43] Furthermore, ubiquitin staining of the intranuclear inclusions becomes evident soon after the huntingtin inclusions become visible.[43] On the basis of these studies in transgenic animals, human brains have been reexamined with antibodies to huntingtin and ubiquitin, and both are positive.[45] No other overt neuropathologic damage is obvious except for the diffuse atrophy. More selective studies of cellular dysfunction and alteration of specific gene expression in these animals are being conducted, as are studies of therapeutic strategies to delay onset and progression of the disease. Additional animal models are being designed that include transgenes driven by different promoters, transgenes driven by conditional promoters that can be turned on and off by exogenous drugs, and so-called knock-ins, in which a stretch of expanded CAG repeats is introduced into the mouse homolog of the affected gene.

Clinical Issues

The identification of the mutations in at least six CAG repeat disorders has led to the ability to diagnose these genetic defects through a simple blood test.[77] The test results, however, have implications well beyond the simple diagnosis of one individual's genetic risk.[78] Each new diagnosis brings a burden of risk of an untreatable illness to the extended family, along with a risk of genetic discrimination. These issues need to be addressed at the state and federal levels, and testing programs should be carefully conducted and monitored.

Testing

Symptomatic

Every movement disorders specialist and most general neurologists have patients referred to them with involuntary movements and cognitive decline but no family history. It is now possible to have a blood sample screened for the HD mutations and other CAG repeat diseases. In the past, patients were told that a definitive diagnosis was not possible but that, after ruling out metabolic and structural causes, HD was most likely. Direct genetic testing is now possible. The clinician, however, should be cognizant of the implications of genetic testing. Consultation with the patient's spouse, children, and other family members is recommended to inform them of the implications of the testing and the current lack of treatment for the disease. Preferably, families should visit with a genetic counselor before having the test performed. Tests should not be performed on asymptomatic minors or others unable to give informed consent.

Presymptomatic

Presymptomatic testing protocols have been recommended by the International Huntington's Association and the World Federation of Neurology Huntington's Research Group.[79] Minors and mentally or cognitively impaired persons should be excluded. Persons undergoing presymptomatic testing should receive at least two counseling sessions with a genetic counselor or the equivalent. They should be accompanied by a family member or significant other who will participate in the entire testing process. Symptomatic disease should be ruled out by a neurologic examination, and each potential test subject should visit a psychologist or psychiatrist who can monitor the person after testing is completed. A waiting period of one month for the test subject to contemplate the decision is recommended before the blood test is sent to the laboratory for analysis. The results should be given to the subject face to face and not by telephone or mail. Close follow-up is recommended because even those with a negative result can become depressed and need counseling.

Treatment

Symptomatic

Current therapies for the CAG repeat disorders are purely symptomatic and are tailored to the specific symptom of concern.[80] In general, the fewer medications the better because the side effects are often more deleterious than the medicines are beneficial. Depression, psychosis, behavioral disorders, and seizures are all amenable to therapy. Ataxia and gait, speech, and swallowing disorders are not amenable to pharmacological treatment, but speech and physical therapy can help optimize function. In the early stages of HD, the chorea and rigidity can be alleviated somewhat by dopamine antagonists and agonists, respectively, but these substances become less helpful later in the illness. High caloric intake may be necessary to avoid dramatic weight loss.

Preventive

The real hope for these diseases is the development of preventive therapies that will delay the onset and retard the progression of the disease.[81] Trials of glutamate antagonists and coenzyme Q in HD are in progress, and the availability of transgenic models will allow more rapid and systematic studies of additional strategies. If the onset of disease could be delayed 10 to 20 years and the rate of progression slowed by 50 percent, the personal and social impact of the illness would be vastly reduced.

Summary

The past five years have brought extraordinary advances in our understanding of the genetics of several devastating neurologic diseases. With the advent of new molecular biologic approaches, the pathophysiological mechanisms causing neuronal dysfunction in these diseases should be elucidated. Such breakthroughs should allow the development of effective new therapies. A breakthrough in understanding any of the CAG repeat disorders is likely to have relevance for understanding the pathophysiology of all the others, and each should bring new and exciting insight into how the brain normally functions.

References

1. Young AB: Huntington's disease: lessons from and for molecular genetics. *Neuroscientist* 1:30, 1994

2. Gusella JF, Wexler NS, Conneally PM, et al: A polymorphic DNA marker genetically linked to Huntington's disease. *Nature* 306:234, 1983

3. Huntington's Disease Collaborative Research Group: A novel gene containing a trinucleotide repeat that is expanded and unstable on Huntington's disease chromosomes. *Cell* 72:971, 1993

4. Warren ST: The expanding world of trinucleotide repeats. *Science* 271:1374, 1996

5. Paulson HL, Fischbeck KH: Trinucleotide repeats in neurogenetic disorders. *Annu Rev Neurosci* 19:79, 1996

6. Ross CA: When more is less: pathogenesis of glutamine repeat neurodegenerative diseases. *Cell* 15:493, 1995

7. Brook JD, McDurrach ME, Harley HG, et al: Molecular basis of myotonic dystrophy: expansion of a trinucleotide repeat at the 3' end of a transcript encoding a protein kinase family member. *Cell* 68:799, 1992

8. Brooks BP, Fischbeck KT: Spinal and bulbar muscular atrophy: a trinucleotide-repeat expansion neurodegenerative disease. *Trends Neurosci* 18:459, 1995

9. Priller J, Scherzer CR, Faber PW, et al: Frataxin gene of Friedreich's ataxia is targeted to mitochondria. *Ann Neurol* 42:265, 1997

10. Gusella JF, Persichetti F, MacDonald ME: The genetic defect causing Huntington's disease: repeated in other contexts? *Mol Med* 3:238, 1997

11. Zhuchenko O, Bailey J, Bonnen P, et al: Autosomal dominant cerebellar ataxia (SCA6) associated with small polyglutamine expansions in the α1A-voltage-dependent calcium channel. *Nat Genet* 15:62, 1997

12. Martin JB: Pathogenesis of neurodegenerative disorders: the role of dynamic mutations. *NeuroReport* 8:1, 1996

13. Harper PS: The epidemiology of Huntington's disease. *Hum Genet* 89:365, 1992

14. Imbert G, Saudou F, Yvert G, et al: Cloning of the gene for spinocerebellar ataxia 2 reveals a locus with high sensitivity to expanded CAG/glutamine repeats. *Nat Genet* 14:285, 1996

15. Myers RH, MacDonald ME, Koroshetz WJ, et al: De novo expansion of a (CAG)n repeat in sporadic Huntington's disease. *Nat Genet* 5:168, 1993

16. Macdonald ME, Barnes G, Srinidhi J, et al: Gametic but not somatic instability of CAG repeat length in Huntington's disease. *J Med Genet* 30:982, 1993

17. Leeflang EP, Zhang L, Tavare S, et al: Single sperm analysis of the trinucleotide repeats in the Huntington's disease gene: quantitation of the mutation frequency spectrum. *Hum Mol Genet* 4:1519, 1995

18. Davis MB, Bateman D, Quinn NP, et al: Mutation analysis in patients with possible but apparently sporadic Huntington's disease. *Lancet* 344:714, 1994

19. Durr A, Dode C, Hahn V, et al: Diagnosis of "sporadic" Huntington's disease. *J Neurol Sci* 129:51, 1995

20. Goldberg YP, Kremer B, Andrew SE, et al: Molecular analysis of new mutations for Huntington's disease: intermediate alleles and sex of origin effects. *Nat Genet* 5:174, 1993

21. Rubinstein DC, Leggo J, Coles R, et al: Phenotypic characterization of individuals with 30-40 CAG repeats in the Huntington disease (HD) gene reveals HD cases with 36 repeats and apparently normal elderly individuals with 36-39 repeats. *Am J Hum Genet* 59:16, 1996

22. Vonsattel JP, Myers RH, Stevens TJ, et al: Neuropathological classification of Huntington's disease. *J Neuropathol Exp Neurol* 44:559, 1985

23. de la Monte SM, Vonsattel JP, Richardson EP, et al: Morphometric demonstration of atrophic changes in cerebral cortex, white matter, and neostriatum in Huntington's disease. *J Neuropathol Exp Neurol* 47:516, 1988

24. Hedreen JC, Peyser CE, Folstein SE, et al: Neuronal loss in layers V and VI of cerebral cortex in Huntington's disease. *Neurosci Lett* 133:257, 1991

25. Myers RH, Vonsattel JP, Stevens TJ, et al: Clinical and neuropathological assessment of severity in Huntington's disease. *Neurology* 38:341, 1991.

26. Tomoda A, Ikezawa M, Ohtani Y, et al: Progressive myoclonic epilepsy: dentato-rubro-pallido-luysian atrophy (DRPLA) in childhood. *Brain Dev* 13:266, 1991

27. Sudarsky L, Coutinho P: Machado-Joseph disease. *Clin Neurosci* 3:17, 1995

28. Albin RL, Young AB, Penney JB: Functional anatomy of the basal ganglia. *Trends Neurosci* 13:1, 1989

29. Reiner A, Albin RL, Anderson KD, et al: Differential loss of striatal projection neurons in Huntington disease. *Proc Natl Acad Sci USA* 85:5733, 1988

30. Graveland GA, Williams RS, DiFiglia M: Evidence for degenerative and regenerative changes in neostriatal spiny neurons in Huntington's disease. *Science* 227:770, 1985

31. Portera-Cailliau C, Hedreen JC, Price DL, et al: Evidence for apoptotic cell death in Huntington disease and excitotoxic animal models. *J Neurosci* 15:3775, 1995

32. Hedreen JC, Folstein SE: Early loss of neostriatal striosome neurons in Huntington's disease. *J Neuropath Exp Neurol* 54:105, 1995

33. Ferrante RJ, Kowall NW, Beal MF, et al: Selective sparing of a class of striatal neurons in Huntington's disease. *Science* 230:561, 1985

34. Dawbarn D, Dequidt ME, Emson PC: Survival of basal ganglia neuropeptide Y-somatostatin neurones in Huntington's disease. *Brain Res* 340:251, 1985

35. Cicchetti F, Gould PV, Parent A: Sparing of striatal neurons coexpressing calretinin and substance P (NK1) receptor in Huntington's disease. *Brain Res* 730:232, 1996.

36. DiFiglia M, Sapp E, Chase K, et al: Huntingtin is a cytoplasmic protein associated with vesicles in human and rat brain neurons. *Neuron* 14:1075, 1995

37. Gutekunst CA, Levey AI, Heilman DJ, et al: Identification and localization of huntingtin in brain and human lymphoblastoid cell lines with anti-fusion protein antibodies. *Proc Natl Acad Sci USA* 92:8710, 1995

38. Sharp AH, Loev SJ, Schilling G, et al: Widespread expression of Huntington's disease gene (IT15) protein product. *Neuron* 14:1065, 1995

39. Trottier Y, Lutz Y, Stevanin G, et al: Polyglutamine expansion as a common pathological epitope detected in Huntington's disease, in spinocerebellar ataxia 1 and 3 and two additional autosomal dominant cerebellar ataxias. *Nature* 378:403, 1995

40. Kosinski CM, Cha J-H, Young AB, et al: Huntingtin immunoreactivity in the rat neostriatum: differential accumulation in projection neurons and interneurons. *Exp Neurol* 144: 239, 1997

41. Ferrante RJ, Gutekunst C-A, Persichetti F, et al: Heterogeneous topographic and cellular distribution of huntingtin expression in the normal human neostriatum. *J Neurosci* 17:3052, 1997

42. Aronin N, Chase K, Young C, et al: CAG expansion affects the expression of mutant huntingtin in the Huntington's disease brain. *Neuron* 15:1, 1995

43. Davies SW, Turmaine M, Cozens BA, et al: Formation of neuronal intranuclear inclusions (NII) underlies the neurological dysfunction in mice transgenic for the HD mutation. *Cell* 90:537, 1997

44. Roizin L, Stellar S, Liu JC: Neuronal nuclear-cytoplasmic changes in Huntington's chorea: electron microscope investigations. *Advances in Neurology 23, Huntington's Disease.* Chase TN, Wexler NS, Barbeau A, Eds. Raven Press, New York, 1979, p 95

45. DiFiglia M, Sapp E, Chase KO, et al: Aggregation of huntingtin in neuronal intranuclear inclusions and dystrophic neurites in brain. *Science* 277:1990, 1997

46. Paulson HL, Perez MK, Trottier Y, et al: Intranuclear inclusions of expanded polyglutamine protein in spinocerebellar ataxia type 3. *Neuron* 19:333,1997

47. Perutz MF: Glutamine repeats and inherited neurodegenerative diseases: molecular aspects. *Curr Opin Struct Biol* 6:848,1996

48. Perutz MF: Glutamine repeats as polar zippers: their possible role in inherited neurodegenerative diseases. *Proc Natl Acad Sci USA* 91:5355, 1994

49. Green H: Human genetic diseases due to codon reiteration: relationship to an evolutionary mechanism. *Cell* 74:955, 1993

50. Kahlem P, Terre C, Green H, Djian P: Peptides containing glutamine repeats as substrates for transglutaminase-catalyzed cross-linking: relevance to disease of the nervous system. *Proc Natl Acad Sci USA* 93:14580, 1996

51. Burke JR, Enghild JJ, Martin ME, et al: Huntingtin and DRPLA proteins selectively interact with the enzyme GAPDH. *Nat Med* 2:347, 1996

52. Kalchman MA, Graham RK, Xia G, et al: Huntingtin is ubiquitinated and interacts with a specific ubiquitin-conjugating enzyme. *J Biol Chem* 271:19385, 1996

53. Li X-J, Sharp AH, Li S-H, et al: Huntingtin-associated protein (HAP1): discrete neuronal localizations in the brain resemble those of neuronal nitric oxide synthase. *Proc Natl Acad Sci USA* 93:4839, 1996

54. Li X-J, Li S-H, Sharp AH, et al: A huntingtin-associated protein enriched in brain with implications for pathology. *Nature* 378:398, 1995

55. Bao J, Sharp AH, Wagster MV, et al: Expansion of polyglutamine repeat in huntingtin leads to abnormal protein interactions involving calmodulin. *Proc Natl Acad Sci USA* 93:5037, 1996

56. Goldberg YP, Nicholson DW, Rasper DM, et al: Cleavage of huntingtin by apopain, a proapoptotic cysteine protease, is modulated by the polyglutamine tract. *Nat Genet* 13:442, 1996

57. Coyle JT, Schwarcz R: Lesion of striatal neurones with kainic acid provides a model for Huntington's chorea. *Nature* 263:244, 1976

58. McGeer EG, McGeer PL: Duplication of biochemical changes of Huntington's chorea by intrastriatal injection of glutamic and kainic acids. *Nature* 263:517, 1976

59. Beal MF, Kowall NW, Ellison DW, et al: Replication of the neurochemical characteristics of Huntington's disease by quinolinic acid. *Nature* 321:168, 1986

60. Young AB, Greenamyre JT, Hollingsworth Z, et al: NMDA receptor losses in putamen from patients with Huntington's disease. *Science* 241:981, 1988

61. Albin RL, Young AB, Penney JB, et al: Abnormalities of striatal projection neurons and N-methyl-D-aspartate receptors in presymptomatic Huntington's disease. *N Engl J Med* 322:1293, 1990

62. Myers RH, Sax DS, Koroshetz WJ, et al: Factors associated with slow progression in HD. *Arch Neurol* 48:800, 1981

63. Beal MF, Hyman BT, Koroshetz WJ: Do defects in mitochondrial energy metabolism underly the pathology of neurodegenerative diseases? *Trends Neurosci* 16:125, 1993

64. Jenkins BG, Koroshetz WJ, Beal MF, et al: Evidence for impairment of energy metabolism in vivo in Hungtington's disease using localized 1H NMR spectroscopy. *Neurology* 43:2689, 1993

65. Koroshetz WJ, Jenkins BG, Rosen BR, et al: Energy metabolism defects in Huntington's disease and effects of coenzyme Q10. *Ann Neurol* 41:160, 1997

66. Gu M, Gash MT, Mann VM, et al: Mitochondrial defect in Huntington's disease caudate nucleus. *Ann Neurol* 39:385, 1996

67. Browne SE, Bowling AC, MacGarvey U, et al: Oxidative damage and metabolic dysfunction in Huntington's disease: selective vulnerability of the basal ganglia. *Ann Neurol* 41:646, 1997

68. Beal MF, Brouillet E, Jenkins BG, et al: Neurochemical and histologic characterization of striatal excitotoxic lesions produced by the mitochondrial toxin 3-nitropropionic acid. *J Neurosci* 13:4191, 1993

69. Brouillet E, Hantraye P, Ferrante RJ, et al: Chronic mitochondrial energy impairment produces selective striatal degeneration and abnormal choreiform movements in primates. *Proc Natl Acad Sci USA* 92:7105, 1995

70. Matsuishi T, Sakai T, Naito E, et al: Elevated cerebrospinal fluid lactate/pyruvate ratio in Machado-Joseph disease. *Acta Neurol Scand* 93:72, 1996

71. Beal MF: Effects of mitochondrial toxins in animals and man. *Mitochondrial Dysfunction and Oxidative Damage in Neurodegenerative Diseases*, R.G. Landes Co, Austin, TX, 1995, p 35

72. Rosenow F, Herholz K, Lanfermann H, et al: Neurological sequelae of cyanide intoxication: the patterns of clinical magnetic resonance imaging, and position emission tomography findings. *Ann Neurol* 38:825, 1995

73. Baraban CD: Central neurotoxic effects of intraperitoneally administered 3-acetylpyridine, harmaline and niacinamide in Sprague-Dawley and Long-Evans rats: a critical review of central 3-acetylpyridine neurotoxicity. *Brain Res Rev* 9:21, 1985

74. Schatz G, Dobberstein B: Common principles of protein translocation across membranes. *Science* 271:1519, 1996

75. Burright EN, Clark HB, Servadio A, et al: SCA1 transgenic mice: model for neurodegeneration caused by an expanded CAG trinucleotide repeat. *Cell* 82:937, 1995

76. Mangiarini L, Sathasivam K, Seller M, et al: Exon 1 of the HD gene with an expanded CAG repeat is sufficient to cause a progressive neurological phenotype in transgenic mice. *Cell* 87:493, 1996

77. Harper PS, Houlihan G, Tyler A: Genetic counselling in Huntington's disease. *Huntington's Disease*. Harper PS, Ed. WB Saunders Co, London, 1996, p 359

78. Harper PS, Soldan J, Tyler A: Predictive tests in Huntington's disease. *Huntington's Disease*. Harper PS, Ed. WB Saunders Co, London, 1996, p 395

79. International Huntington's Association and World Federation of Neurology: Guidelines for the molecular genetic predictive test in Huntington's disease. *J Med Genet* 31:555, 1994

80. Folstein S: *Huntington's Disease: A Disorder of Families*. Johns Hopkins University Press, Baltimore, 1989

81. Shoulson I, Kieburtz K: Neuroprotective therapy for Huntington's disease. *Neuroprotection in CNS Diseases*. Bar PR, Beal FM, Eds. Marcel Dekker, New York, 1997, p 457

Acknowledgments

Figure 1   Marcia Kammerer.

Figures 2 and 3   Seward Hung.

Figure 4   Anne B. Young.

CHAPTER 4

The Molecular Genetics of Alzheimer's Disease

Rudolph E. Tanzi, Ph.D.

Alzheimer's disease (AD) accounts for up to 70 percent of all late-onset cases of dementia and claims more than 100,000 lives a year in the United States. AD, a progressive neurodegenerative disorder of the central nervous system, is clinically characterized by personality changes; language and visuospatial problems; and global cognitive decline, including deficits in memory, reasoning, orientation, and judgment. A valid diagnosis of AD requires the presence of specific neuropathologic features, including abundant amounts of neurofibrillary tangles (NFTs) and β-amyloid (senile) plaques in the brain.[1] NFTs are found inside dying neurons and as threads in the neuropil. They consist of twisted filaments made up primarily of the cytoskeletal protein tau, which is associated with microtubules. Senile or neuritic plaques are extracellular lesions that contain a dense central core of β-amyloid and are often surrounded by degenerating nerve terminals and activated glial cells. The major component of β-amyloid is a 39 to 43 amino acid peptide[2] called Aβ, which is produced by the processing of a larger protein called the β-amyloid precursor protein (APP). β-amyloid is found in cerebral blood vessels and in the senile plaques that litter the cerebral cortex, hippocampus, and temporal lobe regions of AD patients' brains. The molecular dissection of the neuropathologic lesions defining AD have led to the greatest advances regarding the etiology and genetics of this disease.

Epidemiology and Genetics of Familial Alzheimer's Disease

The progressive development of AD neuropathology, including NFTs and Aβ deposits, is also observed in association with advanced age but to a lesser extent than in is observed in patients with AD.[3] Given the inevitable progression of these neuropathologic lesions with increased age in virtually all individuals and the fact that AD is diagnosed based on the abundance not the simple presence of these lesions, the argument can be made that everyone would develop AD if they simply lived long enough. Along these lines, it is conceivable that environmental risk factors together with one's genetic constitution do not determine onset of AD, but rather the age at which the disease will strike and whether it will occur within the span of an average lifetime. The two most important risk factors for AD are increasing age and family history. Regardless of differences in life span, the risk of AD in women has been reported to be greater than in men.[4] Other proposed risk factors include head injury and a lack of higher education.[5,6]

Whereas most cases of AD occur relatively late in life (> 60 years), a subset (approximately five percent) occurs in persons younger than 60 years and frequently clusters in families. These cases are termed *early-onset familial Alzheimer's disease* (FAD). Early-onset FAD is inherited in an autosomal dominant fashion and provides the greatest evidence for a so-called *deterministic* genetic component in AD. Early-onset FAD is a genetically heterogeneous disorder that is caused by defects in at least three different genes located on chromosomes 1, 14, and 21. Clinical and neuropathologic features, as well as the severity of AD caused by the various mutations in the three known FAD genes, are relatively uniform. Familial clustering is also observed in late-onset AD but to a much lesser extent. Part of the reason for this is that late-onset FAD is often difficult to determine because of limitations in determining the family history; affected family members may have died of other age-related illnesses before showing symptoms of late-onset AD. The only reported clinical difference between the early- and late-onset cases of AD has been the presentation of myoclonus uniquely in early-onset AD.[7]

Genetic Clues from β-Amyloid

The first genetic clues to the etiology of AD began with Glenner and Wong's biochemical analysis of Aβ extracted from blood vessels of a patient with Down syndrome (trisomy 21).[2] By middle age, the brains of Down syndrome patients inevitably display the neuropathologic lesions of AD as well as some features of AD-related dementia. Glenner and Wong found that the major component of Aβ is a 4 kd peptide, which they termed the amyloid β-protein because of its β-pleated sheet structure. In their seminal paper reporting the sequence of the first 28 amino acids of the amyloid β-protein, Glenner and Wong also correctly predicted that an AD gene resides on chromosome 21.[2] By 1987, four different groups were able to utilize Glenner and Wong's Aβ amino acid sequence to clone the gene encoding the precursor of the 4 kd amyloid β-protein APP and map it to chromosome 21.[8-11] The amyloid β-protein, which later became known as the Aβ peptide, was found to be only a small domain within APP, an integral type I single membrane spanning protein. The Aβ domain has its C-terminal region embedded in the predicted membrane portion of APP and its N-terminus in the predicted ectodomain of the protein [*see Figure 1*]. How Aβ is actually excised from APP to give rise to Aβ remains a mystery and is the topic of intense investigation.

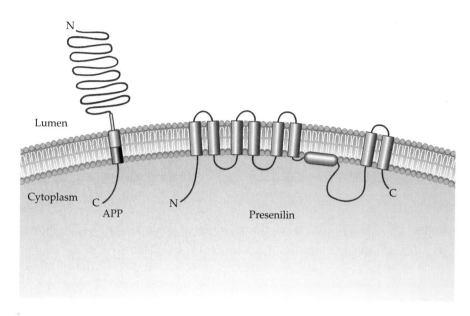

Figure 1 *The β-amyloid precursor protein (APP) has a single transmembrane-spanning domain and is a type I integral membrane spanning protein. The C-terminal portion of APP is cytoplasmic in orientation. The Aβ domain (red) is inserted partly in the membrane and partly in the lumen. Presenilin is shown in the eight-transmembrane model with a topology in which the N-terminal (N), C-terminal (C), and large hydrophilic loop regions face the cytoplasm.*

Concurrent with the cloning and mapping of the *APP* gene to chromosome 21, independent genetic linkage studies of four large, multigenerational, early-onset FAD pedigrees led to the suggestion of an FAD gene on chromosome 21.[12] Ultimately, the linkage of these four families to DNA markers on chromosome 21 turned out to be a spurious result, and it was subsequently shown that the *APP* gene was not the site of the FAD gene defect in these four families.[13,14] In fact, it was subsequently shown that these four families were actually linked to a gene on chromosome 14.[15] However, the combination of these two findings ignited a worldwide race to identify the first gene defect for FAD.

After APP was cloned, researchers around the world scoured the *APP* sequence for FAD mutations. The first *APP* gene mutation was reported by Levy and colleagues in 1990[16] but not for patients with FAD. The first *APP* gene mutation was found in a case of hereditary cerebral hemorrhage with amyloidosis. In this disorder, which was found in a Dutch family, accumulation of Aβ in cerebral blood vessels leads to stroke and often death usually by the fifth decade of life. The missense mutation in this family resided in exon 17 of the *APP* gene, which encodes a portion of the Aβ domain. Soon thereafter, a similar screening protocol was employed to show that this same exon of the *APP* gene contained what would turn out to be the first reported FAD gene mutation, V717I (substitution of isoleucine for valine at amino acid 717[17]). Subsequently, four more mutations were reported for the *APP* gene,[18-21] all of which were adjacent to or within the Aβ domain of the *APP* gene, and all of which would be shown to influence the generation of Aβ in brain tissue [*see Table 1*]. All of these mutations were shown to be nearly 100 percent penetrant and were thus considered causative for early-onset FAD. However, the *APP* gene mutations were later shown to be responsible for only a small proportion (roughly two percent) of all published cases of FAD and roughly five to seven percent of early-onset FAD.[14] In fact, there are currently only 20 known families worldwide that segregate FAD mutations in the APP gene.

Table 1   Effects of FAD Mutations on APP and Aβ Production

Gene Mutation	Effect on APP	Effect on Aβ
APP-717 (London)	Differential γ-secretase cut	Aβ42:Aβ40 ratio increased
APP-670/671 (Swedish)	Increased β-secretase cut	Aβ40 and Aβ42 increased
APP-692 (Flemish)	Decreased α-secretase cut?	Aβ42:Aβ40 ratio increased
APP-693 (Dutch)	Unclear	Increased Aβ aggregation
PS1-FAD mutations	Differential γ-secretase cut	Aβ42:Aβ40 ratio increased
PS2-FAD mutations	Differential γ-secretase cut	Aβ42:Aβ40 ratio increased
Trisomy 21 (Down syndrome)	Increased APP expression	Aβ40 and Aβ42 increased
Apolipoprotein E4	Competes for LDL receptor–related protein (LRP; 103)	Increased Aβ aggregation

The pathogenicity of the FAD mutations in APP was subsequently supported by the reports that transgenic mice expressing the APP717V→F mutation[22] or APP-Swedish mutation[23] produce Aβ deposition in the brain. However, no neuronal loss or NFTs are observed in these mice, suggesting that the production of Aβ in mice is not sufficient to cause definitive AD neuropathology. Other components, such as human forms of the tau protein or a humanized microglial-astroglial system, may also be needed to obtain an animal model for AD that extends beyond Aβ deposition. The role of Aβ in AD neuropathology has also been bolstered by reports that Aβ aggregates can be neurotoxic to neurons.[23-25] However, it is not yet clear whether the aggregated peptide directly induces neuronal cell death in brain or initiates secondary consequences, such as inflammatory responses involving glial and microglial cell activation, free radical generation, and complement activation, which lead to neuronal cell death.

The Genetic Risk Factor: Apolipoprotein E

The apolipoprotein E gene (*APOE4*) on chromosome 19 is associated with late-onset AD (onset older than age 60 to 65 years[26]). APOE is a major lipoprotein in the serum that plays a role in cholesterol storage, transport, and metabolism. The *APOE* gene has three alleles, ε2, ε3, and ε4, resulting from amino acid substitutions at codons 112 and 158. The ε4 (*APOE4*) allele has been associated with AD based on the finding that it was overrepresented in FAD patients as well as in sporadic AD cases relative to the unaffected population.[26,27] *APOE4* is claimed by Corder and colleagues to account for up to 50 percent of the genetic risk for AD.[27] However, it should be noted that the *APOE4* polymorphism is a common or so-called public polymorphism that is present in at least one copy in approximately one in three individuals. Thus, in many cases of the co-occurrence of AD and the *APOE4* allele, random coincidence cannot be ruled out. The risk for AD conferred by *APOE4* is dose-dependent,[27] with two doses of *APOE4* conferring significantly greater risk than one dose. The earliest age of onset of AD attributed to *APOE4* is associated with *APOE4* homozygosity, with the median age of onset being younger than 70 years. However, less than two percent of the general population are *APOE4* homozygotes. The risk for AD conferred by a single dose of *APOE4* is considerably lower and the age of onset distribution is significantly higher. In contrast, the inheritance of the *APOE2* allele is associated with decreased risk for AD[28]; however, this has not been universally observed.[29]

Studies with much larger and thus more reliable patient samples have increased our knowledge of the true fraction of AD attributable to *APOE4* and the age-dependent risk for AD conferred by *APOE4*.[30-32] In a population-based sample of 1,899 individuals aged 65 and older, multiple regression analyses demonstrated an influence of *APOE2* and *APOE4* on performance in a delayed-recall task over a four- to seven-year period.[30] However, the magnitude of the result was modest, with an odds ratio for developing impairment equal to 1.37 for *APOE4* and 0.53 for *APOE2*. In that same study, 85 percent of elderly *APOE4* homozygotes (average age, 81 years) were unimpaired on the mental status test. Hence, a substantial number of individuals who are *APOE4* homozygotes can reach old age without cognitive impairment or dementia.

In another study, aimed at examining the risk for AD conferred by *APOE* status in 578 residents of East Boston aged 65 years and older, persons who were *APOE4* positive had only 2.27 times the risk of incident AD compared with persons who harbored the highly common *APOE3/APOE3* genotype.[31] Interestingly, the *APOE4* allele was found to account for only a small fraction of AD in that same study. The authors concluded that removal of the risk for AD conferred by *APOE4* lowers the incidence of AD by only 13.7 percent.

In a third study, the impact of *APOE4* on age of onset of AD was assessed in 310 families containing at least two siblings or first-degree relatives with AD.[32] The goal of this study was to determine the age range at which *APOE4* conferred the greatest risk for AD. Whereas *APOE4* was found to be associated with AD in this study, the risk for AD conferred by this allele was greatest in the group with onset between 61 and 65 years of age, followed by those with onset between 65 and 70 years of age. After age 70, the influence of *APOE4* as a risk factor for AD dramatically declined. After age 75, the allele frequency of the *APOE4* allele did not significantly differ among affected individuals and their unaffected siblings, suggesting that from age 76 onward, *APOE4* is not specifically associated with AD in the families studied. Individuals with two copies of *APOE4* had a significantly lower age of onset than those with one or no copies of the allele (66.4 versus 72.0; $P < 0.001$), but individuals with one copy of *APOE4* displayed no differences in age of onset from those with none.

Collectively, these data suggest that whereas *APOE4* is associated with AD, the inheritance of two copies of the allele (two percent of the population) appears to have a substantially greater impact as a risk factor for AD than the inheritance of one copy. Several individuals in these studies who were *APOE4* homozygotes and older than 85 years suffered no cognitive impairment. Thus, even *APOE4* homozygosity is not sufficient to cause AD. These findings suggest that it remains critical to search for additional AD genes and genetic risk factors, especially for the development of AD after the age of 70 (approximately 60 percent of AD cases[32]). Most importantly, these findings strongly argue against using *APOE4* genotyping as a predictive test for AD[33,34] and emphasize the need for more large-scale population studies to assess the true impact of *APOE4* as a risk factor for AD.

Putative novel AD genes for cases with age of onset older than 70 years have been reported on chromosomes 3 and 12 (unpublished data) [*see Epilogue*]. Roses and colleagues have orally presented preliminary data suggesting the existence of a late-onset gene on chromosome 12 in AD families with affected individuals with minimal or no association with *APOE4*. We have recently generated data suggesting the existence of a novel late-onset AD gene on chromosome 3q25-26 that appears to be interactive with the *APOE4* allele (unpublished data). One compelling candidate gene in this region of chromosome 3 is the butyl cholinesterase gene, a polymorphism in which the K-variant appears to exhibit a mild association with AD in kindreds with an average age of onset older than 70 years (unpublished data).

The Presenilin Genes

In 1992, mutations in the *APP* gene were shown to be responsible for only a tiny fraction of FAD based on a large-scale assessment of FAD families.[14] In that same year, evidence for a second early-onset FAD gene was found on the long arm of chromosome 14.[15,35,36] The presenilin 1 gene (*PS1*) on chromosome 14 was identified in 1995 by positional cloning[37] and shown to contain mutations in the original four FAD families that St. George-Hyslop and colleagues[12] had previously found to be linked to chromosome 21. These families did not harbor APP mutations; the linkage results on chromosome 21 appear to have been false positives. With the discovery of the chromosome 14 gene linkage, it became apparent that *PS1* was the major early-onset FAD gene. Subsequent to the cloning of *PS1*, a homologous gene, presenilin 2 (*PS2*) was isolated and mapped to the region of chromosome 1 that had been linked to FAD in a set of kindreds of Volga German Origin.[38] *PS2* was found to contain a single amino acid substitution (missense) in the affected members of various Volga German FAD families.

To date, the *PS1* gene has been found to harbor 45 different FAD mutations in over 75 families of various ethnic origins, whereas *PS2* has been found to contain only two different FAD mutations [*see Table 2*].[39-41] All except one of these are missense mutations that result in single amino acid substitutions. The exception is a mutation that deletes exon 9 from *PS1* in three different FAD families.[41,42] The reason for the large difference in the number of FAD families with mutations in *PS1* and *PS2* is not clear. The mean age of onset in *PS1*-linked FAD pedigrees is approximately 45 years (range, 28 to 62 years), whereas the average age of onset in the Volga German families with the N141I *PS2* mutation is 52 years, with individual onset ages ranging from 40 to 85 years. Hence, it may be worthwhile to search for *PS2* mutations in FAD kindreds with later onset than those traditionally used to search for mutations in early-onset FAD genes. The mutations in *PS1* appear to be fully penetrant, with only one reported exception.[43] Thirty-four of the 45 reported FAD mutations in *PS1* occured in single families, making them genetically private mutations. Studies to date suggest that *APOE* genotype has no effect on the age of onset or phenotype of FAD in patients with *PS1* mutations.[44] In contrast, *APOE* genotype is reported to affect the age of onset and degree of amyloid burden in mutations of the *APP* gene.[45]

Mutations in the *APP* and the presenilin genes do not account for all cases of early-onset FAD,[46] and it is estimated that approximately 50 percent of early-onset FAD is genetically unaccounted for by these mutations.[39,40] Hence, genetic diagnosis of new cases of FAD based on the knowledge of existing *PS1* FAD mutations is not feasible, because the prediction is that over

Table 2 Presenilin Mutations

Mutation	Number of Families	Ethnicity	Domain/Exon	Mutation	Number of Families	Ethnicity	Domain/Exon
PS1				L250S	1	British	TM-6/7
A79V	1	Caucasian	TM-1/4	A260V	1	Japanese	TM-6/8
V82L	1	Caucasian	TM-1/4	L262F	1	Swedish	TM-6/8
V96F	1	Caucasian	TM-1/4	C263R	1	Caucasian	HL-6/8
Y115H	1	Caucasian	HL-1/5	P264L	2	Caucasian	HL-6/8
Y115C	1	Caucasian	HL-1/5	P267S	1	United Kingdom	HL-6/8
E120K	1	British	HL-1/5	R269G	1	Caucasian	HL-6/8
E120D	1	Caucasian	HL-1/5	R269H	1	Caucasian	HL-6/8
N135D	1	Caucasian	TM-2/5	R278T	1	Australian	HL-6/8
M139V	4	United Kingdom/ Germany	TM-2/5	E280G	2	United Kingdom	HL-6/8
				E280A	5	Columbian/Japanese	HL-6/8
M139T	1	Caucasian	TM-2/5	A285V	1	Japanese	HL-6/8
M139I	1	Caucasian	TM-2/5	L286V	2	German/Israeli	HL-6/8
I143T	1	Belgian	TM-2/5	S290C	3	British/Australian/ Finnish	HL-6/9
I143F	1	British	TM-2/5	(Δex9)			
M146L	10	Italian	TM-2/5	E318G	1	Swedish	HL-6/9
M146V	3	Italian/United Kingdom/Finnish	TM-2/5	G378E	1	French	TM-7/11
				G384A	3	Caucasian/Belgian/ Japanese	TM-7/11
H163Y	1	Swedish	HL-2/6	L392V	1	Italian	TM-7/11
H163R	6	American/Canadian/ Japanese/Swedish	HL-2/6	C410Y	2	Ashkenazi Jewish	TM-8/11
G209V	1	Caucasian	TM-4/7	A426P	1	Caucasian	TM-8/12
I213Y	1	Caucasian	TM-4/7	P436S	1	Caucasian	C-terminus/ 12
A231T	1	Caucasian	TM-5/7				
A231V	1	Caucasian	TM-5/7				
M233T	1	Australian	TM-5/7	*PS2*			
L235P	1	Caucasian	TM-5/7	N141I	7	Volga German	TM-2
A246E	1	Nova Scotian	TM-6/7	M239V	1	Italian	TM-5

HL, hydrophilic loop domain; TM, transmembrane domain.

70 percent of new cases of early-onset FAD in the age range consistent with *PS1* mutations would be novel mutations even if they occurred in *PS1*. Therefore, a simple genetic test for presenilin gene mutations as a means for diagnosing new cases of early-onset FAD is not practical at the present time.

Mechanisms of FAD Gene Defects: Toward a Common Pathogenic Pathway

The most valuable clues to unraveling the genetics and pathogenic mechanisms in AD have come from studies of Aβ deposition. Regardless of whether the formation of Aβ is causative for AD or simply a consequence of the neuropathogenic cascade, tracking the route to Aβ deposition has led to the identification of both *APP* and *APOE4* as FAD genes and has delineated some important biochemical pathways in this disease. Following maturation in the endoplasmic reticulum and Golgi,[47] the precursor of Aβ, APP, can undergo processing by the α-secretase pathway, in which APP is cleaved within the Aβ domain, preventing formation of β–amyloid and leading to the secretion of the extracellular portion of APP.[47-49] Alternatively, APP can be processed by the β-secretase pathway, which leads to a proteolytic clip at the N-terminus of the Aβ domain, yielding a C-terminal APP fragment containing the intact Aβ domain [*see Figure 2*]. Following β-secretase cleavage, an additional proteolytic clip of the Aβ-containing C-terminus of APP by γ-secretase then excises Aβ from APP,[50-52] allowing Aβ deposition to occur.

The FAD mutations in the *APP* gene alter secretion of Aβ in cultured cells in AD patient fibroblasts in different ways [*see Table 1*].[53-55] Aβ peptides ranging in size from 39 to 43 amino acids can be extracted from senile plaques and are known to be secreted normally by a variety of cells, including neurons.[52] The longer versions of Aβ (42 or 43 amino acids) are more prone to aggregate[56] than the shorter (40 amino acid) and more common (90 percent in the brain) form of the peptide. Aβ42 has been shown to be the principal component of senile plaques. Codon 717 mutations of *APP* have been shown to lead to an increase in the ratio of Aβ42:Aβ40. The Swedish APP FAD mutant leads to a threefold increase in overall Aβ secretion [*see Table 1*].

Current evidence supports a common pathogenic mechanism by which the early-onset FAD mutations cause AD by increasing the ratio of Aβ42:Aβ40 in the brain. In support of this hypothesis, plasma and fibroblasts from patients and at-risk carriers for the presenilin mutations contain increased amounts of the longer, more amyloidogenic form of Aβ, Aβ42.[57] Similar increases in the Aβ42:Aβ40 ratio have been observed in transfected cell lines and transgenic animals expressing mutant forms of *PS1* and *PS2*.[58-61] Increased deposition of Aβ42[62,63] and Aβ40[63] are also found in the brains of patients with *PS1* mutations. These data argue for a potentially central role of Aβ, and more specifically Aβ42, in the generation and deposition of β-amyloid in the AD brain. This hypothesis is strengthened by the observation that the *APOE4* genotype of the protein is associated with significantly increased amyloid burden in AD and Down syndrome patients.[64] Whereas mutations in the *APP* gene and the presenilins both appear to influence the processing of APP leading to more secretion of Aβ42, the risk factor, *APOE4*, appears to influence the rate of aggregation of Aβ and subsequent Aβ deposition[39] [*see Table 1*]. Despite these findings,

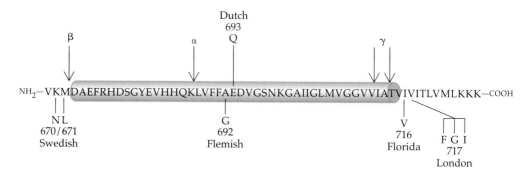

Figure 2 *The known mutations in APP [see Table 1] are shown along with a novel FAD mutation (Florida) at position 716. Also shown are the α-, β-, and γ-secretase cleavage sites. Two γ-secretase sites are shown to indicate the sites that yield either Aβ40 (first γ site) or Aβ42 (second γ site) when combined with a clip at the β-secretase site. Cleavage at the α-secretase site yields the normal N-terminal (NH₂) ectodomain fragment of APP, which is secreted, and the nonamyloidogenic C-terminal (COOH) fragment of APP, which remains embedded in the membrane until it is internalized via a clathrin-coated pit internalization signal on the cytoplasmic domain of APP (not shown). The transmembrane portion of APP terminates with the triple lysine motif (KKK) sequence at the C-terminal end.*

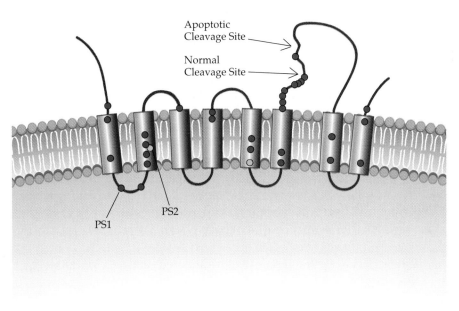

Figure 3 *In the eight-transmembrane model of presenilin, the N-terminal (N), C-terminal (C), and large hydrophilic loop regions face the cytoplasm. FAD mutations in PS1 are indicated by red circles and the two PS2 FAD mutations by yellow circles. The yellow circle in the second transmembrane region represents the Volga German FAD mutation N141I. The normal and apoptotic endoproteolytic cleavage sites are indicated by arrows.*

others have maintained that the generation of Aβ is merely a secondary consequence of AD neuropathology.[65]

The Consequences of Presenilin FAD Mutations

The *PS1* and *PS2* genes encode proteins predicted to be 463 and 448 amino acids in size, respectively. The presenilin proteins exhibit a serpentine-like topology because they are predicted to wind in and out of the membrane multiple times; each is predicted to contain six to nine transmembrane (TM) domains (seven TM[37,38]; nine TM[66]; eight TM[67,68]; six TM[67,69]). The currently favored model has eight TM domains[67,68] [*see Figure 1*]. APP is a single membrane-spanning protein. Whereas *PS1* and *PS2* share 67 percent amino acid identity, two nonhomologous regions are found in the N-terminus and the large hydrophilic loop lying between predicted TM6 and TM7. Two clusters of FAD mutations in *PS1* are observed in exons 5 and 8 [*see Figure 3 and Table 2*]. Together, these exons contain roughly 60 percent of the known *PS1* mutations. The average age of onset of symptoms observed for the mutations in these clusters is younger than that of other mutations in *PS1*. For the FAD mutations in exon 5, the mean age of onset is 40 years, and for exon 8 it is 43 years, compared with a mean age of onset of 47 years for all other mutations in *PS1*. Thus these mutation hot spots appear to be in critical functional or conformational domains in the presenilins.

Like *APP*, both *PS1* and *PS2* are ubiquitously expressed throughout the body.[37,38] In the brain, they are expressed predominantly in neurons[70]; thus, FAD resulting from *PS1/PS2* mutations most likely begins in neuronal populations. *PS1* is expressed more highly in brain areas that are vulnerable to AD neuropathologic changes relative to those that are spared in AD.[71] However, the expression pattern of *PS1* alone is not sufficient to account for selective vulnerability of certain sets of neurons in the AD brain.[71]

At the subcellular level, both PS1 and PS2 are primarily localized to intracellular membranes in the endoplasmic reticulum and, to a lesser extent, the

Golgi.[70] However, most groups have been unable to localize PS1 or PS2 to the plasma membrane. Cook and colleagues[72] report that PS1 is localized to dendrites in human neuronal cell lines. Although the biologic role of the presenilins is not known, valuable clues have been derived from the observation that both presenilins share homology with two proteins in *Caenorhabditis elegans* (*C. elegans*), SEL-12 (50 percent identity[73]) and SPE-4 (25 percent identity[74]). It is likely that the presenilin homologue, SEL-12 functions in the nematode as a co-receptor for the nematode Notch receptor LIN-12. On the other hand, it may facilitate the downstream signal events initiated by the LIN-12 Notch receptor or play a role in the trafficking or recycling of LIN-12 in the cell. In any event, because LIN-12 is needed for development and cell differentiation, one would expect the presenilins to be important during development. In support of this prediction, null mice in which the *PS1* gene has been knocked out die in utero or shortly after birth and display axial skeleton defects, somite segmentation defects, cerebral hemorrhage,[75] and neuronal loss.[76] Mutations in the other nematode homologue, *spe-4*, have been shown to disrupt protein transport during spermatogenesis. These findings, along with the subcellular localization of the presenilins to the endoplasmic reticulum and Golgi, suggest that the presenilins may play a role in intracellular trafficking and transport critical for proper development. In one of the *PS1* knockout studies, *PS1* was shown to be required for the expression of *Notch1* and *Dlll* (a delta-like Notch receptor ligand) in the paraxial mesoderm.[75] Hence, presenilins may also play a primary or secondary role in modulating gene expression. Two groups have recently shown that human PS1 and PS2 could rescue all aspects of the *sel-12* mutant phenotype in *C. elegans*.[77,78] However, FAD mutant versions of PS1 were considerably less effective in this task. These data provide further evidence that the presenilin genes may play a role in Notch signaling either as downstream effector molecules or in the trafficking and recycling of the Notch receptors. FAD mutations in *PS1* and *PS2* also lead to increased Aβ42:Aβ40 ratio, suggesting that normal intracellular transport and processing of APP is adversely affected by such mutations. However, effects of presenilin FAD mutations on APP expression have yet to be ruled out.

Presenilin Processing, Apoptosis, and FAD

Elucidation of the processing and degradation of PS1 and PS2 proteins in human neuronal cell lines transfected with presenilin genes reveal PS2 as a 53 to 54 kd protein[79,80] and PS1 as a 48 kd species.[59,60,81-84] In brain and untransfected native cell lines, little or no full-length PS1 and PS2 are found. The bulk of the presenilins appear to undergo proteolytic processing, which is visualized in the form of these two proteolytic fragments. In native human neuroglioma (H4) cells, the endogenous cleavage fragments of PS1 protein are 27 kd (N-terminal fragment) and 21 kd (C-terminal fragment). The fragments are the result of an endoproteolytic cleavage within the large hydrophilic loop between TM6 and TM7 [*see Figure 4*]. The fragments produced by this clip are saturable and are generated in a highly regulated manner.[81] PS2 protein also undergoes regulated endoproteolytic cleavage to give rise a 30 kd N-terminal fragment and a 25 kd C-terminal fragment.[79,80]

An important clue to the neuropathogenesis of AD has come from the observation that the presenilins can also undergo alternative endoproteolytic processing that is associated with both programmed cell death (apoptosis) and the FAD mutations.[83] Apoptotic cell death is reported to be one pathological feature of the AD brain, although the exact role and importance of apoptosis in the pathogenesis of AD remain unclear.[24,85,86] Apoptotic characteristics such as cell shrinkage, increased DNA fragmentation, and altered morphology of the nuclei of neurons in the AD brain have all been observed. It is now known that overexpression of the *PS2* gene in a variety of cell lines induces apoptosis. This is particularly evident in the brains of patients with the Volga German *PS2*-N141I mutation.[87,88] During apoptosis, PS2 protein also appears to undergo altered endoproteolytic processing at a site in the large hydrophilic loop that lies distal to the normal cleavage site[83] [*see Figure 4*]. The result is the production of two alternative endoproteolytic cleavage fragments: a 34 kd N-terminal fragment and a 20 kd C-terminal frag-

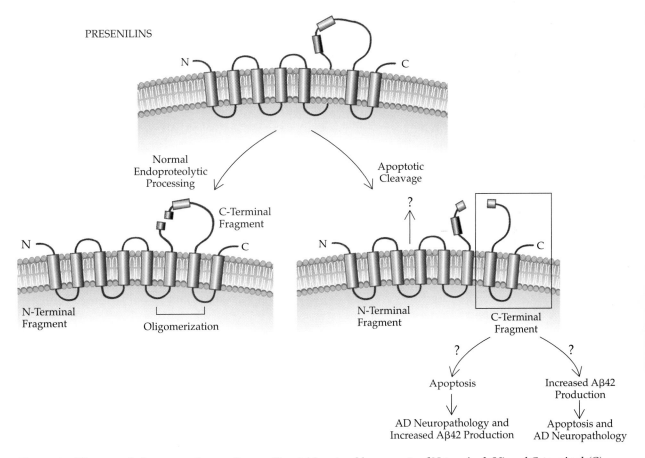

Figure 4   *The normal cleavage pathway of presenilin yields saturable amounts of N-terminal (N) and C-terminal (C) fragments. The normal clip site (red) is contained within a portion of the PS1 gene encoded by exon 9. Once generated, the two normal cleavage products can undergo oligomerization. During apoptosis, when the presenilins are overexpressed or proteasomal degradation is blocked, the presenilins can be diverted into the apoptotic cleavage pathway in which cleavage occurs at a distal CPP32-like cleavage site (dark purple) and gives rise to a larger N-terminal fragment and a smaller, detergent-resistant (insoluble) C-terminal fragment. Because the Volga German FAD mutation PS2-N141I leads to increased ratios of Aβ42:Aβ40, it is postulated that the generation of the alternative presenilin cleavage fragments may contribute to abnormal processing of APP and FAD pathogenesis. Once the apoptotic cleavage products are generated, it is not clear whether they induce apoptosis followed by AD neuropathology and increased Aβ42 production or whether they first induce increased production of Aβ42 followed by apoptotic changes and AD neuropathology.*

ment.[39,40,79,80,83] Likewise for PS1 protein, apoptosis resulting from overexpression of this protein in neuronal cell lines leads to two alternative fragments because of a more distal endoproteolytic cleavage event.[83]

The apoptosis-associated presenilin fragments are generated as a result of cleavage by the member of the family of apoptotic proteases known as CPP32 or caspase-3.[83] Induction and activation of cell death–specific proteases such as interleukin-1β converting enzyme and caspase-3 are instrumental steps in the apoptotic cell death process.[89-95]

The apoptotic cleavage of the presenilins by caspase-3 occurs endogenously when native human neuronal cells are treated with reagents that induce apoptosis (e.g., staurosporine and etoposide). Thus, the presenilins appear to be cell death substrates that are cleaved by a protease in the caspase-3 family during apoptosis.

These findings raise the question of the role played by the alternative presenilin cleavage products in the apoptotic cell death process. The biochemical properties of the apoptotic presenilin fragments are quite different from those of the normal cleavage products. Unlike the normal C-terminal fragments of PS1 and PS2 proteins, the apoptotic versions are detergent resistant and show a very slow rate of turnover.[39-40,79,80,83] In contrast, the normal presenilin cleavage products are soluble. The localization of the alternative presenilin C-terminal fragments to the detergent-resistant fraction suggests that they may be associated with cytoskeletal elements. Alternatively, they may be able to self-aggregate into a detergent (triton)-resistant protein complex/aggregate. In either case, the generation of such fragments could lead to profound disturbances in the intracellular compartments in which they are generated. The most likely compartment is the endoplasmic reticulum, in which APP also undergoes its first round of maturation and from which Aβ42 has recently been reported to originate (Virginia Lee, personal communication, 1997).

PS2 protein has previously been shown be degraded by the ubiquitin-proteasome pathway.[39,40,79,80] It is therefore conceivable that proteasomal degradation of full-length PS2 protein serves as a means for regulating the endoproteolytic cleavage of the full-length protein [*see Figure 4*]. Interestingly, when proteasome degradation is blocked, a greater amount of the apoptotic C-terminal fragment is observed. Thus, it appears that when excess levels of full-length PS2 protein override the regulated cleavage pathway, it is diverted into the alternative (apoptotic) pathway.

Perhaps the biggest question is how apoptosis-associated cleavage of the presenilins is related to the presenilin FAD mutations and associated increases in the ratio of Aβ42:Aβ40. PS2 genes carrying the Volga German FAD mutation N141I have been shown to lead to an approximately 3.8-fold increase in the ratio of Aβ42:Aβ40.[61] Interestingly, a similar increase (3.5-fold) has been observed for the ratio of the apoptotic:normal PS2 protein C-terminal cleavage fragments in a comparison of human neuroglioma cell lines expressing either wild-type PS2 or PS2-N141I proteins.[39,40,79,80,83] Increased generation of apoptotic PS2 protein cleavage fragments may play a role in the neuropathogenic mechanism by which the *PS2*-N141I gene mutation leads to an increased Aβ42:Aβ40 ratio and AD neuropathology [*see Figure 4*]. Along similar lines, an FAD mutation that leads to the deletion of exon 9 in the *PS1* gene[42] would remove the normal cleavage site while preserving the more distal apoptotic clip site, perhaps leading to increased apoptotic cleavage of this molecule as well.

Alternative PS1 and PS2 protein fragments generated under apoptotic conditions could potentially serve to alter the apoptotic threshold and make cells more vulnerable to apoptosis. The increase in apoptotic PS2 protein fragments in cells expressing the *PS2*-N141I FAD mutant raises the possibility that this fragment may contribute to the enhanced susceptibility to apoptosis associated with this mutation in transiently PS2 protein–transfected cell lines.[87] Treatment of a variety of neuronal cell lines with Aβ has previously been reported to induce apoptotic neuronal cell death both in vitro and in vivo[96-98] and to down-regulate the expression of the antiapoptotic gene *bcl-2* in primary neurons.[99] Given these observations, at least two potential pathogenic pathways can be envisaged for the presenilin FAD mutations [*see Figure 4*]. One possibility is that the apoptotic presenilin cleavage products may directly participate in inducing programmed cell death, which in turn leads to pathogenic changes associated with FAD (e.g., neuronal cell death, synaptic degeneration, and the generation of Aβ42). Alternatively, the apoptotic presenilin cleavage products may initially induce the increased production of Aβ42, which in turn induces programmed cell death and neuronal degeneration. It is still not clear how the enhanced generation of the insoluble, apoptotic presenilin fragments and the increased generation of Aβ42 resulting from FAD mutant forms of the presenilins interplay to bring about AD neuropathogenesis. However, this should be an important topic for future studies.

Pathogenesis of NFTs

In addition to β-amyloid deposition, NFTs are another prime hallmark of AD neuropathology, although they also occur in a number of other neurologic disorders, including progressive supranuclear palsy, dementia pugilistica, corticobasal degeneration, and Pick's disease.[100] NFTs consist primarily of abnormal cytoskeletal elements termed *paired helical filaments* (PHFs). The major component of PHF is the microtubule-associated protein tau, which is abnormally phosphorylated in the NFT.[100] Although normal tau is a relatively flexible protein, hyperphosphorylation causes the protein to become more rigid. As a result, microtubule function could be disordered, leading to impaired axonal transport and neurodegeneration. Hyperphosphorylation may also lead to an inability of tau to bind to microtubules, allowing it to accumulate in neurons. The relationship between NFT formation and the generation of Aβ is not clear. β-amyloid may induce tau directly to form PHF, may activate kinases that hyperphosphorylate tau, or adversely affect the intracellular milieu to promote NFT formation. Whatever the case, a true understanding of the etiology and pathogenesis of AD must account for NFT formation.

Therapeutic Options

Numerous industrial and academic laboratories have been attempting over the past decade to develop compounds that can retard or forestall the neuropathogenesis of AD.[101] These include a wide range of drugs, although the anticholinesterases (e.g., tacrine hydrochloride [Cognex] and donepezil [Aricept]) are currently the most popular treatment of choice. Other drugs under development include antioxidants, neurotrophic factor agonists and nootropic agents, muscarinic and nicotinic agonists, complement cascade inhibitors, cytokine inhibitors, apoptosis inhibitors, glutamate receptor (e.g., N-methyl-D-aspartate) antagonists, estrogen agonists, acetylcholine release stimulators, and phosphodiesterase inhibitors.

One of the most promising avenues for developing new treatments for AD has been driven by the genetic studies of Alzheimer's disease. This approach involves attempts to identify the various secretase enzymes [*see Figure 2*] that cleave APP to develop therapeutic strategies for blocking the production of β-amyloid. A number of pharmaceutical companies are close to testing inhibitors of the β and γ secretases. Other studies have been targeted at identifying factors and proteins that modulate and facilitate posttranslational modification and trafficking of APP. In this regard, a family of novel proteins known as the FE65-like proteins have been shown to bind to a portion of the cytoplasmic domain of APP that contains a clathrin-internalization motif utilized for endocytosis of APP.[102] FE65-like proteins could conceivably regulate the function and the trafficking and internalization of APP. Similar results have been obtained for the mammalian APP homologues known as the APP-like proteins, APLP1 and APLP2, which do not contain Aβ domains.[102-104] The development of compounds that stimulate activity of proteins like FE65 might be useful for promoting the nonamyloidogenic processing of APP.

Summary

Four different genes have now been identified as FAD loci. Pathogenic mutations in the genes *APP*, *PS1*, and *PS2* account for approximately 50 percent of early-onset FAD. Over 50 different mutations in these genes are able to cause AD with virtually 100 percent penetrance. A larger subset of AD cases with age of onset older than 60 years has been associated with the inheritance of the *APOE4* allele, which confers increased risk for AD but is not sufficient to cause the disease. Risk conferred for AD by *APOE4* is a dose-dependent phenomenon that is maximized in cases with onset between 61 and 70 years of age.

The common molecular event in AD neuropathogenesis that has been associated with all of the known FAD genes is the excessive generation or aggregation of the Aβ (and particularly Aβ42) peptide and the deposition of Aβ in the brain. Whereas the *APP* and presenilin gene mutations most likely adversely affect the processing and transport of APP to promote the enhanced production of Aβ42, *APOE4* appears to affect the aggregation rate of Aβ as evidenced by increased amyloid load in brains of *APOE4*-positive versus *APOE4*-negative AD patients.

Apoptotic cell death has previously been reported to be a pathological feature of AD, although the extent to which apoptosis contributes to the pathogenesis of Alzheimer's disease is not known. PS1 and PS2 proteins have also been implicated in the apoptotic pathway. Overexpression of PS2 protein has been shown to lead to apoptosis in transiently transfected cells. During apoptosis, the presenilins are diverted from their normal regulated endoproteolytic cleavage pathway into an alternative processing pathway in which they are cleaved by an apoptosis-activated protease in the caspase-3 family. It is therefore possible that the alternative PS1 and PS2 protein cleavage products generated under apoptotic conditions may serve as proapoptotic effectors that may alter the apoptotic threshold and make cells more susceptible to apoptosis. The increase in apoptotic PS2 protein fragments associated with the Volga German FAD mutant N141I suggests that these polypeptides may contribute to the report of enhanced susceptibility to apoptosis caused by this mutation. Proteasomal inhibitors cause enhanced generation of the PS2 protein apoptotic fragments and have also been reported to induce apoptosis in human monoblast cells[105] and cause increased secretion of Aβ.[106] Exactly how the aberrant cellular metabolism of the presenilins, the induction of apoptosis, the generation of apoptosis-associated resistant presenilin cleavage fragments, and the generation of Aβ interplay to contribute to AD neuropathogenesis clearly requires further study. Whereas the exact mechanisms by which the known FAD gene changes lead to the onset of AD remain unclear, novel therapies aimed at these targets may carry the greatest potential for developing effective treatments of this formidable disease.

References

1. Terry RD, Katzman R: Senile dementia of the Alzheimer type. *Ann Neurol* 14:497, 1983
2. Glenner GG, Wong CW: Alzheimer's disease: initial report of the purification and characterization of a novel cerebrovascular amyloid protein. *Biochem Biophys Res Commun* 120:885, 1984
3. Morris JC: The neurology of aging: normal versus pathologic change. *Geriatrics* 46:47, 1991
4. Bachman DI, Wolf PA, Linn R, et al: Prevalence of dementia and probable senile dementia of the Alzheimer's type in the Framingham Study. *Neurology* 42:1115, 1992
5. Snowdon D, Kemper S, Mortimer J, et al: Linguistic ability in early life and cognitive function and Alzheimer's disease in late life: findings from the Nun Study. *JAMA* 275:528, 1996
6. Van Duijn C, Tanja T, Haaxma R, et al: Head trauma and the risk of Alzheimer's disease. *Am J Epidemiol* 135:775, 1992
7. Lampe TH, Bird TD, Nochlin D, et al: Phenotype of chromosome 14-linked Alzheimer's disease: genetic and clinicopathological description. *Ann Neurol* 34:368, 1994
8. Tanzi RE, Gusella JF, Watkins PC, et al: Amyloid β protein gene: cDNA, mRNA distribution and genetic linkage near the Alzheimer locus. *Science* 235:880, 1987
9. Kang J, Lemaire H, Unterbeck A, et al: The precursor of Alzheimer's disease amyloid A4 protein resembles a cell-surface receptor. *Nature* 325:733, 1987
10. Goldgaber D, Lerman JI, McBride OW, et al: Characterization and chromosomal localization of a cDNA encoding brain amyloid of fibril protein. *Science* 235:877, 1987
11. Robakis NK, Ramakrishna N, Wolfe G, et al: Molecular cloning and characterization of a cDNA encoding the cerebrovascular and the neuritic plaque amyloid peptides. *Proc Natl Acad Sci USA* 84:4190, 1987
12. St. George-Hyslop PH, Tanzi RE, Polinsky RJ, et al: The genetic defect causing familial Alzheimer's disease maps on chromosome 21. *Science* 235:885, 1987
13. Tanzi RE, St. George-Hyslop PH, Haines JL, et al: The genetic defect in familial Alzheimer's disease is not tightly linked to the amyloid B-protein gene. *Nature* 329:156, 1987
14. Tanzi RE, Vaula G, Romano DM, et al: Assessment of amyloid ß protein precursor gene mutations in a large set of familial and sporadic Alzheimer disease cases. *Am J Hum*

Genet 51:273, 1992

15. St. George-Hyslop PH, Haines J, Rogaev E, et al: Genetic evidence for a novel familial Alzheimer's disease locus on chromosome 14. *Nat Genet* 2:330, 1992

16. Levy E, Carman MD, Fernandez-Madrid IJ, et al: Mutation of the Alzheimer's disease amyloid gene in hereditary cerebral hemorrhage, Dutch type. *Science* 248:1124, 1990

17. Goate A, Chartier-Harlin M, Mullan M, et al: Segregation of a missense mutation in the amyloid precursor protein gene with familial Alzheimer's disease. *Nature* 349:704, 1991

18. Chartier-Harlin M, Crawford F, Houlden H, et al: Early-onset Alzheimer's disease caused by mutations at codon 717 of the β-amyloid precursor protein gene. *Nature* 353:844, 1991

19. Hendriks L, van Duijn CM, Cras P, et al: Presenile dementia and cerebral haemorrhage linked to a mutation at codon 692 of the β-amyloid precursor protein gene. *Nat Genet* 1:218, 1992

20. Mullan M, Crawford F, Axelman K, et al: A pathogenic mutation for probable Alzheimer's disease in the N-terminus of β-amyloid. *Nat Genet* 1:345, 1992

21. Murrell J, Farlow M, Ghetti B, et al: A mutation in the amyloid precursor protein associated with hereditary Alzheimer disease. *Science* 254:97, 1991

22. Games D, Adams D, Alessandrini R, et al: Alzheimer-type neuropathology in transgenic mice overexpressing V717F β-amyloid precursor protein. *Nature* 373:523, 1995

23. Hsiao K, Chapman P, Nilsen S, et al: Correlative memory deficits, Aβ elevation and amyloid plaques in transgenic mice. *Science* 274:99, 1996

24. Cottman CW, Bridges R, Pike C, et al: Mechanisms of neuronal cell death in Alzheimer's disease. *Alzheimer's Disease: Advances in Clinical and Basic Research.* Corain B, Ed. John Wiley and Sons, New York, 1993, p 281

25. Pike CJ, Burdick D, Walencewicz AJ, et al: Neurodegeneration induced by β-amyloid peptides in vitro: the role of peptide assembly state. *J Neurosci* 13:1676, 1993

26. Saunders AM, Strittmatter WJ, Schmechel D, et al: Association of apolipoprotein E allele e4 with late-onset familial and sporadic Alzheimer's disease. *Neurology* 43:1467, 1993

27. Corder E, Saunders A, Strittmatter W, et al: Gene dose of apolipoprotein E type 4 allele and the risk of Alzheimer's disease in late onset families. *Science* 261:921, 1993

28. Corder E, Saunders A, Risch N, et al: Protective effect of apolipoprotein E type 2 for late onset Alzheimer disease. *Nat Genet* 7:180, 1994

29. van Duijn C, de Knijff P, Wehnert A, et al: The apolipoprotein E e2 allele is associated with increased risk of early onset Alzheimer's disease and a reduced survival. *Ann Neurol* 37:605, 1995

30. Hyman BT, Gomez-Isla T, Brigg M, et al: Apolipoprotein E and cognitive change in an elderly population. *Ann Neurol* 40:55, 1996

31. Evans DA, Beckett LA, Field TS, et al: Apolipoprotein E e4 and incidence of Alzheimer disease in a community population of older persons. *JAMA* 277:822, 1997

32. Blacker D, Haines JL, Rodes L, et al: APOE-4 and age of onset of Alzheimer's disease: the NIMH Genetics Initiative. *Neurology* 48:139, 1997

33. National Institute of Aging/Alzheimer's Association Working Group: Apolipoprotein genotyping in Alzheimer's disease position statement. *Lancet* 347:1091, 1996

34. Statement on use of apolipoprotein E testing for Alzheimer disease. American College of Medical Genetics/American Society of Human Genetics Working Group on ApoE and Alzheimer's disease. *JAMA* 274:1627, 1995

35. Schellenberg GD, Bird TD, Wijsman EM, et al: Genetic linkage evidence for a familial Alzheimer's disease locus on chromosome 14. *Science* 258:668, 1992

36. Van Broeckhoven C, Backhovens H, Cruts M, et al: Mapping of a gene predisposing to early-onset Alzheimer's disease to chromosome 14q24.3. *Nat Genet* 2:334, 1992

37. Sherrington R, Rogaev EI, Liang Y, et al: Cloning of a novel gene bearing missense mutations in early onset familial Alzheimer disease. *Nature* 375:754, 1995

38. Levy-Lahad E, Wasco W, Poorkaj P, et al: Candidate gene for the chromosome 1 familial Alzheimer's disease locus. *Science*, 269:973, 1995

39. Tanzi RE, Kovacs DM, Kim T-W, et al: The gene defects responsible for familial Alzheimer's disease. *Neurobiol Dis* 3:159, 1996

40. Tanzi RE, Kovacs DM, Kim T-W, et al: The presenilin genes and their role in early-onset familial Alzheimer's disease. *Alzheimer Dis Rev* 1:91, 1996

41. Hardy J: Amyloid, the presenilins and Alzheimer's disease. *Trends Neurosci* 20:154, 1997

42. Perez-Tur J, Froelich S, Prihar G, et al: A mutation in Alzheimer's disease destroying a splice acceptor site in the presenilin-1 gene. *Neuroreport* 7:297, 1996

43. Rossor MN, Fox NC, Beck J, et al: Incomplete penetrance of familial Alzheimer's disease in a pedigree with a novel presenilin-1 gene mutation. *Lancet* 347:1560, 1996

44. Van Broeckhoven C, Backhovens H, Cruts M, et al: APOE genotype does not modulate age of onset in families with chromosome 14 encoded Alzheimer's disease. *Neurosci Lett* 169:179, 1994

45. Sorbi S, Nacmias B, Forleo P, et al: Epistatic effect of APP717 mutation and apolipoprotein E genotype in familial Alzheimer's disease. *Ann Neurol* 38:124, 1993

46. Cruts M, Hendriks L, Van Broeckhoven C: The presenilin genes: a new gene family involved in Alzheimer disease pathology. *Hum Mol Genet* 5:1449, 1996

47. Weidemann A, Konig G, Bunke D, et al: Identification, biogenesis and localization of precursors of Alzheimer's disease A4 amyloid protein. *Cell* 57:115, 1989

48. Sisodia SS, Koo EH, Beyreuther K, et al: Evidence that β-amyloid protein in Alzheimer's disease is not derived by normal processing. *Science* 248:492, 1990

49. Esch FS, Keim PS, Beattie EC, et al: Cleavage of amyloid β peptide during constitutive processing of its precursor. *Science* 248:1122, 1990

50. Gandy SE, Greengard P: Processing of Aβ-amyloid precursor protein: cell biology, regulation and role in Alzheimer's disease. *Int Rev Neurobiol* 36:29, 1994

51. Shoji M, Golde TE, Ghiso J, et al: Production of the Alzheimer amyloid β protein by normal proteolytic processing. *Science* 258:126, 1992

52. Haass C, Schlossmacher MG, Hung AY, et al: Amyloid β-peptide is produced by cultured cells during normal metabolism. *Nature* 359:322, 1992

53. Cai X-D, Golde TE, Younkin SG: Release of excess amyloid β protein from a mutant amyloid β protein precursor. *Science* 259:514, 1993

54. Suzuki N, Cheung TT, Cai XD, et al: An increased percentage of long amyloid beta protein secreted by familial amyloid beta protein precursor (beta APP717) mutants. *Science* 264:1336, 1994

55. Citron M, Oltersdorf T, Haass C, et al: Mutation of the β-amyloid precursor protein in familial Alzheimer's disease increases β-protein production. *Nature* 360:672, 1991

56. Jarrett JT, Berger EP, Lansbury PT: The carboxy terminus of the beta amyloid protein is critical for the seeding of amyloid formation: implications for the pathogenesis of Alzheimer's disease. *Biochemistry* 32:4693, 1993

57. Scheuner D, Eckman C, Jensen M, et al: Aβ42(43) is increased in vivo by the PS1/2 and APP mutations linked to familial Alzheimer's disease. *Nat Med* 2:864, 1996

58. Borchelt DR, Thinakaran G, Eckman CB, et al: Familial Alzheimer's disease-linked presenilin 1 variants elevate Aβ1-42/1-40 ratio in vitro and in vivo. *Neuron* 17:1005, 1996

59. Citron M, Westaway D, Xia W, et al: Mutant presenilins of Alzheimer's disease increase production of 42-residue amyloid beta-protein in both transfected cells and transgenic mice. *Nat Med* 3:67, 1992

60. Duff K, Eckman C, Zehr C, et al: Increased amyloid-β42(43) in brains of mice expressing mutant presenilin 1. *Nature* 383:710, 1996

61. Tomita T, Maruyama K, Takaomi CS, et al: The presenilin 2 mutation (N141I) linked to familial Alzheimer disease (Volga German families) increases the secretion of amyloid ß protein ending at the 42nd (or 43rd) residue. *Proc Natl Acad Sci USA* 94:2025, 1997

62. Lemere CA, Lopera F, Koski KS, et al: The E280A presenilin 1 Alzheimer mutation produces increased Aβ42 deposition and severe cerebellar pathology. *Nat Med* 2:1146, 1996

63. Gomez-Isla T, Wasco W, Pettingell WP, et al: Novel presenilin 1 gene mutation: increased β-amyloid and neurofibrillary changes. *Ann Neurol* 41:809, 1997

64. Hyman BT, West HL, Rebeck GW, et al: Quantitative analysis of senile plaques in Alzheimer's disease: observation of log-normal size distribution and molecular epidemiology of differences associated with ApoE genotype and trisomy 21 (Down syndrome). *Proc Natl Acad Sci USA* 92:3586, 1995

65. Roses AD: Apolipoprotein E affects the rate of Alzheimer disease expression: β amyloid burden is a secondary consequence dependent on APOE genotype and duration of disease. *J Neuropathol Exp Neurol* 53:429, 1994

66. Slunt HH, Thinakaran G, Lee MK, et al: Nucleotide sequence of the chromosome 14-encoded S182 cDNA and revised secondary structure prediction. *Int J Exp Clin Invest* 2:188, 1995

67. Doan A, Thinakaran G, Borchelt DR, et al: Protein topology of presenilin 1. *Neuron* 17:1023, 1996

68. Li X, Greenwald I: Membrane topology of the *C. elegans* SEL-12 presenilin. *Neuron* 17:1015, 1996

69. Lehmann S, Chiesa R, Harris DA: Evidence for a six-transmembrane domain structure

of presenilin 1. *J Biol Chem* 272:12047, 1997

70. Kovacs DM, Fausett HJ, Page KJ, et al: Alzheimer associated presenilins 1 and 2: neuronal expression in brain and localization to intracellular membranes in mammalian cells. *Nat Med* 2: 224, 1996

71. Page K, Hollister R, Tanzi RE, et al: In situ hybridization of presenilin 1 mRNA in Alzheimer's disease and in lesioned rat brain. *Proc Natl Acad Sci USA* 93:14020, 1996

72. Cook DG, Sung JC, Golde TE, et al: Expression and analysis of presenilin 1 in a human neuronal system: localization in cell bodies and dendrites. *Proc Natl Acad Sci USA* 93: 9223, 1996

73. Levitan D, Greenwald I: Facilitation of lin-12-mediated signalling by sel-12, a *Caenorhabditis elegans* S182 Alzheimer's disease gene. *Nature* 377:351, 1995

74. L'Hernault SW, Arduengo PM: Mutation of a putative sperm membrane protein in *Caenorhabditis elegans* prevents sperm differentiation but not its associated meiotic-divisions. *J Cell Biol* 119:55, 1992

75. Wong P, Zheng H, Chen H, et al: Presenilin 1 is required for Notch1 and Dll1 expression in the paraxial mesoderm. *Nature* 387:288, 1997

76. Shen J, Bronson R, Feng Chen D, et al: Skeletal and CNS defects in presenilin-1 deficient mice. *Cell* 89:629, 1997

77. Levitan D, Doyle TG, Brousseau D, et al: Assessment of normal and mutant human PS function in *Caenorhabditis elegans*. *Proc Natl Acad Sci USA* 93:14940, 1996

78. Baumeister R, Leimer U, Zweckbronner I, et al: Human presenilin-1, but not familial Alzheimer's disease (FAD) mutants, facilitate *Caenorhabditis elegans* notch signalling independently of proteolytic processing. *Genes Funct* 1:149, 1997

79. Kim T-W, Hallmark OG, Pettingell W, et al: Proteolytic processing and ubiquitin-proteasomal degradation of wild-type and mutant forms of presenilin 2. *Alzheimer's Disease: Biology, Diagnosis, and Therapeutics*. Iqbal K, Winblad B, Nishimura T, et al, Eds. John Wiley and Sons, West Sussex, England, 1997, p 576

80. Kim T-W, Hallmark OG, Pettingell WH, et al: Endoproteolytic cleavage and proteasomal degradation of presenilin 2 in transfected cells. *J Biol Chem* 272:1106, 1997

81. Thinakaran G, Borchelt D, Lee M, et al: Endoproteolysis of presenilin 1 and accumulation of processed derivatives in vivo. *Neuron* 17:181, 1996

82. Mercken M, Takahashi H, Honda T, et al: Characterization of human presenilin 1 using N-terminal specific monoclonal antibodies: evidence that Alzheimer mutations affect proteolytic processing. *FEBS Lett* 389:297, 1996

83. Kim T-W, Pettingel WH, Jung YK, et al: Aberrant cleavage of Alzheimer-associated presenilins during apoptosis by a caspase-3 family protease. *Science* 277:373, 1997

84. Walter J, Grünberg J, Capell A, et al: Proteolytic processing of the Alzheimer's disease associated presenilin-1 generates an *in vivo* substrate for protein kinase C. *Proc Natl Acad Sci USA*, 94:5349, 1997

85. LeBlanc A: Apotosis and Alzheimer's disease. *Molecular Mechanism of Dementia*. Wasco W, Tanzi RE, Eds. Humana Press, Totowa, NJ, 1996, p 57

86. Su J, Anderson A, Cummings B, et al: Immunohistochemical evidence for apotosis in Alzheimer's disease. *NeuroReport* 5:2529, 1994

87. Wolozin B, Iwasaki K, Vito P, et al: PS2 participates in cellular apoptosis: constitutive activity conferred by Alzheimer mutation. *Science* 274:1710, 1996

88. Deng G, Pike CJ, Cotman CW: Alzheimer-associated presenilin-2 confers increased sensitivity to apoptosis in PC12 cells. *FEBS Lett* 397:50, 1996

89. Chinnaiyan AM, Dixit VM: The cell-death machine. *Curr Biol* 6:555, 1991

90. Fraser A, Evan G: A license to kill. *Cell* 85:781, 1996

91. Jacobson MD, Weil M, Raff MC: Role of Ced-3/ICE-family proteases in staurosporine-induced programmed cell death. *J Cell Biol* 133:1041, 1996

92. Jacobson MD, Weil M, Raff MC: Programmed cell death in animal development. *Cell* 88:347, 1997

93. Kaufmann SC, Desnoyers S, Ottaviano Y, et al: Specific proteolytic cleavage of poly(ADP-ribose) polymerase: an early marker of chemotherapy-induced apoptosis. *Cancer Res* 53:3976, 1993

94. Martin SJ, Green DR: Protease activation during apotosis: death by a thousand cuts? *Cell* 82:349, 1995

95. Nicholson DW, Ali A, Thornberry NA, et al: Identification and inhibition of the ICE/Ced-3 protease necessary for mammalian apoptosis. *Nature* 376:37, 1995

96. Forloni G, Chiesa R, Smiroldo S, et al: Apoptosis mediated neurotoxicity induced by

chronic application of ß amyloid fragment 25-35. *NeuroReport* 4:523, 1993

97. LeBlanc A: Increased production of 4 kDa amyloid β-peptide in serum deprived human primary neuron cultures: possible involvement of apoptosis. *J Neurosci* 15:7837, 1995

98. Loo D, Copani A, Pike CJ, et al: Apoptosis is induced by β-amyloid in cultured central nervous system neurons. *Proc Natl Acad Sci USA* 90:7951, 1993

99. Paradis E, Douillard H, Koutroumanis M, et al: Amyloid β peptide of Alzheimer's disease downregulates Bcl-2 and upregulates Bax expression in human neurons. *J Neurosci* 16:7533, 1996

100. Goedert M, Trojanowski JQ, Lee VM-Y: Tau protein and the neurofibrillary pathology of Alzheimer's disease. *Mechanisms of Dementia*. Wasco W, Tanzi R, Eds. Humana Press, Totowa, NJ, 1997, p 199

101. Alzheimer's disease-drug status update in ID research alerts. *Alzheimer Dis* 2:383, 1997

102. Guenette SY, Chen J, Jondro PD, Tanzi RE: Association of a novel human FE65-like protein with the cytoplasmic domain of the β-amyloid precursor protein. *Proc Natl Acad Sci USA* 93:10832, 1996

103. Wasco W, Bupp K, Magendantz M, et al: Identification of a mouse brain cDNA that encodes a protein related to the Alzheimer disease-associated amyloid β protein precursor. *Proc Natl Acad Sci USA* 89:10758, 1992

104. Wasco W, Gurubhagavatula S, Paradis MD, et al: Isolation and characterization of APLP2 encoding a homologue of the Alzheimer's associated amyloid β protein precursor. *Nat Genet* 5:95, 1993

105. Imajoh-Ohmi S, Kawaguchi T, Suguyama S, et al: Lactacystin, a specific inhibitor of the proteasome induces apoptosis in human monoblast U937 cells. *Biochem Biophys Res Commun* 217:1070, 1995

106. Marambaud P, Chevallier N, Barelli H, et al: Proteasome contributes to the α-secretase pathway of amyloid precursor protein in human cells. *J Neurochem* 68:698, 1997

Acknowledgments

This work was supported by grants from the National Institute on Aging, National Institute of Neurological Disorders and Stroke, and The Metropolitan Life Foundation.

Figures 1 through 4   Kathy Konkle.

CHAPTER 5

Genetics of Epilepsy

James O. McNamara, M.D.

The epilepsies constitute a common, devastating, and remarkably diverse collection of disorders that affect roughly one percent of the population in the United States.[1] Current therapies consist of pharmacological agents that inhibit seizures, yet at least 30 percent of affected individuals have persistent seizures despite optimal drug therapy.[2] No means of preventing or curing epilepsy are available apart from invasive neurosurgical procedures that are suitable for only a minority of patients.

Insight into the mechanisms of the epilepsies in cellular and molecular terms may lead to novel therapeutic approaches. Progress in understanding the underlying mechanisms has emerged in the past two decades,[3] but has not translated into improved therapies. The fundamental problem limiting these investigations is the enormous complexity of the mammalian brain.

The intent here is to outline the extraordinary progress of genetic investigations of the epilepsies in the past several years. To provide a framework for understanding genetic studies, the terminology and classification of the epilepsies are outlined, with particular emphasis on the role of genetic determinants in the various syndromes. Successes with identifying the chromosomal localization of mutant genes underlying several epilepsy syndromes are then reviewed, and six epilepsy syndromes in which the mutant gene has been identified are considered. Finally, the issue of how these discoveries might lead to a better understanding of the pathogenesis of the disorder and ultimately more effective therapy is addressed.

Terminology and Classification of Epileptic Seizures

The term *seizure* refers to a transient alteration of behavior caused by the disordered, synchronous, and rhythmic firing of populations of central nervous system neurons. The term *epilepsy* refers to a chronic disorder of brain function characterized by recurrent seizures. Seizures can be termed *nonepileptic* when they are evoked in a normal brain by treatments such as electroshock or chemical convulsants, or *epileptic* when they occur without apparent provocation in the brain of a patient with epilepsy.

Seizures are thought to arise from the cerebral cortex. Epileptic seizures have been classified into *partial* seizures, those beginning focally in a small region of the cortex, and *generalized* seizures, those characterized by widespread involvement of both hemispheres from the outset.[4] The clinical manifestations of a seizure are determined by the functions normally served by the cortical region giving rise to the seizure. For example, a seizure involving the motor cortex is associated with clonic jerking of the body part controlled by this region of the cortex. Examples of generalized seizures include *myoclonic, absence,* and *tonic-clonic.* A myoclonic seizure consists of sudden, brief, lightning-like jerks that may be generalized or limited to one or more muscle groups, ordinarily without loss of consciousness.[5] An absence seizure is characterized by the abrupt interruption of ongoing activities associated with a blank stare and impaired responsiveness, typically lasting a few seconds. A tonic-clonic seizure consists of tonic contraction of muscles throughout the body followed by clonic convulsive movements; the seizure is accompanied by loss of consciousness, typically lasts 40 to 60 seconds, and is followed by impaired consciousness lasting variable periods of time. Partial seizures are subdivided into *simple* or *complex,* the former associated with preservation of consciousness and the latter with impairment of consciousness. Both simple and complex partial seizures can remain confined to a localized region of the cortex or propagate widely in the nervous system, resulting in secondarily generalized tonic-clonic seizures. A given patient often exhibits more than one kind of seizure. In addition to this *seizure* classification, a more recent classification specifies *epileptic syndromes,* which refers to a cluster of symptoms frequently occurring together and includes seizure types, etiology, age of onset, and other factors.[6] The type of seizure specifies the optimal antiseizure drug, whereas the type of epilepsy reflects the etiology and prognosis.

Evidence of the enormous diversity of the epilepsies lies in the fact that more than 40 distinct epilepsy syndromes have been identified.[6] A subset of the many epilepsy syndromes in which an inherited pattern is prominent is presented [*see Table 1*]. This classification will require continual updating as the mechanisms of the epilepsies are better understood. One basis for classification of epileptic syndromes is whether the epileptic seizure is partial (localization-related) or generalized (initial clinical manifestation indicates involvement of both cerebral hemispheres) at onset. Another basis is whether the epileptic syndrome is symptomatic (has an identified cause) or idiopathic (lacks an identified cause). Progress in the genetics of epilepsies is one way of updating this classification system because genetic etiologies are being ascertained for some of the idiopathic and symptomatic epilepsies.

Genetic Determinants of the Epilepsies

A brief overview of some of the genetic determinants of epilepsy [*see Table 1*] emphasizes that the majority of epileptic syndromes for which a mutant gene has been identified are symptomatic, generalized-onset epilepsies. These diseases include epileptic seizures as a prominent feature but are associated with additional prominent abnormalities of CNS function and often structure. Patients with these disorders experience epileptic seizures but also exhibit profound and/or progressive cognitive impairments and other neurologic symptoms and signs, alerting the clinician to the presence of a genetic disturbance (or other disorder). Indeed, the sole exception to this pattern thus far is the autosomal dominant frontal lobe epilepsy syndrome, which is classified as a localization-related, idiopathic form; epileptic seizures constitute the only nervous system abnormality evident in these individuals. Chromosomal linkage has been established for another epilepsy in this category, partial epilepsy with auditory symptoms, but the mutant gene has yet to be identified.

Table 1   Epilepsy Syndromes

Type	Syndrome	Pattern of Inheritance	Locus	Gene
Idiopathic Partial-Onset Epilepsy	Autosomal dominant frontal lobe epilepsy	Autosomal dominant	20q13.2–q13.3	Nicotinic acetylcholine receptor subunit α_4
	Partial epilepsy with auditory symptoms	Autosomal dominant	10q22–q24	Unknown
Symptomatic Partial-Onset Epilepsy	Periventricular heterotopia	X-linked dominant	Xq28	Unknown
	Northern epilepsy syndrome	Autosomal recessive	8p22–p23.1	Unknown
Idiopathic Generalized-Onset Epilepsy	Benign familial neonatal convulsions	Mendelian	20q/8q	Unknown
	Childhood absence epilepsy	Complex	Unknown	Unknown
	Juvenile myoclonic epilepsy	Complex	6p	Unknown
	Juvenile absence epilepsy	Complex	Unknown	Unknown
	Generalized tonic-clonic epilepsy	Complex	Unknown	Unknown
Symptomatic Generalized-Onset Epilepsy	Miller-Dieker syndrome	De novo	17q13.3	Platelet-activating factor acetylhydrolase
	Northern epilepsy syndrome	Autosomal recessive	8p22–p23.1	Unknown
	Unverricht-Lundborg disease	Autosomal recessive	21q22.3	Cystatin B
	Myoclonic epilepsy with ragged red fiber disease	Mitochondrial	Mitochondrial	tRNA^{lys}
	Batten disease	Autosomal recessive	16p	CLN3
	Angelman syndrome	Mostly de novo	15q11–q13	UBE3A
	Lafora's disease	Autosomal recessive	6q23–q25	Unknown

One feature of obvious clinical relevance is the proportion of patients with epilepsy afflicted with each of the epilepsy syndromes [*see Table 1*]. Among those syndromes for which the mutant gene has been identified, each is rare and clearly accounts for less than one percent of patients afflicted with epilepsy. Indeed, every epilepsy syndrome is extremely rare, with the exception of four: juvenile myoclonic epilepsy, childhood absence epilepsy, juvenile absence epilepsy, and generalized tonic-clonic epilepsy. These are the most common forms of idiopathic generalized epilepsies. The idiopathic generalized epilepsy syndromes as a whole account for roughly 40 to 50 percent of all epilepsies.[1] Difficulties in accurate diagnosis preclude unambiguously accurate assessments of the proportion of patients with epilepsy exhibiting each of the idiopathic generalized epilepsy syndromes; nonetheless, it appears that juvenile myoclonic epilepsy is an especially common form, accounting for approximately 10 percent of all epilepsies.[7]

A number of other epileptic syndromes account for the remaining percentage of epilepsy, namely, symptomatic, partial-onset epilepsies such as temporal lobe epilepsies. These lack an overt genetic determinant and account for the majority of adult-onset epilepsies.[2,8] Typical causes are neoplasms or vascular malformations or complications months to years after brain injury or stroke. Some evidence indicates that genetic determinants may contribute to susceptibility to epilepsy arising after a brain insult.[9]

The patterns of inheritance of different epilepsies and the proportion of the epilepsies inherited in these distinct patterns bears a striking resemblance to other human diseases, such as Alzheimer's disease (AD). Some are inherited in mendelian patterns (e.g., autosomal dominant and recessive) and others are inherited as complex disorders [*see Table 1*]. Mutant genes are identified for six rare forms of epilepsy, each of which is inherited in a mendelian or mitochondrial pattern. In fact, the most common forms of genetically determined epilepsy show a complex pattern of inheritance and are called idiopathic generalized-onset epilepsies. These inheritance patterns are reminiscent of those observed in AD in which several mutant genes underlying rare forms of AD have been determined, including the amyloid precursor protein[10,11] and the presenilins[12,13] [*see Chapter 4*]. Each of these subsets of AD is inherited in a mendelian pattern but accounts for only a tiny fraction of all AD (less than one percent). By contrast, the most common form of AD, late-onset AD, appears to be a complex genetic disease with multiple genetic determinants, one of which is conferred by the apolipoprotein ε4 isotype.[14,15] Some of the lessons learned in the successful hunt for genes conferring susceptibility to common forms of AD may facilitate elucidation of the genes conferring susceptibility to common forms of epilepsy.

Progress with Linkage Analysis

The first step in localizing a mutant gene underlying an epileptic syndrome is accomplished by astute clinicians, through recognition that a particular form of epilepsy is familial. Clinicians are crucial for defining the phenotype in a large pedigree and in recruitment of family members for genetic study. Accurate characterization of the clinical phenotype includes a history, a physical examination, an electroencephalogram (EEG), and a blood sample for DNA isolation. The collaborating molecular geneticist uses this information together with polymorphic DNA markers distributed throughout the genome to establish a statistical relationship or linkage between the phenotype and DNA markers in a specific region of a specific chromosome [*see Chapter 1*]. Positional cloning and identification of candidate genes leads to characterization of the mutant gene responsible for the disease.

Benign Familial Neonatal Convulsions

Benign familial neonatal convulsions constitute a rare disorder characterized by unprovoked tonic seizures, ocular symptoms, apnea, and other autonomic features. Approximately half of the affected individuals experience the first seizure by the third day of life, and roughly two thirds experience remission by six weeks of age. Intellectual development is usually normal, but 10 to 20 percent acquire epilepsy later in life.[16] The segregation pattern of benign familial neonatal convulsions is consistent with an autosomal dominant pattern of inheritance with high penetrance. Leppert and colleagues[17] established tight linkage (maximum logarithm of the odds [LOD] score of 5.64) between two polymorphic markers on chromosome 20q and the disease locus in a single pedigree containing 19 affected individuals spanning five generations. Malafosse and colleagues[18] confirmed this localization in additional families. A second locus for benign familial neonatal convulsions was subsequently identified on chromosome 8q[19] in another large pedigree spanning three generations with 14 affected individuals. Identification of two distinct chromosomal regions of mutant genes linked to the same disorder provides strong evidence that mutations of at least two different genes can be responsible for the same phenotype; stated differently, genetic heterogeneity has been established. Additional genetic heterogeneity is suggested by the finding of an autosomal recessive form of this disorder,[20] the chromosomal localization of which is unclear.

Partial Epilepsy with Auditory Symptoms

A pedigree has been identified in which 11 members were classified as affected with partial seizures with onset between eight and 19 years of age.[21] Inheritance was consistent with an autosomal dominant pattern. Seizure types included simple partial, complex partial, and tonic-clonic. At least half of the affected individuals reported nonspecific auditory disturbances such as hearing a ringing noise (tinnitus) and machine-like humming at seizure onset. Affected individuals exhibited normal intelligence and no abnormalities on neurologic examination. The seizures appeared to be responsive to phenytoin. Linkage between markers on chromosome 10q22-q24 and the disease locus was established with a maximum two-point LOD score of 3.99. Key recombinants permitted narrowing the susceptibility locus to a 10 centimorgan (cM) interval.

Northern Epilepsy Syndrome

Northern epilepsy syndrome has been identified in northern Finland and is characterized by the onset of tonic-clonic seizures in otherwise healthy children between five and 10 years of age.[22] Impairments of cognitive function begin two to five years after seizure onset and progress rapidly. All patients have tonic-clonic seizures and roughly one third also exhibit complex partial seizures. Seizures occur with very high frequency despite optimal antiseizure treatment, with seizure frequency typically peaking at one to two per week in early adulthood and diminishing thereafter. Many patients become seizure free after 35 years of age. Inheritance is consistent with an autosomal recessive pattern. Linkage analysis permitted assignment of the mutant gene to a region of 4 cM on chromosome 8p22-p23.1.[23]

Lafora's Disease

Lafora's disease is an autosomal recessive form of progressive myoclonic epilepsy characterized by the onset of generalized tonic-clonic seizures or drop attacks in the second decade of life. Shortly thereafter, massive myoclonic jerks occur, followed by a rapidly progressive dementia with apraxia and visual loss. Most affected individuals die within a decade of disease onset. The distinctive histopathology of polyglucosan intracellular inclusion bodies was described by Lafora in the early 1900s. Diagnosis is established by the typical clinical picture combined with identification of the characteristic periodic acid-Schiff–positive cytoplasmic inclusions in myoepithelial cells of the apocrine sweat glands in an axillary skin biopsy. The precise molecular defect responsible for the clinical disorder and polyglucosan accumulation is unknown. Serratosa and colleagues[7] screened the human genome with microsatellite markers at 13 cM intervals and used linkage analysis in nine families together with homozygosity mapping in four consanguineous families to define the Lafora's disease gene region. The locus was narrowed to a 2.5 cM region on chromosome 6q23-q25.

Periventricular Heterotopia

Periventricular heterotopia is a disorder of cortical development characterized by masses of neural cells bordering the lateral ventricles of the cerebral hemispheres [*see Figure 1*]. Affected individuals are at high risk for epilepsy of diverse types, including complex partial and generalized tonic-clonic seizures, but exhibit no other neurologic abnormalities.[24] Approximately 67 percent of affected individuals exhibit epilepsy, with an average age of onset of 14 years. The disorder is inherited in a dominant form with full penetrance as assessed by magnetic resonance imaging or computed tomography. Periventricular heterotopia was shown to be closely linked (maximal multipoint LOD score of 5.37) to markers in distal Xq28.[24] The disease was detected only in females, and these females were obligatory mosaics for the mutation. The mutation is likely lethal to males early in embryogenesis or in the neonatal period.

Juvenile Myoclonic Epilepsy and Related Idiopathic Generalized Epilepsies

Included among the more than 40 specific epileptic syndromes described in humans[6] are four syndromes classified as idiopathic generalized epilepsies. These include juvenile myoclonic epilepsy, childhood absence epilepsy, juvenile absence epilepsy, and generalized tonic-clonic epilepsy. The clinical features of these syndromes are described elsewhere.[6] These syndromes appear to be complex genetic disorders; that is, genetic diseases that do not follow a simple mendelian mode of inheritance.[25] In contrast to the rarity of other genetic forms of epilepsy, such as benign familial neonatal convulsions, these four syndromes together account for the majority of the idiopathic generalized

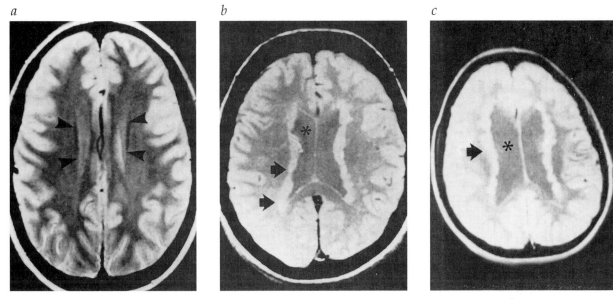

Figure 1 *(a) T1-weighted, magnetic resonance imaging (MRI) scan of a normal brain shows good definition of white matter from gray matter and shows that the periventricular zone (arrowheads) is normally thin and indistinct. (b and c) MRIs of individuals with periventricular heterotopia performed with a similar technique are similar to one another, given differences in reproduction, but are distinctly abnormal. The periventricular zone of both lateral ventricles is distorted and enlarged by increased signal intensity that forms a continuous lining without clear asymmetry (arrows). The heterotopic nodules have MRI signal characteristics identical to gray matter, suggesting that they correspond to neurons. The ventricles (asterisks) are also somewhat enlarged relative to normal. These distinctive MRI abnormalities serve as the basis of phenotyping that permitted identification of the locus at Xq28.*

epilepsies that are thought to represent 30 to 40 percent of all epilepsies.[26] Although none of the susceptibility genes has been identified in these syndromes, the large number of individuals afflicted underscores their importance.

The idea that a genetic determinant underlies these syndromes emerges from two lines of evidence. First, in families of probands with juvenile myoclonic epilepsy, first-degree relatives (i.e., siblings, parents, or offspring, all of whom share half of the genes of the proband) have an increased risk of acquiring juvenile myoclonic epilepsy, childhood absence epilepsy, juvenile absence epilepsy, or generalized tonic-clonic epilepsy.[27-30] The increased risk appears to be approximately 20-fold in that eight percent of first-degree relatives of someone with absence epilepsy or generalized tonic-clonic epilepsy have a form of idiopathic generalized epilepsy,[31] compared with 0.4 percent of the general population. (The prevalence of epilepsy is one percent,[1] and idiopathic generalized epilepsy accounts for roughly 40 percent of epilepsy[26]). The second line of evidence derives from twin studies, in which 65 percent of monozygotic twins but only 24 percent of dizygotic twins were concordant for idiopathic generalized epilepsy.[32] Because both types of twins are likely to experience similar environmental influences, the two to three times greater concordance in the monozygotic twins supports a genetic determinant. The nature or number of genes responsible remains unclear. The lack of 100 percent concordance in monozygotic twins also suggests that environmental factors may contribute to idiopathic generalized epilepsies.

Analyses of idiopathic generalized epilepsy patterns in monozygotic versus dizygotic twins provide clues to the genetic mechanisms. Berkovic and colleagues[32] studied 245 twin pairs in which one or both twins had seizures. Twenty pairs of monozygotic twins were identified in which at least one twin had childhood absence epilepsy, juvenile absence epilepsy, or juvenile myoclonic epilepsy. Thirteen of the 20 twins (65 percent) were concordant for epilepsy. Interestingly, in each instance, the twins exhibited the identical syndrome; stated differently, the epilepsies were syndromically faithful. Each of the seven unaffected twins exhibited an abnormal EEG. Twenty-one pairs of dizygotic twins were identified in which at least one twin had childhood absence epilepsy, juvenile absence epilepsy, or juvenile myoclonic epilepsy. Five of the 21 pairs (24 percent) were concordant for epilepsy. In contrast to the monozygotic twins, four of the five pairs of dizygotic twins concordant for epilepsy exhibited different epilepsy syndromes; that is, the epilepsies in the dizygotic twins were usually syndromically unfaithful. The presence of childhood absence epilepsy and other idiopathic generalized epilepsy syndromes in pedigrees in which the proband exhibited juvenile myoclonic epilepsy[23] is consistent with these observations in dizygotic twins. The studies by Berkovic and colleagues[32] replicate a number of the earlier findings in twin studies.[31,33]

Several conclusions can be drawn from these observations: (1) The high concordance rate in monozygotic versus dizygotic twins implicates a strong genetic determinant. (2) The disorders exhibit a high penetrance because at least 65 percent of monozygotic pairs are concordant for epilepsy. (3) The inheritance of a given disorder does not follow a simple mendelian pattern (e.g., autosomal dominant or recessive, X-linked) in view of a lack of syndromic fidelity among dizygotic twins and the relatively low risk to first-degree relatives. The lack of mendelian patterns of inheritance together with syndromic fidelity among monozygotic but not dizygotic twins leads to the hypothesis that the syndromes reflect variable expression of one or a few common mutant alleles. The variable expression could be attributed to the fact that multiple, functionally related genes underlie each disorder; that is, inheritance of a given gene in any one family predisposes an individual to multiple forms of these disorders, but inheritance of an additional gene (or genes) predisposes an individual to one specific disorder. Additionally, interactions between genetic predisposition and environmental factors could contribute to the penetrance and expression of these disorders.

The fact that juvenile myoclonic epilepsy and related forms of idiopathic generalized epilepsy are the most common forms of familial epilepsy led to investigation of the chromosomal site of a presumed mutant allele. Delgado-Escueta and colleagues[34] initially reported linkage to a locus predisposing to juvenile myoclonic epilepsy in the HLA region of chromosome 6p. Individuals were characterized as affected if other idiopathic generalized epilepsy syndromes, such as childhood absence epilepsy, juvenile absence epilepsy, and generalized tonic-clonic epilepsy, were present. The maximum LOD score obtained was 3.78 using a recombination fraction of 0.01, assuming autosomal dominant inheritance and 90 percent penetrance. Confirmation of these findings was obtained in distinct populations by Weissbecker and colleagues in Berlin using HLA serologic markers[35] and by Greenberg and colleagues in New York using polymorphic DNA markers.[36] Meanwhile, Delgado-Escueta and colleagues[37] again reported linkage (LOD score above 7) between juvenile myoclonic epilepsy or an abnormal EEG and a locus on chromosome

6p, in this instance 6p11-p21.2, using an autosomal dominant model with 70 percent penetrance. The extent to which this locus of chromosome 6p11-p21.2 overlaps with the original HLA locus reported by this group is unclear. In contrast to the conclusions of this work, a study of 25 families with juvenile myoclonic epilepsy identified in the United Kingdom and Sweden[38] found no evidence for linkage on 6p despite using different models for linkage analysis (pairwise or multipoint), different patterns of inheritance (autosomal recessive or dominant), or variable penetrance (age-dependent high or low penetrance). This discrepancy suggests that the disorder is genetically heterogeneous.

Mutant Genes Underlying Epileptic Syndromes

Gene hunters for the epilepsies have succeeded in detecting the mutant genes underlying six forms of epilepsy in the past seven years, beginning with identification of a mitochondrial gene mutation in 1990. Identification of the precise genetic defect enables the clinician to establish a molecular diagnosis that is useful both for prognosis of the affected individual and for genetic counseling. Because three of these syndromes exhibit similar clinical manifestations (e.g., distinct forms of progressive myoclonic epilepsies[5]), establishing a molecular diagnosis can simplify the clinician's task. As outlined below, identification of the mutant gene provides vital information for elucidating the molecular pathogenesis of a disorder, an understanding that could lead to effective treatment or prevention of a disease.

Myoclonic Epilepsy and Ragged Red Fiber Disease

Myoclonic epilepsy and ragged red fiber disease (MERRF)[39-43] represents a rare familial disease with exclusively maternal inheritance. The age of onset is remarkably variable, ranging from three to 63 years. Epileptic seizures are a prominent feature; the seizures are typically myoclonic, but tonic-clonic, elementary partial, and complex partial seizures also occur.[42] The myopathy is termed ragged red fiber because of a distinctive disordered appearance of occasional muscle fibers. Paradoxically, the clinical manifestations of the myopathy are often mild or absent. Additional clinical features include ataxia, dementia, and hearing loss. The hallmark of the disease is the extraordinary variability in its severity, rate of progression, and manifestations, even among members of the same family[42] [*see Chapter 15*].

The maternal inheritance affecting both male and female offspring pointed to a mitochondrial gene because the oocyte but not the sperm contributes mitochondria to the zygote. Shoffner and colleagues[44] discovered that a point mutation, an A to G mutation at nucleotide pair 8344, of a mitochondrial gene encoding a tRNA^{lys} is responsible for the disease. The disease is associated with defects in mitochondrial oxidative phosphorylation complexes enriched in proteins encoded by mitochondrial genes, namely complexes I and IV. The variability of the oxidative phosphorylation defect among different tissues is consistent with a random distribution of mitochondria containing wild-type or mutant DNA during organogenesis early in embryonic development. Because mitochondrial translation products derived from lymphoblastoid cell lines of MERRF patients exhibited a reduction in labeling of high relative to low molecular weight polypeptides, a defect in protein synthesis was suspected. Thus, among the potential mitochondrial genes under consideration, a mutation of a mitochondrial DNA (mtDNA) ribosomal RNA (rRNA) or transfer RNA (tRNA) gene seemed likely.[45] Shoffner and colleagues[44] demonstrated the presence of the mutation in three MERRF pedigrees and its absence in 75 controls and showed that the mutation was heteroplasmic; that is, the proportion of wild-type and mutant mtDNA varied within a given tissue among affected relatives.

Batten Disease

Batten disease is one of five recognized inherited neurodegenerative disorders characterized by the accumulation of autofluorescent pigments, ceroid and lipofuscin, in neurons and other cell types.[46] The five subtypes of these disorders, collectively termed the neuronal ceroid lipofuscinoses, are distinguished by age of onset, chromosomal site, and both clinical and pathological features. Juvenile-onset Batten disease is inherited in an autosomal recessive pattern. It is the most common neurodegenerative disease of childhood, with an estimated incidence of one per 25,000 births.[47] The onset of the disease is between the ages of five and 10 years and typically begins with visual impairment. Subsequently, tonic-clonic and myoclonic seizures develop along with cognitive impairments that progress inexorably and culminate in death in the second or third decade. The International Batten Disease Consortium[48] used a positional cloning approach to identify *CLN3* as the mutant gene underlying Batten disease. The gene was disrupted by small but distinct deletions in Finnish and Moroccan patients, and a point mutation altering a splice donor site was identified in an additional family. *CLN3* encodes a 428 amino acid protein of unknown function.

Unverricht-Lundborg Disease

Progressive myoclonic epilepsy of the Unverricht-Lundborg type is a rare disorder that typically begins between the ages of 8 and 13 years.[5] The earliest manifestations include myoclonic and tonic-clonic seizures. Eventually, dysarthria, ataxia, and intention tremor develop. Intellectual decline proceeds in an indolent fashion, with a loss of roughly 10 IQ points per decade. Survival into adulthood is common, and some individuals reach the sixth decade. In contrast to Batten disease, no storage material has been found. Inherited in an autosomal recessive pattern, linkage analysis localized the gene underlying progressive myoclonic epilepsy of the Unverricht-Lundborg type to a region of two million base pairs on chromosome 21q22.3.[49] Genetic and physical mapping information refined the location of the gene to a 175 kilobase (kb) region. Using a direct complementary DNA (cDNA) selection technique, Pennacchio and colleagues identified a cDNA that encoded cystatin B, a protein inhibitor of cysteine proteases.[50] This gene is expressed in all tissues examined. Lymphoblastoid cell lines were used from affected and unaffected individuals to compare the expression of cystatin B messenger RNA (mRNA). Affected individuals from four families exhibited undetectable levels of cystatin B mRNA in comparison to asymptomatic carriers and unaffected individuals. This suggested the presence of a mutation of the cystatin B gene that would result in decreased amounts of the mRNA. A point mutation was

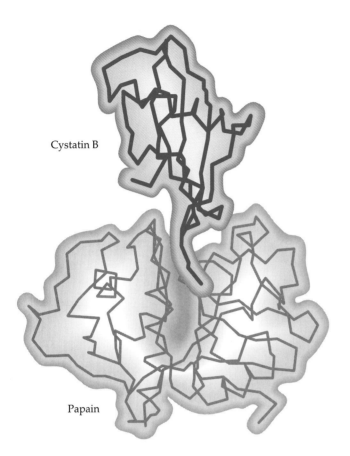

Cystatin B

Papain

Figure 2　*The crystal structure of the cysteine protease inhibitor cystatin B complexed with the cysteine protease papain. Although detailed information regarding the structure and function of the cysteine protease inhibitor is available, the chain of molecular and cellular events by which the mutations of the cystatin B gene result in a phenotype of progressive cognitive impairment and myoclonic epilepsy is unknown.*

found in a 3′ splice site of intron 1 in affected members of one family; the highly conserved nature of nucleotides in this position in all introns underscores its functional significance as a cause of the disease. A second mutation was found in two additional families at amino acid position 68 generating a stop codon. A mutation was subsequently identified that accounts for the majority of progressive myoclonic epilepsy of the Unverricht-Lundborg–type patients; namely, an unstable 15 to 18 nucleotide minisatellite repeat expansion ranging in size from 0.5 to 1.5 kb in the putative promoter region of the cystatin B gene[51,52] [*see Figure 2*]. In contrast to the highly unstable trinucleotide microsatellite repeat expansions associated with other neurologic disorders, including Huntington's disease, the mutation associated with the Unverricht-Lundborg type appears to be less unstable and has not been associated with anticipation (the progressively earlier onset and severity of a disorder throughout generations of a pedigree).

Miller-Dieker Syndrome

The Miller-Dieker syndrome, a form of type 1 lissencephaly, is a brain malformation characterized by a smooth surface of cerebral cortex; microscopic analyses disclose evidence of a four-layer cortex in contrast to the normal six-layer structure.[53] Affected individuals exhibit an abnormal facial appearance, profound mental retardation, and multiple neurologic abnormalities, including hypotonia and eventually spastic quadriplegia.[53] Although seizures may occur in the first week of life, more commonly they begin later in the first year and include myoclonic, tonic, and tonic-clonic seizures. Most cases of Miller-

Dieker syndrome arise as a consequence of a de novo deletion or translocation involving chromosome 17p13.3. The presence of microdeletions in this region led to positional cloning of a gene, *LIS-1*,[54] that is deleted in these patients. *LIS-1* was subsequently demonstrated to encode a 45 kd subunit of the brain platelet-activating factor acetylhydrolase.[55] Platelet-activating factor is a phospholipid that has been implicated in both synaptic plasticity and gene expression.[56,57] Platelet-activating factor acetylhydrolase inactivates platelet-activating factor by removing the acetyl group at the *sn-2* position.

Autosomal Dominant Nocturnal Frontal Lobe Epilepsy

Autosomal dominant nocturnal frontal lobe epilepsy is a recently recognized form of partial epilepsy characterized by frequent, brief, and often violent seizures that usually occur at night. Onset typically is in childhood. The seizures are often mistaken for a sleep disturbance such as night terrors and nightmares. Affected individuals exhibit no neurologic symptoms or signs other than epilepsy. Berkovic and colleagues identified a large Australian kindred and used linkage analysis with polymorphic DNA markers to assign the gene to chromosome 20q13.2-q13.3.[58] Because the neuronal nicotinic acetylcholine receptor α_4-subunit (CHRNA4) had been mapped to this region of chromosome 20 and the gene is expressed in frontal cortex, this particular gene was examined further. A missense mutation was identified in which serine replaces phenylalanine at residue 248 in the putative second transmembrane domain of this receptor [*see Figure 3*].[59] The presence of the mutation in all 21 affected family members and in four obligate carriers but not in 333 healthy control subjects supported a causal link to the phenotype.

Angelman Syndrome

Angelman syndrome is a rare genetic disorder characterized by epileptic seizures, severe mental retardation, gait disturbance, and microcephaly evident in infants.[60] Tonic-clonic, myoclonic, and absence seizures are present.[61] The disorder is also termed the happy puppet syndrome because of the puppet-like appearance due to the abnormality of motor control together with the easily provoked smiling and laughter in affected individuals. A diversity of genetic mechanisms underlie Angelman syndrome, but the unifying theme is that a single gene active preferentially or exclusively inherited from the mother on chromosome 15 is necessary to prevent the disorder. Approximately 70 percent of Angelman syndrome cases are caused by spontaneously arising deletions of bands q11–q13 of chromosome 15 inherited from the mother. Approximately two percent are caused by inheriting both copies of chromosome 15 from the father, a phenomenon termed *uniparental disomy*. Another two percent of Angelman syndrome cases are caused by *imprinting mutations* in which the gene expression of the maternally inherited chromosome 15 follows paternal rather than maternal patterns. The remaining 25 percent represent the familial cases. A chromosomal rearrangement transmitted from a normal mother to her Angelman syndrome daughter provided the critical opportunity for Kishino and colleagues[62] to identify a mutant gene responsible for some cases of Angelman syndrome. The mother transmitted to her daughter a copy of chromosome 15 containing a balanced inversion of part of chromosome 15q. This inversion disrupted the 5′ end of a

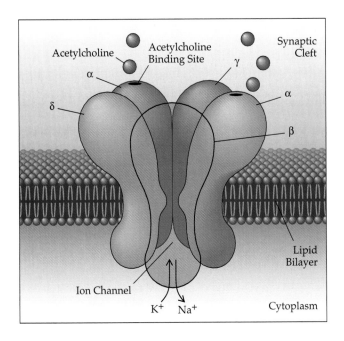

Figure 3 *The nicotinic acetylcholine receptor (nAChR) of the electric fish* Torpedo marmorata *provides a framework for understanding the human neuronal nAChR. The nAChR is a multisubunit protein (two α, one β, one γ, and one δ) that forms binding sites for acetylcholine and a channel for ions to traverse the cell membrane. Each subunit contains four domains that traverse the cell membrane; the second domain of each subunit contributes to the wall of the ion channel.*

gene termed *UBE3A* or E6-associated protein (E6-AP).[62] They also discovered distinct mutations of this same gene leading to premature termination of translation in members of two other families with Angelman syndrome; similar observations were made simultaneously by Matsuura and colleagues.[63] The *UBE3A* gene encodes a protein that functions in proteolysis, particularly in ubiquitin targeting of proteins for degradation in the proteasome.[64]

Epilepsy Genes: What a Surprise!

In contrast to a number of other diseases of the CNS, the availability of a diversity of both in vivo and in vitro models of epilepsy permits study of the underlying mechanisms. Extensive investigations[3,65] have begun to shed light on mechanisms of epileptogenesis. Experimental work has simultaneously elucidated plausible mechanisms for suppressive activity of antiseizure drugs.[66,67] Taken together, this information has suggested categories of candidate genes, mutations of which might culminate in an epileptic phenotype. For example, insight into molecular determinants of synaptic and intrinsic properties controlling neuronal excitability suggests a broad diversity of candidates that might engender increased excitability and an epileptic phenotype. The identification of the α_4-subunit of an nAChR as a mutant gene underlying autosomal dominant nocturnal frontal lobe epilepsy came as a surprise because scientists had struggled to define whether and how ACh activates neuronal nAChRs in the CNS.[68,69] This enigma about nAChRs can be contrasted with the easily demonstrated roles of glutamate and γ-aminobutyric acid (GABA) receptors in synaptic function in the mammalian nervous system. The magnitude of the surprise that a mutant nAChR gene can cause epilepsy is even less than that evoked by discovery that mutation of the gene for cysteine protease inhibitor cystatin B is linked to a form of progressive myoclonic epilepsy. Similar surprise attended recognition that the genes encoding CLN3, UBE3A, or tRNA^{lys} can cause epilepsy. The beauty of these discoveries is that they necessitate a broader view of mechanisms of epileptogenesis. The challenge for scientists now is to begin to decipher how these mutant genes culminate in epilepsy.

How Does the Genotype Produce the Phenotype?

The principal issue for clinicians and their patients is how discovery of a mutant gene translates into prevention or more effective treatment of the epilepsy. Identification of a mutant gene has immediate implications for genetic counseling as well as predicting prognosis of the affected individual. Development of more effective therapeutic strategies will almost certainly require an understanding of how the mutant gene culminates in the phenotype of epilepsy.

Defining the functional consequences of the mutation in an in vitro expression system would be a logical first step toward understanding how the mutant gene results in increased neuronal excitability and epilepsy. This in turn may provide the first clues to therapeutic approaches. In vitro expression systems are available (*Xenopus* oocytes or a cell line such as human embryonic kidney 293 cells) that permit characterization of differences in the physiologic and biochemical properties of the mutant and wild-type gene products. To understand how such an analysis can be informative, consider the mutant gene, encoding the α_4-subunit of the nicotinic ACh receptor, found to underlie autosomal dominant nocturnal frontal lobe epilepsy. This protein is a subunit of a heteromeric ligand-gated ion channel receptor for the neurotransmitter, ACh [*see Figure 3*]. Neuronal nicotinic ACh receptors are thought to consist of five proteins assembled in an ion channel.[70] These pentamers consist of various combinations of roughly 10 subunits (α_2-α_8; β_2-β_4) that are differentially expressed throughout the nervous system and form pharmacologically and physiologically distinct receptors. In mammalian brain the nAChR subtype, consisting of α_4-β_2–subunits, appears to account for 90 percent of all nAChRs. Each of these subunits exhibits four transmembrane domains, the second of which (M2) is thought to contribute to the walls of the ACh-gated channel. ACh is thought to open the channel by triggering a conformational change that alters the association of amino acid residues of the M2 segment. The amino acid residue at position 248 of the normal α_4-subunit is serine; by contrast, the mutant α_4-subunit in autosomal dominant nocturnal frontal lobe epilepsy contains a phenylalanine at position 248 as a result of the point mutation[58] [*see Figure 4*]. Interestingly, serine 248 residues in the five subunits forming the AChR are thought to constitute a critical narrow ring in the transmembrane ion channel. What is the impact on channel function of substituting a phenylalanine for serine at 248? Does this substitution modify the permeability of any or all ions through the channel? Does it modify the sensitivity to drugs that activate or inhibit the channel? Preliminary studies of the mutant $\alpha_4$ser248phe-subunit coexpressed in *Xenopus* oocytes with β_2-subunits[71] disclosed that the mutant receptor desensitized more rapidly and recovered from desensitization more slowly than the wild-type receptor. By contrast, the magnitude of the macroscopic currents in the oocyte and the potency of ACh did not appear to differ between the wild-type and mutant receptors. One might expect that enhanced desensitization of the mutant receptor would result in a net diminution of activity of synaptic mutant

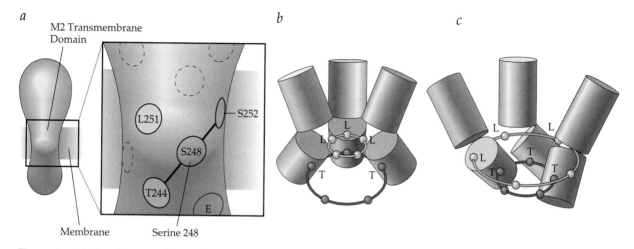

Figure 4   *Electron images of* Torpedo marmorata *postsynaptic membranes activated by brief mixing with acetylcholine. (a) The amino acid residues viewed from inside the pore; note that serine 248 faces the pore of the channel. (b and c) The closed and open configurations of the second transmembrane domains. Serine 248 occupies a pivotal position in the region bordering the pore in these configurations.*

nAChRs. The properties of desensitization of the mutant receptor in oocytes are characterized by a time course of tens of seconds; precisely how these findings affect the kinetics of synaptic nAChRs operating at a millisecond level of resolution is unclear. More detailed study of the kinetic properties of the receptor under conditions that permit rapid exchange of drugs and analysis of individual channels will be required to more fully understand the impact of the mutation on channel function.

Let us assume that the sole abnormality in function of the mutant nAChR is a more rapid desensitization and delayed recovery from desensitization. This raises the question as to how such properties culminate in the phenotype of frontal lobe epilepsy. To begin to address this question, engineering mice expressing the mutant nAChR in a pattern similar to the expression in humans could provide a powerful tool. However, engineering such mice is a much more difficult undertaking than expressing the mutant nAChR in an in vitro expression system. One approach is to use recombinant DNA techniques to engineer a gene consisting of the promoter of the α_4-subunit driving expression of the mutant α_4-subunit and to inject this transgene into a fertilized egg; the resulting transgenic mouse expresses both wild-type alleles in addition to the transgene. This approach has been successfully used to produce an animal model of amyotrophic lateral sclerosis in which a mutant form of superoxide dismutase is expressed as a transgene[72] [*see Chapter 12*]. Alternatively, strategies such as gene targeting could be exploited to engineer mice expressing the mutant gene in place of the wild-type or normal gene. This enables the investigator to address a host of questions. How does the altered function of the mutant AChR affect the physiology of ACh synapses? How does the mutant gene product expressed in a subset of CNS neurons produce hyperexcitability of small neuronal networks and in the ensemble of interconnected neurons in vivo? Can pharmacological intervention prevent the onset or ameliorate the progression of the epileptic disorder? Answers to these and other questions should provide clues to rational therapeutic approaches to this disorder based upon knowledge of the underlying mechanisms.

References

1. Hauser WA, Hesdorffer DC: *Epilepsy: Frequency, Causes and Consequences*. Demos Press, New York, 1990

2. Mattson RH, Cramer JA, Collins JF, et al: Comparison of carbamazepine, phenobarbital, phenytoin, and primidone in partial and secondarily generalized tonic-clonic seizures. *N Engl J Med* 313:145, 1985

3. McNamara JO: Cellular and molecular basis of epilepsy. *J Neurosci* 14:3413, 1994

4. Commission on Classification and Terminology of the International League Against Epilepsy: Proposal for revised clinical and electroencephalographic classification of epileptic seizures. *Epilepsia* 22:489, 1981

5. Berkovic SF, Andermann FA, Carpenter S, et al: Progressive myoclonus epilepsies: specific causes and diagnoses. *N Engl J Med* 315:296, 1986

6. Commission on Classification and Terminology of the International League Against Epilepsy: Proposal for revised classification of epilepsies and epileptic syndromes. *Epilepsia* 30:389, 1989

7. Serratosa JM, Delgado-Escueta AV, Posada I, et al: The gene for progressive myoclonus epilepsy of the Lafora type maps to chromosome 6q. *Hum Mol Genet* 4:1657, 1995

8. Hauser WA, Kurland LT: The epidemiology of epilepsy in Rochester, Minnesota, 1935 through 1967. *Epilepsia* 16:1, 1975

9. Andermann E: Multifactorial inheritance of generalized and focal epilepsy. *Genetic Basis of the Epilepsies*. Anderson VE, Hauser WA, Penry JK, et al, Eds. Raven Press, New York, 1982, p 355

10. Goate A, Chartier-Harlin MC, Mullan M, et al: Segregation of a missense mutation in the amyloid precursor protein gene with familial Alzheimer's disease. *Nature* 349:704, 1991

11. Murrell J, Farlow M, Ghetti B, et al: A mutation in the amyloid precursor protein associated with hereditary Alzheimer's disease. *Science* 254:97, 1991

12. Sherrington R, Rogaev EI, Liang Y, et al: Cloning of a gene bearing missense mutations in early-onset familial Alzheimer's disease. *Nature* 375:754, 1995

13. Rogaev EI, Sherrington R, Rogaeva EA, et al: Familial Alzheimer's disease in kindreds with missense mutations in a gene on chromosome 1 related to the Alzheimer's disease type 3 gene. *Nature* 376:775, 1995

14. Corder EH, Saunders AM, Strittmatter WJ, et al: Gene dose of apolipoprotein E type 4 allele and the risk of Alzheimer's disease in late onset families. *Science* 261:921, 1993

15. Saunders AM, Strittmatter WJ, Schmechel D, et al: Association of apolipoprotein E allele Σ4 with late-onset familial and sporadic Alzheimer's disease. *Neurology* 43:1467, 1993

16. Ronen GM, Rosales TO, Connolly M, et al: Seizure characteristics in chromosome 20 benign familial neonatal convulsions. *Neurology* 43:1355, 1993

17. Leppert M, Anderson VE, Quattlebaum T, et al: Benign familial neonatal convulsions linked to genetic markers on chromosome 20. *Nature* 337:647, 1989

18. Malafosse A, Leboyer M, Dulac O, et al: Confirmation of linkage of benign familial neonatal convulsions to D20S19 and D20S20. *Hum Genet* 89:54, 1992

19. Lewis TB, Leach RJ, Ward K, et al: Genetic heterogeneity in benign familial neonatal convulsions: identification of a new locus on chromosome 8q. *Am J Hum Genet* 53:670, 1993

20. Schiffman R, Shapira Y, Ryan SG: An autosomal recessive form of benign familial neonatal seizures. *Clin Genet* 40:467, 1991

21. Ottman R, Risch N, Hauser WA, et al: Localization of a gene for partial epilepsy to chromosome 10q. *Nat Genet* 10:56, 1995

22. Tahvanainen E, Ranta S, Hirvasniemi E, et al: The gene for a recessively inherited human childhood progressive epilepsy with mental retardation maps to the distal short arm of chromosome 8. *Proc Natl Acad Sci USA* 91:7267, 1994

23. Ranta S, Lehesjoki AE, Hirvasniemi A, et al: Genetic and physical mapping of the progressive epilepsy with mental retardation (EPMR) locus on chromosome 8p. *Genome Res* 6:351, 1996

24. Eksioglu YZ, Scheffer IE, Cardenas P, et al: Periventricular heterotopia: an X-linked dominant epilepsy locus causing aberrant cerebral cortical development. *Neuron* 16:77, 1996

25. Ott J: *Analysis of Human Genetic Linkage*. Johns Hopkins University Press, Baltimore, 1991

26. Greenberg DA, Delgado-Escueta AV: The chromosome 6p epilepsy locus: exploring

mode of inheritance and heterogeneity through linkage analysis. *Epilepsia* 34:12, 1993

27. Janz D, Christian W: Impulsive petit mal. *J Neurol* 176:346, 1957

28. Tsuboi T, Christian W: On the genetics of the primary generalized epilepsy with sporadic myoclonias of impulsive petit mal type: a clinical and electroencephalographic study of 399 probands. *Humangenetik* 19:155, 1973

29. Janz D, Durner M, Beck-Mannagetta G, et al: Family studies on the genetics of juvenile myoclonic epilepsy (epilepsy with impulsive petit mal). *Genetics of the Epilepsies*. Beck-Mannagetta G, Anderson VE, Doose H, et al, Eds. Springer-Verlag, Berlin, 1989, p 43

30. Delgado-Escueta AV, Greenberg D, Weissbecker K, et al: Gene mapping in the idiopathic generalized epilepsies: juvenile myoclonic epilepsy, childhood absence epilepsy, epilepsy with grand mal seizures, and early childhood myoclonic epilepsy. *Epilepsia* 31:S19, 1990

31. Metrakos K, Metrakos J: Genetics of convulsive disorders. II: Genetic and electroencephalographic studies in centrencephalic epilepsy. *Neurology* 11:464, 1961

32. Berkovic SF, Howell RA, Hopper JL: Twin study of epilepsy syndromes. *Epilepsia* 34:38, 1993

33. Lennox W: The heredity of epilepsy as told by relatives and twins. *JAMA* 146:529, 1951

34. Greenberg DA, Delgado-Escueta AV, Widelitz H, et al: Juvenile myoclonic epilepsy (JME) may be linked to the BF and HLA loci on human chromosome 6. *Am J Med Genet* 31:185, 1988

35. Weissbecker KA, Durner M, Janz D, et al: Confirmation of linkage between juvenile myoclonic epilepsy locus and the HLA region on chromosome 6. *Am J Med Genet* 38:32, 1991

36. Greenberg DA, Durner M, Shinnar S, et al: Association of HLA class II alleles in patients with juvenile myoclonic epilepsy compared with patients with other forms of adolescent-onset generalized epilepsy. *Neurology* 47:750, 1996

37. Liu AW, Delgado-Escueta AV, Serratosa JM, et al: Juvenile myoclonic epilepsy locus in chromosome 6p21.2-p11: linkage to convulsions and electroencephalography trait. *Am J Hum Genet* 57:368, 1995

38. Whitehouse WP, Rees M, Curtis D, et al: Linkage analysis of idiopathic generalized epilepsy (IGE) and marker loci on chromosome 6p in families of patients with juvenile myoclonic epilepsy: no evidence for an epilepsy locus in the HLA region. *Am J Hum Genet* 53:652, 1993

39. Tsairis P, Engel WK, Kark P: Familial myoclonic epilepsy syndrome associated with skeletal muscle mitochondrial abnormalities. *Neurology (Minneapolis)* 23:408, 1973

40. Fukuhara N, Tokiguchi S, Shirakawa K, et al: Myoclonus epilepsy associated with ragged-red fibres (mitochondrial abnormalities): disease entity or a syndrome? *J Neurol Sci* 47:117, 1980

41. DiMauro S, Bonilla E, Zeviani M, et al: Mitochondrial myopathies. *Ann Neurol* 17:521, 1985

42. Berkovic SF, Carpenter S, Evans A, et al: Myoclonus epilepsy and ragged-red fibres (MERRF). *Brain* 112:1231, 1989

43. DeVivo DC: The expanding clinical spectrum of mitochondrial diseases. *Brain Dev* 15:1, 1993

44. Shoffner JM, Lott MT, Lezza AMS, et al: Myoclonic epilepsy and ragged-red fiber disease (MERRF) is associated with mitochondrial DNA tRNA^{Lys} mutation. *Cell* 61:931, 1990

45. Wallace DC, Zheng X, Lott MT, et al: Familial mitochondrial encephalo-myopathy (MERRF): genetic, pathophysiological, and biochemical evidence for a mitochondrial DNA disease. *Cell* 55:601, 1988

46. Dyken PR: Reconsideration of the classification of the neuronal ceroid lipofuscinoses. *Am J Med Genet* 5:69, 1988

47. Zeman W: Studies in the neuronal ceroid lipofuscinosis. *J Neurpathol Exp Neurol* 33:1, 1974

48. International Batten Disease Consortium: Isolation of a novel gene underlying Batten disease, *CLN3*. *Cell* 82:949, 1995

49. Lehesjoki AE, Koskiniemi M, Sistonen P, et al: Localization of a gene for progressive myoclonus epilepsy to chromosome 21q22. *Proc Natl Acad Sci USA* 88:3696, 1991

50. Pennacchio LA, Lehesjoki AE, Stone NE, et al: Mutations in the gene encoding cystatin B in progressive myoclonus epilepsy (EPM1). *Science* 271:1731, 1996

51. Lafreniere RG, Rochefort DL, Chretien N, et al: Unstable insertion in the 5' flanking region of the cystatin B gene is the most common mutation in progressive myoclonus epilepsy type 1, EPM1. *Nature Genet* 15:298, 1997

52. Virtaneva K, D'Amato E, Miao J, et al: Unstable minisatellite expansion causing recessively inherited myoclonus epilepsy, EPM1. *Nat Genet* 15:393, 1997

53. Dobyns WB, Reiner O, Carozzo R, et al: Lissencephaly: a human brain malformation associated with deletion of the LIS1 gene located at chromosome 17p13. *JAMA* 270:2838, 1993

54. Reiner O, Carozzo R, Shen Y, et al: Isolation of a Miller-Dieker lissencephaly gene containing G protein beta-subunit-like repeats. *Nature* 364:717, 1993

55. Hattori M, Adachi H, Tsujimoto M, et al: Miller-Dieker lissencephaly gene encodes a subunit of brain platelet-activating factor. *Nature* 370:216, 1994

56. Bazan NG, Allan G: Platelet-activating factor in the modulation of excitatory amino acid neurotransmitter release and of gene expression. *J Lipid Mediat Cell Signal* 14:321, 1996

57. Kato K, Clark GD, Bazan NG, et al: Platelet-activating factor as a potential retrograde messenger in CA1 hippocampal long-term potentiation. *Nature* 367:175, 1994

58. Phillips HA, Scheffer IE, Berkovic SF, et al: Localization of a gene for autosomal dominant nocturnal frontal lobe epilepsy to chromosome 20q13.2. *Nat Genet* 10:117, 1995

59. Steinlein OK, Mulley JC, Propping P, et al: A missense mutation in the neuronal nicotinic acetylcholine receptor α_4 subunit is associated with autosomal dominant nocturnal frontal lobe epilepsy. *Nat Genet* 11:201, 1995

60. Williams CA, Zori RT, Hendrickson J, et al: Angelman syndrome. *Curr Probl Pediatr* 25:216, 1995

61. Guerrini R, DeLorey TM, Bonanni P, et al: Cortical myoclonus in Angelman syndrome. *Ann Neurol* 40:39, 1996

62. Kishino T, Lalande M, Wagstaff J: UBE3A/E6-AP mutations cause Angelman syndrome. *Nat Genet* 15:70, 1997

63. Matsuura T, Sutcliffe JS, Fang P, et al: De novo truncating mutations in E6-AP ubiquitin-protein ligase gene (UBE3A) in Angelman syndrome. *Nat Genet* 15:74, 1997

64. Huigebretse J, Scheffner M, Howley PM: Cloning and expression of the cDNA for E6-AP, a protein that mediates the interaction of the human papillomavirus E6 oncoprotein with p53. *Mol Cell Biol* 13:775, 1993

65. Dichter MA, Ayala GF: Cellular mechanisms of epilepsy: a status report. *Science* 237:157, 1987

66. Macdonald RL, Kelly KM: Antiepileptic drug mechanisms of action. *Epilepsia* 34:S1, 1994

67. McNamara JO: Drugs effective in the therapy of the epilepsies. *Goodman & Gilman's Pharmacological Basis of Therapeutics,* 9th ed. Hardman JG, Limbird LE, Eds. McGraw Hill, New York, 1996, p 461

68. Sivilotti L, Colquhoun D: Acetylcholine receptors: too many channels, too few functions. *Science* 269:1681, 1995

69. Zhang ZW, Coggan JS, Berg DK: Synaptic currents generated by neuronal acetylcholine receptors sensitive to α-bungarotoxin. *Neuron* 17:1231, 1996

70. Lindstrom J, Anand R, Peng X, et al: Neuronal nicotinic receptor subtypes. *Ann NY Acad Sci* 757:100, 1995

71. Weiland S, Witzemann V, Villaroel A, et al: An amino acid exchange in the second transmembrane segment of a neuronal nicotinic receptor causes partial epilepsy by altering its desensitization kinetics. *FEBS Lett* 398:91, 1996

72. Gurney ME, Pu H, Chiu AY, et al: Motor neuron degeneration in mice that express a human Cu,Zn superoxide dismutase mutation. *Science* 264:1772, 1994

73. Stubbs MT, Laber M, Bode W, et al: The refined 2.4 A x-ray crystal structure of recombinant human stefin B in complex with the cysteine proteinase papain: a novel type of proteinase inhibitor interaction. *EMBO J* 9:1939, 1990

74. Unwin N: Acetylcholine receptor channel imaged in the open state. *Nature* 373:37, 1995

Acknowledgments

This work was supported by a grant from the Department of Veterans Affairs. The author also wishes to thank Mark Routbort and Rod Radtke for their careful reading and comments on this manuscript.

Figure 1   From "Periventricular Heterotopia: An X-linked Dominant Epilepsy Locus Causing Aberrant Cerebral Cortical Development," by Y.Z. Eksioglu, I.E. Scheffer, P. Cardenas, et al., in *Neuron* 16:77, 1996. Used by permission.

Figure 2   Seward Hung. From "The Refined 2.4 A X-ray Crystal Structure of Recombinant Human Stefin B in Complex with the Cysteine Proteinase Papain: A Novel Type of Proteinase Inhibitor Interaction," by M.T. Stubbs, M. Laber, W. Bode, et al, in *EMBO J* 9:1939, 1990. Used by permission.

Figure 3   Seward Hung.

Figure 4   Adapted by Seward Hung from "Acetylcholine Receptor Channel Imaged in the Open State," by N. Unwin, in *Nature* 373:37, 1995. Used by permission. © 1995 MacMillan Magazines Ltd.

CHAPTER 6

Molecular Genetics of Brain Tumors

Mark A. Israel, M.D.

Primary tumors of the central nervous system (CNS) occurred in approximately 18,000 individuals in 1996, accounting for an estimated 13,300 deaths in the United States, with a mortality rate of 2.6 per 100,000. Overall, CNS cancer constitutes about one percent of newly diagnosed malignant tumors. In children and adolescents, brain tumors are a major health problem, second only to leukemia as a cause of cancer-related deaths; they are the third leading cause of cancer-related deaths between 15 and 34 years of age.[1]

Brain tumors are a diverse group of neoplasms. They may be benign (e.g., meningiomas, schwannomas, low-grade gliomas) or malignant (e.g., astrocytomas, glioblastomas, ependymomas). Central nervous system malignant tumors are usually confined to the CNS and rarely metastasize outside the brain and spinal cord. Tumors that arise in the brain occur with varying frequency in patients of different ages and sex, and each type has characteristic clinical manifestations. Initially, CNS tumors were classified by surgeons and pathologists on the basis of their location and histologic appearance, which provided important information for treatment with conventional modalities such as surgery, X-irradiation, and chemotherapy and contributed to prognostication. Because our ability to cure patients with malignant brain tumors has not improved significantly in the past several decades, there is now considerable interest in utilizing an understanding of the molecular pathology underlying malignant characteristics of these tumors to stimulate the development of more precise nosologies and more specific, less toxic antineoplastic therapies.

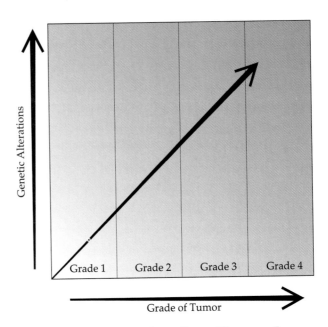

Figure 1 *Multistep progression of genetic changes associated with astrocytic tumors. Enhanced degrees of malignancy in tumors arising in various central nervous system tissues are associated with the accumulation of genetic alterations.*

Brain Tumors as Genetic Disorders: Tumor Suppressor Genes and Oncogenes Contribute to the Development of Brain Tumors

Malignant brain tumors, like other cancers, arise in association with the accumulation of structural alterations of specific genes within individual cells [*see Figure 1*].[2] These mutations cause the cell to proliferate in an uncontrolled manner. Such inappropriate growth is accompanied by other features of malignancy, such as loss of differentiated characteristics of the cell type of origin, acquisition of the ability to invade surrounding normal tissues and to metastasize to other areas of the neuraxis, and resistance to antineoplastic therapies.

Tumor Suppressor Genes

Several types of mutations that mediate the development of malignancy are associated with chromosomal changes that include either a loss or a gain of function [*see Table 1*]. During the development of a tumor, loss-of-function mutations lead to an impairment in cellular activities that function physiologically to restrain cell proliferation. The genes that encode these growth-restraining activities are known as *tumor suppressor genes*, and inactivating mu-

Table 1   Cytogenetic and Genetic Alterations in Brain Tumors

Tumor Type	Chromosomal Alteration
Astrocytoma	1p⁻, 9p⁻, del10, 13q⁻, 17p⁻, 17q⁻, 19q, 22q⁻, DMs*
Medulloblastoma	17p⁻, 6q⁻, 16q, DMs
Meningioma	22q⁻
Oligodendroglioma	1p⁻, 9p⁻, 19q⁻

*DMs–double minute chromosomes.

Table 2 Tumor Suppressor Genes and Proto-oncogenes
Implicated in Human Brain Tumors

Tumor Suppressor Gene		Proto-oncogene	
Gene or Locus	Chromosomal Location	Gene or Locus	Chromosomal Location
p53	17p13.1	EGFR	7p12
NF2	22q12	PDGFR-A	5q31-q32
NF1	17q11.2	ros-1	6q22
RB1	13q14	H-ras, N-ras	11p15; 1p13
APC	9q31	c-myc, N-myc	8q24; 2p23-p24
MEN1	11q13	gli	12q13-q14
PTEN	10q23		

tations of such genes have been recognized to contribute to the development of many different tumor types, including brain tumors [*see Table 2*]. A single functioning copy of a tumor suppressor gene is often sufficient to maintain normal function, so loss of both the maternally and the paternally derived allele is necessary to promote neoplasia.[3] The loss of these two alleles usually occurs as the result of two independent mutational events. Because the loss of a segment of chromosomal DNA containing one allele and a point mutation in the other allele is the mechanism by which this most commonly occurs, a frequently used technique to implicate a tumor suppressor gene in the pathogenesis of any particular tumor is evaluation of tumor DNA with polymorphic markers for a site within or near the gene of interest. If two distinct alleles can be detected in control DNA isolated from a patient's fibroblasts or blood cells but only a single allele can be detected in tumor DNA, investigators frequently assume that this so-called loss of heterozygosity marks the involvement of the tumor suppressor gene located near the polymorphic site. At the present time, it is generally agreed that of the known tumor suppressor genes, those most likely to be important for the development of malignant brain tumors are *p53*, *RB*, and *PTEN (or MMAC1)*.

The *p53* gene is located on chromosome 17p and encodes the p53 protein, which acts as a transcription factor to induce or repress transcription of other genes through sequence-specific interaction with DNA. p53 expression can influence multiple cellular functions, including progression through the cell cycle,[4] genomic stability,[5] and induction of programmed cell death (apoptosis).[6] *p53* mutations are common molecular genetic alterations in malignant astrocytomas.[7,8] However, *p53* mutations occur as commonly in low-grade astrocytomas as in high-grade glioblastomas.[9] This may indicate that the mutation of *p53* is associated principally with the change from normal tissue to low-grade neoplasia, rather than with progression from low- to high-grade malignancy. In one reported series of low-grade astrocytomas that recurred as more anaplastic gliomas, *p53* mutations were found in nearly half of the original tumors, and there were no new mutations in the high-grade recurrences.[10] Rarely, acquisition of a *p53* mutation occurs during the progression of an astrocytic tumor to a higher degree of malignancy.[11] Hence, *p53* mutations may be an early change in some tumors and a late event in others.

The retinoblastoma gene, *RB1*, is a tumor suppressor gene located on chromosome 13q14. Patients with hereditary retinoblastoma have a germ-line mutation in one *RB1* allele that either is inherited or occurs de novo during gametogenesis. Both copies of the *RB1* gene are mutated or deleted in tumor tissue from these patients, as well as in retinoblastomas that occur sporadically in individuals who do not carry the germ-line mutation.[12] Patients with a germ-line mutation of *RB1* demonstrate enhanced propensity to develop retinoblastoma because one *RB1* allele is inactivated in every body cell from birth [*see Figure 1*]. Typically, such patients develop multifocal and bilateral tumors. It is not known why these patients usually develop tumors of the retina. They also have increased risk for development of other tumors,[13] such as brain tumors[14]; *RB1* deletions are also found in sporadically occurring gliomas. Importantly, other genes in the proliferation control pathway, where *RB1* functions, are frequently altered in tumors that have an apparently normal *RB1*. The *cdk4* gene product, which inhibits RB1 function, is amplified and overexpressed in some of these tumors,[15] and *cdkN2*, which encodes a 16 kd protein (p16) initially isolated as an inhibitor of cdk4, has been reported to be deleted in a high proportion of human glial cell lines[16] and in some uncultured primary glial tumors [*see Figure 2*].[17] These findings provide strong evidence for involvement of the *RB1* growth regulatory pathway in the pathogenesis of glial tumors.

Another tumor suppressor gene, *PTEN*[18] (or *MMAC1*),[19] has been identified at chromosome 10q23-q24, which is commonly mutated or deleted in glioma tumor cell lines. It is of particular interest because the region of chromosome 10 in which *PTEN* is located is deleted in approximately 90 percent of cases of the highest-grade astrocytoma, glioblastoma multiforme. Although the function of the protein encoded by this locus is not known, the predicted protein sequence suggests that it may possess phosphatase activity, which might be antagonistic to the many kinases associated with growth stimulatory mechanisms. Other tumor suppressor genes likely to be of importance for the development of brain tumors include *NF1, NF2, TSC1, TSC2,* and *VHL*.

Oncogenes

The second class of mutations known to contribute to the development of malignant tumors occurs in genes that are typically involved in cellular proliferation [*see Table 2*]. These genes, known as proto-oncogenes, encode proteins that are growth factors or growth factor receptors, mediators of signaling pathways, or regulators of gene expression. Activating mutations convert proto-oncogenes to *oncogenes*, which either encode a structurally different protein with novel biologic activities or encode a normal protein that is expressed at inappropriately high levels or at inappropriate times.

Remarkably, only a few oncogenes have been identified that are likely to contribute to the development of primary brain tumors, and these are best characterized in astrocytic tumors. The prototype oncogene with a purported role in brain tumor oncogenesis is the epidermal growth factor receptor gene (*EGFR*)[20] [*see Figure 3*]. EGFR is a transmembrane protein containing an extracellular domain that binds EGF or transforming growth factor–α (TGF-α), a transmembrane domain, and an intracellular domain with tyrosine kinase activity. Activation of *EGFR* function may occur either by amplification of gene transcription or by somatic mutations that enhance the function of

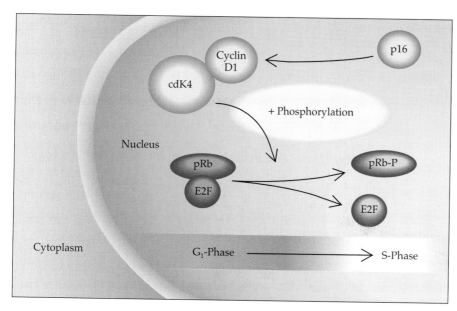

Figure 2 *Alteration in the regulators of cell cycle progression are found in gliomas. Progression into the S phase from the G_1 phase of the cell cycle is regulated by phosphorylation of pRb, the product of the tumor suppressor gene, RB. Conversion of pRB to its phosphorylated form, pRb-P, results in its dissociation from the transcription factor E2F and allows E2F to mediate expression of S phase–specific genes. Phosphorylation of pRb is mediated in part by the cdk4/cyclin D1 holoenzyme, which can be inhibited by cdk inhibitors, such as p16.*

the protein. Amplification is invariably associated with enhanced, constitutive expression of EGFR. A large number of mutations in *EGFR* genes have been identified, which confer a growth advantage on the tumor and are selected during the progressive expansion of the neoplastic tissue. *EGFR* amplification is found in about 36 percent of glioblastomas, seven percent of anaplastic astrocytomas, and three percent of astrocytomas.[21] Whereas many different mutations in amplified *EGFR* genes have been identified, a remarkably common alteration results from an 801 base pair (bp) deletion. This deletion results in a mutant protein that includes a novel epitope at the deletion site and is expressed on the extracellular surface of tumor cells.[22] Monoclonal antibodies against this particular epitope conjugated to radioisotopes are being tested for antineoplastic activity.

Other proto-oncogenes have been reported to be amplified or overexpressed in human gliomas or glial tumor cell lines. Amplifications of N-*ras*, N-*myc*, *myc*, *MET*, *cdk4*, *MDM2*, *PDGFR-A*, and *gli* genes have been reported

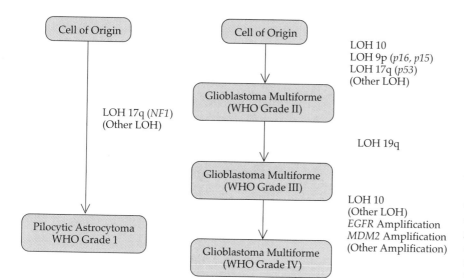

Figure 3 *Astrocytomas with increasing malignant potential are associated with the accumulation of known genetic alterations. LOH is loss of heterozygosity. WHO is the World Health Organization.*

in astrocytic tumor specimens. Overexpression of the H-*ras* and N-*ras* protein products, cell-membrane–associated GTP-binding proteins, have been reported in gliomas (71 percent of high-grade tumors versus zero percent of low-grade tumors).[23] Other genes, including *ROS*, *TGF-β_1*, and *TGF-β_2*, are also highly expressed and may contribute to the malignant behavior of these tumors, although there is no evidence of somatic mutation activating these genes or the proteins they encode.

Genetic Predispositions to the Development of Brain Tumors

Cancer syndromes associated with an inherited predisposition to the development of brain tumors are important clinical entities that provide insights into the pathogenesis of different tumor types [*see Table 3*]. Our current understanding of cancer predisposition syndromes, for which hereditary retinoblastoma is the prototype, indicates that individuals carry a germ-line mutation in one of the two alleles of a tumor suppressor gene important in the development of that tumor type. Whereas some of the hereditary syndromes that carry a predisposition for the development of brain tumors are very rare, others, such as neurofibromatosis, are relatively common, affecting approximately one in 4,000 individuals in the United States. Individuals with von Recklinghausen's neurofibromatosis (NF1) inherit a defective copy of the *NF1* gene. The protein product of the *NF1* gene (neurofibromin) functions to antagonize the activity of p21^{ras}, a known growth stimulatory protein. Patients with NF1 are predisposed to develop pilocytic astrocytomas of the optic nerve and, rarely, malignant astrocytomas within the cerebral hemispheres. A few somatic mutations of the *NF1* gene are reported in sporadic glial tumors.[24]

Another such syndrome is the inherited Li-Fraumeni syndrome, in which germ line *p53* mutations are found.[25] Family members are also predisposed to other tumor types, of which brain tumors are among the most common, including astrocytomas, medulloblastomas, and choroid plexus tumors.[26] It is suggested that glioma patients with multifocal tumors, a second primary

Table 3 Hereditary Syndromes, Including Brain Tumors

Syndrome	*Gene*	*Chromosomal Location*	*Associated CNS Tumors*
Neurofibromatosis type 1	*NF1*	17q	Neuroma, schwannoma, meningioma, optic glioma
Neurofibromatosis type 2	*NF2*	22q12	Acoustic schwannoma, glial tumors
Retinoblastoma	*RB*	13q14	Retinoblastoma, pinealoblastoma
Li-Fraumeni syndrome	*p53*	17p13	Glioma, medulloblastoma
von Hippel-Lindau syndrome	*VHL*	3p	Retinal angioma, cerebellar hemangioblastoma
Multiple endocrine neoplasia	*MEN1*	11q13	Pituitary adenoma, malignant schwannoma
Tuberous sclerosis	*TSC1* *TSC2*	9q 16p13	Giant cell subependymal astrocytoma, astrocytoma, ependymoma
Turcot syndrome	*APC*	5q21	Astrocytoma, medulloblastoma

malignancy, or a family history of cancer are at increased risk for carrying a germ line *p53* mutation.[26] The finding of multifocal brain tumors in Li-Fraumeni patients suggests that multiple tumors develop more frequently in patients affected by inherited predispositions than in the general population. In contrast to *NF1* mutations, *p53* mutations are also found frequently in sporadically occurring astrocytic tumors.

Mutation of the *NF2* gene on chromosome 22q causes neurofibromatosis type 2 (NF2), which is characterized by the development of bilateral acoustic schwannomas that present with progressive deafness early in the third decade of life in over 90 percent of affected patients. Individuals with NF2 also have a predisposition for the development of meningiomas, gliomas, and schwannomas of nerves other than the eighth cranial nerve. The *NF2* gene encodes a protein known as neurofibromin 2, merlin, or schwannomin, which has extensive homology to a family of cytoskeletal proteins.[27] Tuberous sclerosis (Bourneville's disease) is another condition associated with cutaneous lesions, hydrocephalus, seizures, and mental retardation. Subependymal nodules, visible on neuroimaging studies, are characteristic. This syndrome is associated with mutations at chromosomes 9q (*TSC-1*) and 16p (*TSC-2*), and each of these genes encodes tuberins, proteins that modulate the GTPase activity of other signaling proteins.[28] Gene carriers can develop ependymomas and childhood subependymal giant cell astrocytomas, which are benign neoplasms that may be found in the retina or along the border of the lateral ventricles, sometimes obstructing the foramen of Monro and producing hydrocephalus as well as other tumors outside the nervous system.

Von Hippel-Lindau syndrome occurs as the result of a mutation in the *VHL* gene located at chromosome 3p25. This gene encodes a protein of 213 amino acids, and although its function is not yet known, data indicate that it can inhibit the expression of a number of genes that are activated in response to hypoxia, such as the gene encoding vascular endothelial growth factor (VEGF),[29] a gene whose expression is critical for the growth of high-grade astrocytomas. Von Hippel-Lindau syndrome is defined by the presence of retinal angiomas and cerebellar hemangioblastomas, cystic tumors that can present at any age. Hemangioma of the spinal cord; hypernephroma; renal cell carcinoma; pheochromocytoma; and cysts of the kidneys, pancreas, epididymis, or liver may also occur. Recently, the gene responsible for the nevoid basal cell carcinoma syndrome, *PTCH*, was cloned. Patients with this syndrome typically have multiple skin cancers along with benign skin disorders, but medulloblastoma rarely occurs. *PTCH* mutations also occur in sporadic medulloblastomas,[30] raising the possibility that mutations in other genes encoding proteins in the growth regulatory pathway that includes the *PTCH* gene may provide important insights into the pathogenesis of medulloblastoma.

Molecular Pathology of Glial Tumors

Astrocytic Tumors

Tumors derived from astrocytes are the most frequent primary intracranial neoplasms. Their neuropathological appearance is highly variable, and numerous attempts have been made to devise histologic grading systems that accurately predict their clinical course. The most widely used of these is the four-tiered grading system of the World Health Organization. Grade I identifies histologic variants of astrocytoma that have an excellent prognosis following surgical excision. These include juvenile pilocytic astrocytoma, subependymal giant cell astrocytoma (which occurs in patients with tuberous sclerosis), and pleiomorphic xanthoastrocytoma. At the other extreme is grade IV, glioblastoma multiforme, which has a high mitotic rate and invariably invades adjacent tissue. Astrocytoma (grade II) and anaplastic astrocytoma (grade III) are of intermediate histologic grade, and this is reflected in their clinical prognosis. The defining features of aggressive growth are hypercellularity, nuclear and cytoplasmic atypia, endothelial proliferation, mitotic activity, and necrosis. The presence of endothelial proliferation and necrosis, characteristic of glioblastoma multiforme, are regarded as important predictors of a tumor's potential for rapid growth and aggressive invasion of normal surrounding tissue.

Most highly malignant astrocytic tumors, such as glioblastoma multiforme, seem to develop without evidence of a precursor lesion. However, in some cases, these tumors do arise years after diagnosis of a lower grade astrocytic tumor. Although astrocytic oncogenesis is still incompletely understood, some molecular genetic constructs can tentatively be hypothesized to occur at recognizable stages in this progressive process [*see Figure 2*]. Mutation of the *p53* gene is likely to be an early event, associated with the change from normal astrocytes to low-grade neoplasia. This is of particular interest because loss of one of the biologic functions of *p53*, namely its role in stabilizing the genome, may be a crucial step that allows for the accumulation of numerous mutations in a single cell that seem to be frequently associated with the most highly malignant tumors. Loss of heterozygosity for chromosome 22q may be another early event in the oncogenic transformation of astrocytes. Loss of heterozygosity for chromosome 9p, where *p15* and *p16* are located, chromosome 13q, and chromosome 19q, as well as deletion of chromosome 10, where *PTEN* is located, and *EGFR* amplification are associated almost exclusively with astrocytic tumors of higher grade. These changes are therefore assumed to occur later in astrocytic oncogenesis.[31]

All astrocytic tumors do not follow the same sequence of molecular alterations. For example, many glioblastomas display loss of chromosome 10 but show no evidence of *p53* inactivation. These glioblastomas are thought to have followed a different pathway to malignancy than those arising from known low-grade astrocytomas.[31] However, it is not known at present whether these late changes require earlier, as yet uncharacterized mutations for their transforming action to be manifested, or whether these changes alone can produce a glioblastoma from a previously normal cell.

The development of important new technologies has provided considerable information regarding the spectrum of mutational changes that can be found in malignant tumors. Fluorescent in situ hybridization analysis (FISH) of interphase nuclei has made it possible to extend molecular cytogenetics to large numbers of clinical specimens, and comparative genomic hybridization has made it possible to survey the entire genome of individual tumors for genetic changes in a manner that is not biased by assumptions

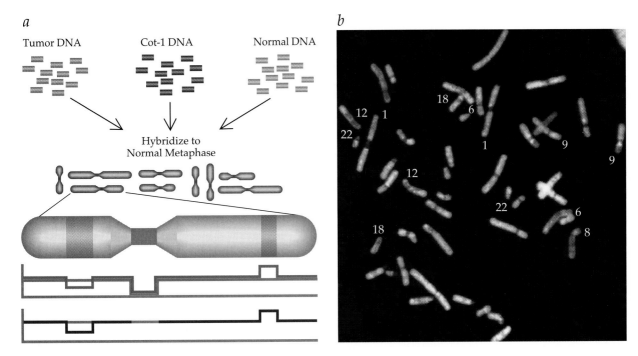

Figure 4 *Comparative genomic hybridization (CGH) makes possible the identification of chromosomal regions of amplification and deletion. (a) DNA extracted from a tumor is labeled with a green fluorescent dye, and DNA from a normal tissue is labeled with a red fluorescent dye. Equal amounts of these DNA probes are mixed with an excess of unlabeled DNA corresponding to highly repeated DNA sequences (Cot-1 DNA) and hybridized to a metaphase spread of chromosomes from normal human tissue. As depicted in the lower portion of (a), areas of amplification appear red, and areas of deletion appear green when viewed under a fluorescent microscope. (b) Photomicrographs of a normal metaphase spread examined under fluorescent light after CGH for the evaluation of chromosomal changes in a childhood brain tumor, medulloblastoma.*

regarding where in the genome such changes might be located [*see Figure 4*]. Comparative genomic hybridization has been used to study a number of different primary brain tumors, but in none has it been more revealing than in the study of high-grade astrocytic tumors [*see Figure 5*].[32] Selected chromosomal regions have been identified by comparative genomic hybridization to be either deleted or amplified in malignant astrocytomas [*see Table 4*]. Although more than 50 putative cancer-related loci have been identified in such analyses, in only a few is the gene likely to be modified by these structural alterations known.

In no case has it been possible to associate defined genetic alterations with a specific phenotypic characteristic of a tumor, such as invasive potential, that is key for progression from a low-grade neoplasm to a highly malignant form. There is also a dearth of information regarding when and how in the pathogenesis of a brain tumor these genetic programs are activated. Understanding of these events is an important target of ongoing research directed toward providing important insights of both pathological and therapeutic importance.

Oligodendrogliomas

The cellular origins of oligodendroglial tumors are poorly understood. There is ongoing confusion over proper diagnostic classifications for these tumors, which either do not display prototypical characteristics of cells arising in the oligodendroglial lineage or may contain multiple components of either oligodendroglial or astrocytic origin. This problem is of considerable clinical significance because oligodendrogliomas, which in most series constitute about 10 percent of glial tumors, are much more responsive than astrocytic tumors to chemotherapy. The physician may be inclined to give patients with brain tumors the most treatable diagnosis, raising the possibility that tumors of uncertain histologic appearance are designated as oligodendrogliomas.

From a molecular perspective, however, there are dramatic differences in the genetic alterations seen in prototypic oligodendrogliomas and astrocytomas. The most important cytogenetic alteration distinguishing oligodendrogliomas from other tumors of the CNS is the consistent presence of deletions of chromosome 1p36[32] and 19q13.3-13.4,[33,34] suggesting that

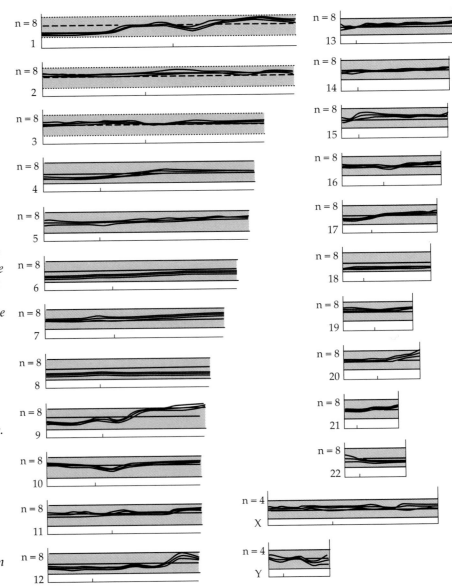

Figure 5 CGH ratio profiles of analyses of a tumor cell line established from an astrocytic brain tumor. On the left side of each plot is the chromosome number and number of individual chromosomes (n) analyzed. The thick and thin lines represent the ratio mean and standard deviation, respectively. The vertical line on the horizontal axis shows the location of the centromere. The hatched line represents a ratio of 1.0, and the dotted lines above and below represent ratios of 1.5 and 0.5, respectively. This tumor demonstrates loss of 1p31–36.3, 4p, 6, 8, 9p, 18, 22, and most of 12. There is gain of chromosomal DNA on 1q, 9q, and distal 12q.

Table 4   Amplified Chromosomal Regions Detected
by Comparative Genomic Hybridization in Malignant Gliomas

1p36.2-pter	8q24 (*myc*)	14q11.2-q13
1p13	9p2	14q22-q23
1q22-q24	10q22	14q31-qter
1q32.1	10q226	15q12-q21
4q12 (*PFGRA*)	11p11.2-p12	17p11.2-p12
5p15.1	11q13	18p
5q31.2	11q14.3-q24	19q13.4
7p12 (*EGFR*)	12p13	20q13.1-q13.3
7q21.1	12q13-q15 (*cdk4, MDM2*)	22q12
7q21.2-q21.3	12q22-qter	
7q31 (*MET*)	13q32-q34	

unidentified tumor suppressor genes are located here. Candidate genes located on chromosome 19q in the region commonly deleted include several genes known to function in DNA repair. The occasional loss of heterozygosity for chromosome 9p loci in oligodendrogliomas suggests the possible loss of the *cdkN2/p16* locus in some oligodendroglial tumors. But once again, mutations in this gene have not been found in tumors that have not lost additional large segments of chromosome 9p, raising the possibility that other genes important for these tumors are located in this chromosomal region. Mutational alterations frequently found in astrocytic tumors (e.g., *p53*) have not been detected in these tumors. Similarly, deletions of portions of chromosome 10 and *EGFR* amplification are found very rarely in these tumors.

The occurrence of mixed tumors that demonstrate regions histologically similar to oligodendroglioma and other areas similar to astrocytoma are being recognized with increased frequency. Molecular studies of carefully dissected tissue from these tumors suggest that both components of such mixed tumors share cytogenetic alterations characteristic of oligodendroglioma, perhaps indicating that they are derived from a common precursor cell.[35] Although it is not known if these histologic changes are the result of local environmental factors within the tumor or arise as the result of stochastic changes in gene expression, it remains possible that these tumors will be sensitive to therapy in a manner similar to that of oligodendrogliomas rather than their more resistant counterpart, astrocytomas.

Ependymomas

Cytogenetic studies of ependymomas reveal deletions of chromosome 22q,[33] the location of the *NF2* gene. Patients with NF2, who carry a mutation in one copy of the *NF2* gene, are known to acquire ependymomas, bilateral vestibular schwannomas, and multiple meningiomas. Sporadic ependymomas have been examined for *NF2* mutations, but to date this mutation has been found infrequently.[36] Similarly, although ependymomas occasionally occur in Li-Fraumeni syndrome patients, who transmit a germline mutation of the *p53* gene, only a single *p53* mutation has been detected in more than 75 reported tumors. Consistent with this observation is the lack of chromosome 10 deletions and *EGFR* amplification, which might have been expected if these malignancies shared a closer pathological relationship with tumors of the astrocytic lineage.

Medulloblastoma

Medulloblastoma is a common malignant childhood brain tumor, accounting for approximately 15 to 20 percent of all brain tumors in this age group. The majority of these tumors are located in the vermis of the cerebellum, where they are thought to arise in the fetal external granular layer. Histopathologically, medulloblastoma is highly cellular, with frequent mitoses and pseudorosette staining patterns. Cytologically, tumor cells have scant cytoplasm and small nuclei that stain darkly, a characteristic of many embryonal childhood tumors. Medulloblastoma has attracted considerable attention because it is sensitive to both radiation and chemotherapy, and the five-year disease-free survival approximates 50 to 70 percent.

Medulloblastoma occurs in three inherited cancer syndromes: Turcot's syndrome, Li-Fraumeni syndrome, and Gorlin's syndrome. In Turcot's syndrome, familial adenomatosis of the colon is combined with CNS tumors that are most frequently medulloblastomas, although astrocytic tumors are also occasionally seen.[37] Patients with familial adenomatous polyposis coli carry a germ-line mutation of the *APC* gene. Mutations in this gene have also been found in patients with Gorlin's syndrome, a disorder characterized by many basal cell nevi and a number of different associated tumors, including medulloblastoma and glioma. However, it has not yet been possible to document loss of the second allele of the *APC* gene in medulloblastoma tumor tissue from such patients. At present, no somatic mutations of the *APC* gene have been detected in sporadically occurring medulloblastoma. In children from Li-Fraumeni families with medulloblastoma, which carry a germ line *p53* mutation, it is possible to detect mutations of the remaining normal *p53* allele in tumor tissue, although again, sporadically occurring medulloblastomas rarely have a mutated *p53* gene. The absence of mutations in *p53* and *APC* in sporadically occurring medulloblastoma is unexpected. Other genes in the same pathways served by *p53* or *APC* may be mutated in medulloblastoma, or perhaps multiple molecular disorders can lead to histopathology typical of the tumor. For these reasons, observations of deletions in the *PTCH* gene in some sporadic medulloblastomas,[30] which is mutated in patients with Gorlin's syndrome, are of particular interest. *PTCH* maps to chromosome 9q22.3-q31, but loss of heterozygosity in this chromosomal region in medulloblastoma tumors is infrequent, raising the possibility that mutations in other genes in the growth regulatory pathway that includes *PTCH* will be found in sporadic medulloblastoma.

Despite the very infrequent occurrence of *p53* mutations in sporadically occurring medulloblastoma, cytogenetic studies have identified chromosome 17p, the region in which *p53* is located, as a frequent site of deletions. These observations raise the possibility of a second important locus on 17p13, telomeric to *p53*.[38] Consistent with this possibility is the observation that deletions at this locus may be more common in tumors that follow an aggressive clinical course. Other molecular genetic changes sometimes associated with medulloblastoma are deletions of chromosomes 6q and 16q and, infrequently, amplification of c-*myc* and N-*myc*. Expression of high levels of messenger RNA (mRNA) encoding the TrkC neurotrophin receptor in medulloblastoma is reported to be associated with favorable outcome in a small series of patients.[39] This may be of pathophysiological importance, because it now appears that most medulloblastomas, although appearing undifferentiated, can express a variety of genes characteristic of a neuronal lineage. If these tumors do express receptors for physiologic regulators of proliferation and differentiation, it might be possible to fashion drugs to modulate their activity.

Meningiomas and Vestibular Schwannomas

Meningiomas arise frequently in patients with NF2. Because deletion of loci on chromosome 22q is a frequent event in both vestibular schwannoma and meningioma when these occur sporadically in otherwise healthy patients, the *NF2* gene (located on chromosome 22q12) is a plausible candidate for inactivation in these tumors. *NF2* mutations and loss of heterozygosity for this locus have been reported by several groups in sporadic meningioma and vestibular schwannoma. There is also evidence for a second tumor suppressor locus important for meningioma formation on chromosome 22q, although no specific gene at this location has been implicated. Deletion of loci on chromosomes 1p, 9q, and 17p are found in malignant meningioma, although mutation of *p53* is infrequent. Gene amplification or other evidence of specific oncogene activation has not yet been identified to occur in meningioma and schwannoma.

Molecular Pathophysiology of Brain Tumors

Invasion of Normal Tissue

When gliomas recur following initial treatment, they virtually always recur locally at the site in which they were first detected.[40] An important reason for this propensity to local growth is the invasive characteristics of these malignant tumors. In contrast to other tumor types, including metastatic lesions, tumors of the CNS demonstrate preferred routes of invasion. These include invasion along white matter tracts and growth defined by the surrounding vasculature.

Pathological activities required for tissue invasion are hypothesized to include attachment of the tumor cell to surrounding extracellular matrix proteins, degradation of the matrix by proteinases produced by tumor cells, and tumor cell migration. The extracellular milieu of the CNS is distinct from other tissues and consists largely of glycosaminoglycans, hyaluronic acid, and chondroitan sulfate, which are detected only in low amounts in other tissues. Although glial tumor cells express a variety of different integrin receptors and recognize many different components in the basement membranes of blood vessels, including laminin and type IV collagen, virtually nothing is known of the manner in which tumor cells recognize white matter tracts to allow for migration along these pathways. Attachment to tissues and specific molecules at the growing edge of tumors is important because it is the key interface between proliferating malignant cells and normal tissue. Among the integrin receptors that glial tumor cells commonly express, $\alpha_V\beta_3$ is of special interest because this receptor is not detectable in normal tissues of the CNS but is detectable in high-grade astrocytic tumors.

Glial tumors express high surface levels of a urokinase-type plasminogen activator (uPA)[41] that binds to urokinase plasminogen receptors (uPAR) that are also highly expressed in the most malignant gliomas.[42] uPAR functions to present activated plasminogen activator to the extracellular matrix, where plasminogen is cleaved to active plasmin. Plasmin, in turn, degrades molecules of the extracellular matrix, such as fibronectin, laminin, and proteoglycans, that are especially prominent in the extracellular matrix of blood vessels. It also activates additional proteases, such as the matrix metalloproteinases and cathepsins, which then act locally to degrade plasmin-resistant molecules of the extracellular matrix. Naturally occurring protease inhibitors specific for plasminogen activators have been described and are produced by some brain tumors. These molecules can inhibit tumor-mediated degradation of extracellular matrix,[43] and their physiologic role is thought to involve maintaining the fine balance required for remodeling of the extracellular matrix during development. Because tumor invasion might be a key requirement for the proliferation of malignant cells in the CNS, disruption of adjacent normal CNS tissue, and establishment of the tumor vasculature, it is possible that therapeutic strategies targeting the invasive activities of glial tumor cells will be effective, even though tumor cells likely have dispersed from the primary tumor mass before the tumor is diagnosed.

Tumor Angiogenesis

Development of a tumor vasculature is a critical event in the establishment of a tumor as well as for its progression.[44] Low-grade brain tumors tend to have a vasculature similar to that observed in normal brain, but high-grade tumors, especially astrocytoma, are highly vascularized with vessels that lack an effective blood-brain barrier and have an enhanced permeability that contributes to tumor-associated vasogenic edema, a cause of considerable morbidity.[45] Because of its potential as a therapeutic target, much attention has been directed toward the tumor vasculature of gliomas and the molecules that mediate tumor neoangiogenesis.

Angiogenesis requires the coordinated interplay of many different cell types and biologic activities, and important candidates for the regulation of angiogenesis in brain tumors have been identified. Among these, VEGF[46] is of particular interest. VEGF is expressed at very high levels in high-grade gliomas,[47] and receptors for this growth stimulatory ligand can be detected on tumor vasculature. Flt-1, the first VEGF receptor recognized,[48] is not expressed in normal brain tissue but in a tumor-stage–dependent manner in the vasculature in astrocytomas.[49] A second VEGF receptor, KDR,[50] is expressed only in high-grade gliomas,[51] although the mechanism of enhanced expression in brain tumors of these two growth factor receptors is unknown. Platelet-derived growth factor and acid and basic fibroblast growth factors are additional glial tumor–produced growth factors for which receptors are expressed on glial tumor endothelium.

The progressive accumulation of genetic alterations during tumor progression may include changes that indirectly modulate cellular pathways important for neoangiogenesis. Consistent with this possibility has been the identification of angiogenic inhibitors whose production is p53 protein dependent.[52,53] The importance of such observations is enhanced by other reports indicating that it is possible to inhibit glial tumor angiogenesis in vivo by transferring a recombinant gene encoding a dominant negative mutant VEGF receptor gene into the vasculature of a rodent glioma tumor model,[54] thereby demonstrating the dependency of glioma tumor growth on neoangiogenesis. Both nonspecific inhibitors of angiogenesis, such as suramin, and specific inhibitors, such as TNP-40, are being evaluated in clinical trials for the treatment of glioblastoma.

Therapeutic Resistance and Apoptosis in Brain Tumors

Primary brain tumors are widely considered particularly resistant to the most commonly used antineoplastic strategies, radiation and chemotherapy. The observation, however, that multiagent chemotherapy can significantly affect the outcome of patients with anaplastic oligodendrogliomas[55] and the recent availability of novel strategies for delivering focal radiation therapy have revived interest in pursuing these modalities for the treatment of primary brain tumors. Whereas considerable controversy persists over the relative contributions of the blood-brain barrier, multidrug resistance transporters, and enhanced mechanisms of DNA repair to therapeutic resistance,[56,57] recent advances in the understanding of the mechanisms of cell death following DNA damage induced by both cytotoxic agents and irradiation may provide new opportunities to enhance tumor cell sensitivity. Foremost among these advances has been recognition of the contribution of apoptosis to the response of various cell types to cytotoxic therapies,[58] including radiotherapy. Apoptosis is a mechanism of active cell death that requires the expression of numerous cellular genes [*see Chapters 4 and 7*]. It is characterized by a rapid loss of plasma membrane integrity and DNA fragmentation.

Primary brain tumors exhibit varying degrees of apoptosis that is recognizable in biopsies obtained at surgery before the initiation of cytotoxic therapy. The successful treatment of an established tumor requires that many logs of tumor cells be killed, and the degree to which apoptosis contributes to the overall response of tumors to therapy is not yet clear. Interestingly, the level of apparently spontaneous apoptosis in these specimens seems highest in those tumors generally regarded as most responsive to therapy, medulloblastoma and oligodendroglioma.[59] The level of apoptosis in prebiopsy specimens may serve as a marker of treatment outcome.[60] It is now possible to demonstrate that after irradiation, medulloblastomas undergo extensive apoptosis that is dependent upon the expression of a functional *p53* gene,[61] whereas only a minimal amount of apoptosis is observed following the irradiation of cell lines from astroglial tumors, and this response is *p53* independent.[62] *BCL2*, a gene whose expression is known to inhibit apoptosis in many different cell types, is highly expressed in some high-grade glial tumors.[63]

Diagnostic and Therapeutic Opportunities

Oncogenetic Alterations as Prognostic Markers

On the basis of our emerging view of cancer as a genetic disorder, it should be possible to identify tumor markers that contribute to the pathophysiology of brain tumors. To date, the potential prognostic importance of EGFR amplification as a marker in glioblastoma has been the most extensively examined. EGFR amplification in tumor cells seems to be associated with a poor outcome.[64] It will also be important to characterize alterations in individual tumors to seek emerging therapies that target specific mutations or cellular pathways.

Knowledge of the genetic basis for tumor development has the potential to improve significantly the management of individuals at increased risk for the development of brain tumors. Early intervention may be possible if inherited predispositions are known. The development of new technologies that make it feasible to routinely screen for germ-line mutations in genes such as *p53*, which are known to enhance the risk of brain tumors, as well as strategies that might detect tumors based on the release into serum of a molecule highly expressed in brain tumors, such as superoxide dismutase,[65] should have a significant impact on these opportunities. Whereas the mechanism of enhanced expression of this cytoplasmic enzyme in astrocytic tumors is unknown, it is possible that cells expressing high levels of this enzyme, which antagonizes the cytotoxic effects of both chemotherapy and irradiation, are genetically selected during the course of tumorigenesis and subsequent antineoplastic therapy.

Targeted Therapy: Monoclonal Antibodies and Gene Therapy

Most therapeutic innovations to date have occurred through improvement of surgical and radiation treatment. Knowledge of brain tumor pathogenesis may presage the development of unanticipated treatment strategies that will be more effective and less toxic. Targeted therapy based on tumor-specific genetic or biochemical alterations is a goal of both conventional drug development seeking small-molecule therapeutics and novel approaches such as administration of antisense oligonucleotides and targeted ribozymes.

Novel strategies that have already reached clinical trials include monoclonal antibodies targeting brain tumor cells and gene therapy. Monoclonal antibodies that localize to tumor tissue and therefore have diagnostic potential are also being developed as treatment modalities. Two particularly interesting strategies involve the use of a radiolabeled antitenascin antibody and an antibody directed against a recurring structural alteration in the EGFR receptor. Tenascin is an extracellular matrix–associated protein found in many high-grade brain tumors but not in the mature CNS.[66] An antibody that effectively targets this molecule has been used to treat a large number of glioma patients with promising results[67] and is being evaluated for the treatment of recurrent cystic gliomas, into which it can be inoculated. Another related strategy is the development of monoclonal antibodies against surface markers that are tumor specific or expressed at high levels on the surface of tumor cells. A radiolabeled antibody against the membrane-associated ganglioside 3'6'-isoLD1 is being evaluated in brain tumor patients,[68] and anti-EGFR monoclonal antibody has been examined for toxicity in brain tumor patients.[69] A novel epitope expressed on the surface of a significant proportion of glioblastomas as a result of a recurring deletion in the EGFR is among the most promising future targets for immunotherapy,[70] and several approaches to target this site are under investigation.

Gene therapy is a novel modality of cancer therapy that involves the therapeutic transfer of new genetic information. To date, treatment of tumors by gene therapy has been greatly limited by the inability of gene transfer strategies to reach more than a small percentage of tumor cells. Nevertheless, a number of strategies specifically designed to overcome this limitation are being pursued. Of these, the two with the greatest impact have been the use of viral vectors to transfer the gene encoding the enzyme thymidine kinase from the Herpes simplex virus (HSV-tk), and attempts to develop tumor vaccines. Gene therapy using transfer of the HSV-tk gene to tumor cells is based on the ability of this enzyme to convert an inactive pro-drug, ganciclovir, into a therapeutically active, phosphorylated metabolite after systemic administration.[71] The tk enzyme of mammalian cells does not have a high affinity for nucleoside analogs such as ganciclovir; hence, this drug is not activated in cells that do not contain the transferred HSV-tk gene. The use of retroviral vectors for gene transfer provides an important measure of therapeutic index for the treatment of brain tumors because the integration of these vectors and stable transfer of the therapeutic gene occurs only in dividing cells. Although the low frequency of gene transfer that even this strategy results in is problematic, it has been shown that after infection with the HSV-tk–containing retrovirus and treatment with ganciclovir, the number of cells that die far exceeds the number of cells into which the HSV-tk gene is transfected. This is referred to as a bystander effect, which is likely to be mediated by both immune mechanisms and the intercellular transfer of phosphorylated ganciclovir metabolites through gap junctions formed between cells that express HSV-tk and uninfected tumor cells.[72]

Attempts are also under way to use gene therapy to transfer cytokine genes to irradiated tumor cells that have been placed in culture following removal at surgery. These cells, which might present both known and unknown tumor cell antigens to the immune system more efficiently than other tumor cells as a result of the high local concentrations of cytokine that result from expression of the transferred gene, are reinoculated subcutaneously into patients. Such an approach to the development of tumor vaccines has been further enhanced by the recognition of tumor-specific antigens and by in vitro strategies for enhancing antigen presentation. The development of vaccines holds promise not only for the primary treatment of tumors but also for prolonging the duration of disease-free remissions.

References

1. American Cancer Society: Cancer Statistics 1997. *CA:A Cancer Journal for Clinicians*, Vol. 47. Lippincott-Raven, New York, January/February 1997

2. James CD, Carlbom E, Dumanski JP, et al: Clonal genomic alterations in glioma malignancy stages. *Cancer Res* 48:5546, 1988

3. Knudson AG: Mutation and cancer: statistical study of retinoblastoma. *Proc Natl Acad Sci USA* 68:820, 1971

4. el-Deiry WS, Tokino T, Velculescu VE, et al: WAF1, a potential mediator of *p53* tumor suppression. *Cell* 75:817, 1993

5. Yin Y, Tainsky MA, Bischoff FZ, et al: Wild-type *p53* restores cell cycle control and inhibits gene amplification in cells with mutant *p53* alleles. *Cell* 70:937, 1992

6. el-Deiry WS, Harper JW, O'Connor PM, et al: WAF1/CIP1 is induced in *p53*-mediated G1 arrest and apoptosis. *Cancer Res* 54:1169, 1994

7. James CD, Olson JJ: Molecular genetics and molecular biology advances in brain tumors. *Curr Opin Oncol* 8:188, 1996

8. Bogler O, Huang HJ, Kleihues P, et al: The *p53* gene and its role in human brain tumors. *Glia* 15:308, 1995

9. Chozick BS, Weicker ME, Pezzullo JC, et al: Pattern of mutant *p53* expression in human astrocytomas suggests the existence of alternate pathways of tumorigenesis. *Cancer* 73:406, 1994

10. Ohgaki H, Eibl RH, Schwab M, et al: Mutations of the *p53* tumor suppressor gene in neoplasms of the human nervous system. *Mol Carcinog* 8:74, 1993

11. Sidransky D, Mikkelsen T, Schwechheimer K, et al: Clonal expansion of *p53* mutant cells is associated with brain tumour progression. *Nature* 355:846, 1992

12. Friend SH, Horowitz JM, Gerber MR, et al: Deletions of a DNA sequence in retinoblastomas and mesenchymal tumors: organization of the sequence and its encoded protein. *Proc Natl Acad Sci USA* 84:9059, 1987

13. Abramson DH, Ellsworth RM, Zimmerman LE: Nonocular cancer in retinoblastoma survivors. *Trans Am Ophthalmol Otolaryngol* 81:454, 1976

14. Draper GJ, Sanders BM, Kingston JE: Second primary neoplasms in patients with retinoblastoma. *Br J Cancer* 53:661, 1986

15. Reifenberger G, Reifenberger J, Ichimura K, et al: Amplification of multiple genes from chromosomal region 12q12-14 in human malignant gliomas: preliminary mapping of the amplicons shows preferential involvement of *CDK4*, *SAS*, and *MDM2*. *Cancer Res* 54:4299, 1994

16. Kamb A, Gruis NA, Weaver-Feldhaus J, et al: A cell cycle regulator potentially involved in genesis of many tumor types. *Science* 264:436, 1994

17. Schmidt EE, Ichimura K, Reifenberger G, et al: *CDKN2* (*p16/MTS1*) gene deletion or *CDK4* amplification occurs in the majority of glioblastomas. *Cancer Res* 54:6321, 1994

18. Li J, Yen C, Liaw D, et al: *PTEN*, a putative protein tyrosine phosphatase gene mutated in human brain, breast, and prostate cancer. *Science* 275:1943, 1997

19. Steck PA, Pershouse MA, Jasser SA, et al: Identification of a candidate tumour suppressor gene, *MMAC1*, at chromosome 10q23.3 that is mutated in multiple advanced cancers. *Nat Genet* 15:356, 1997

20. Collins VP: Gene amplification in human gliomas. *Glia* 15:289, 1995

21. Fuller GN, Bigner SH: Amplified cellular oncogenes in neoplasms of the human central nervous system. *Mutat Res* 276:299, 1992

22. Wong AJ, Ruppert JM, Bigner SH, et al: Structural alterations of the epidermal growth factor receptor gene in human gliomas. *Proc Natl Acad Sci USA* 89:2965, 1992

23. Orian JM, Vasilopoulos K, Yoshida S, et al: Overexpression of multiple oncogenes related to histological grade of astrocytic glioma. *Br J Cancer* 66:106, 1992

24. Hughes RAC: Neurological complications of neurofibromatosis 1. *The Neurofibromatoses: A Pathogenetic and Clinical Overview.* Huson SM, Hughes RAC, Eds. Chapman & Hall Medical, London, 1994, p 204

25. Malkin D, Li FP, Strong LC, et al: Germ line *p53* mutations in a familial syndrome of breast cancer, sarcomas, and other neoplasms. *Science* 250:1233, 1990

26. Kyritsis AP, Bondy ML, Xiao M, et al: Germline *p53* gene mutations in subsets of glioma patients. *J Natl Cancer Inst* 86:344, 1994

27. Trofatter JA, MacCollin MM, Rutter JL, et al: A novel moesin-, ezrin-, radixin-like gene is a candidate for the neurofibromatosis 2 tumor suppressor. *Cell* 72:791, 1993

28. European Chromosome 16 Tuberous Sclerosis Consortium: Identification and characterization of the tuberous sclerosis gene on chromosome 16. *Cell* 75:1305, 1993

29. Iliopoulos O, Levy AP, Jiang C, et al: Negative regulation of hypoxia-inducible genes by the von Hippel-Lindau protein. *Proc Natl Acad Sci USA* 93:10595, 1996

30. Raffel C, Jenkins RB, Frederick L, et al: Sporadic medulloblastomas contain *PTCH* mutations. *Cancer Res* 57:842, 1997

31. von Deimling A, Louis DN, Wiestler OD: Molecular pathways in the formation of gliomas. *Glia* 15:328, 1995

32. Kim DH, Mohapatra G, Bollen A, et al: Chromosomal abnormalities in glioblastoma multiforme tumors and glioma cell lines detected by comparative genomic hybridization. *Int J Cancer* 60:812, 1995

33. Ransom DT, Ritland SR, Kimmel DW, et al: Cytogenetic and loss of heterozygosity studies in ependymomas, pilocytic astrocytomas, and oligodendrogliomas. *Genes Chromosomes Cancer* 5:348, 1992

34. Reifenberger J, Reifenberger G, Liu L, et al: Molecular genetic analysis of oligodendroglial tumors shows preferential allelic deletions on 19q and 1p. *Am J Pathol* 145:1175, 1994

35. Kraus JA, Koopman J, Kaskel P, et al: Shared allelic losses on chromosomes 1p and 19p suggest a common origin of oligodendroglioma and oligoastrocytoma. *J Neuropathol Exp Neurol* 54:91, 1995

36. Rubio M-P, Correa KM, Ramesh V, et al: Analysis of the neurofibromatosis 2 gene in human ependymomas and astrocytomas. *Cancer Res* 54:45, 1994

37. Turcot J, Despres J-P, St Pierre F: Malignant tumors of the central nervous system associated with familial polyposis of the colon: report of two cases. *Dis Colon Rectum* 2:465, 1959

38. McDonald JD, Daneshvar L, Willert JR, et al: Physical mapping of chromosome 17p13.3 in the region of a putative tumor suppressor gene important in medulloblastoma. *Genomics* 23:229, 1994

39. Segal RA, Goumnerova LC, Kwon YK, et al: Expression of the neurotrophin receptor TrkC is linked to a favorable outcome in medulloblastoma. *Proc Natl Acad Sci USA* 91:12867, 1994

40. Burger PC, Dubois PJ, Schold SC, et al: Computerized tomographic and pathologic studies of the untreated, quiescent, and recurrent glioblastoma multiforme. *J Neurosurg* 58:159, 1983

41. Mohanam S, Sawaya RE, Yamamoto M, et al: Proteolysis and invasiveness of brain tumors: role of urokinase-type plasminogen activator receptor. *J Neurooncol* 22:153, 1994

42. Bindal AK, Hammoud M, Shi WM, et al: Prognostic significance of proteolytic enzymes in human brain tumors. *J Neurooncol* 22:101, 1994

43. Cajot JF, Bamat J, Bergonzelli GE, et al: Plasminogen-activator inhibitor type 1 is a potent natural inhibitor of extracellular matrix degradation by fibrosarcoma and colon carcinoma cells. *Proc Natl Acad Sci USA* 87: 6939, 1990

44. Folkman J: The role of angiogenesis in tumor growth. *Semin Cancer Biol* 3:65, 1992

45. Risau W: Molecular biology of blood-brain barrier ontogenesis and function. *Acta Neurochir Suppl (Wien)* 60:109, 1994

46. Plate KH, Breier G, Weich HA, et al: Vascular endothelial growth factor is a potential tumour angiogenesis factor in human gliomas in vivo. *Nature* 359:845, 1992

47. Plate KH, Risau W: Angiogenesis in malignant gliomas. *Glia* 15:339, 1995

48. de Vries C, Escobedo JA, Ueno H, et al: The fma-like tyrosine kinase, a receptor for vascular endothelial growth factor. *Science* 255:989, 1992

49. Plate KH, Breier G, Weich HA, et al: Vascular endothelial growth factor and glioma angiogenosis: coordinate induction of VEGF receptors, distribution of VEGF protein and possible in vivo regulatory mechanisms. *Int J Cancer* 59:520, 1994

50. Terman BC, Dougher-Vermazen M, Carrion ME, et al: Identification of the KDR tyrosine kinase as a receptor for vascular endothelial growth factor. *Biochem Biophys Res Commun* 187:1579, 1992

51. Hatva E, Kaipainen A, Mentula P, et al: Expression of endothelial cell-specific receptor tyrosine kinases and growth factors in human brain tumors. *Am J Pathol* 146:368, 1995

52. Dameron KM, Volpert OV, Tainsky MA, et al: Control of angiogenesis in fibroblasts by p53 regulation of thrombospondin-1. *Science* 265:1582, 1994

53. Van Meir EG, Polverini P, Chazin V, et al: Release of an inhibitor of angiogenesis upon induction of wild-type p53 expression in glioblastoma cells. *Nat Genet* 8:171, 1994

54. Millauer B, Shawer L, Plate KH, et al: Glioblastoma growth inhibited in vivo by a dominant-negative Flk-1 mutant. *Nature* 367:576, 1994

55. Cairncross JG, MacDonald DR: Oligodendroglioma: a new chemosensitive tumor. *J Clin Oncol* 8:2090, 1990

56. Feun LG, Savaraj N, Landy HJ: Drug resistance in brain tumors. *J Neurooncol* 20:165, 1994

57. Stewart DJ: A critique of the role of the blood-brain barrier in the chemotherapy of human brain tumors. *J Neurooncol* 20:121, 1994

58. Lowe SW, Ruley HE, Jacks T, et al: *p53*-dependent apoptosis modulates the cytotoxicity of anticancer agents. *Cell* 74:957, 1993

59. Schiffer D, Cavalla P, Migheli A, et al: Apoptosis and cell proliferation in human neuroepithelial tumors. *Neurosci Lett* 195:81, 1995

60. Wheeler JA, Stephens LC, Tornos C, et al: Apoptosis as a predictor of tumor response to radiation in stage IB cervical carcinoma. *Int J Radiat Oncol Biol Phys* 32:1487, 1995

61. Dee S, Haas-Kogan DA, Israel MA: Inactivation of p53 is associated with decreased levels of radiation-induced apoptosis in medulloblastoma cell lines. *Cell Death Diff* 2:267, 1995

62. Haas-Kogan DA, Dazin P, Hu L, et al: *p53*-independent apoptosis: a mechanism of radiation-induced cell death of glioblastoma cells. *Can J Neurol Sci* 2:113, 1996

63. Krajewski S, Krajewska M, Ehrmann J, et al: Immunohistochemical analysis of *Bcl-2*, *Bcl-X*, *Mcl-1*, and *Bax* in tumors of central and peripheral nervous system origin. *Am J Pathol* 150:805, 1997

64. von Deimling A, von Ammon K, Schoenfeld D, et al: Subsets of glioblastoma multiforme defined by molecular genetic analysis. *Brain Pathol* 3:19, 1993

65. Cobbs CS, Levi DS, Aldape K, et al: Manganese superoxide dismutase (MnSOD) expression in human central nervous system tumors. *Cancer Res* 56:3192, 1996

66. Zagzag D, Friedlander DR, Miller DC, et al: Tenascin expression in astrocytomas correlates with angiogenesis. *Cancer Res* 55:907, 1995

67. Riva P, Arista A, Franceschi G, et al: Local treatment of malignant gliomas by direct infusion of specific monoclonal antibodies labeled with 131I: comparison of the results obtained in recurrent and newly diagnosed tumors. *Cancer Res* 55:5952S, 1995

68. Kurpad SN, Zhao XG, Wikstrand CJ, et al: Tumor antigens in astrocytic gliomas. *Glia* 15:244, 1995

69. Faillot T, Magdelenat H, Mady E, et al: A phase I study of an anti-epidermal growth factor receptor monoclonal antibody for the treatment of malignant gliomas. *Neurosurgery* 39:478, 1996

70. Reist CJ, Archer GE, Kurpad SN, et al: Tumor-specific anti-epidermal growth factor receptor variant III monoclonal antibodies: use of the tyramine-cellobiose radioiodination method enhances cellular retention and uptake in tumor xenografts. *Cancer Res* 55:4375, 1995

71. Martin JC, Dvorak CA, Smee DF, et al: 9-[(1,3-Dihydroxy-2-propoxy)methyl]guanine: a new potent and selective antiherpes agent. *J Med Chem* 26:759, 1983

72. Fick J, Barker FG II, Dazin P, et al: The extent of heterocellular communication mediated by gap junctions is predictive of bystander tumor cytotoxicity in vitro. *Proc Natl Acad Sci USA* 92:11071, 1995

Acknowledgments

Figures 1, 3, 5   Marcia Kammerer.

Figures 2, 4a   Dimitry Schidlovsky.

Figures 4(b) and 5   Courtesy of Dr. B. Feverstien.

CHAPTER 7

Molecular Approaches to Therapy of Stroke

Frank R. Sharp, M.D., Jari Honkaniemi, M.D., Ph.D.,
Stephen M. Massa, M.D., Ph.D.

The brain is particularly vulnerable to ischemic injury, requiring continuous delivery of oxygen and glucose. Decreased oxygen and glucose delivery during ischemia results in rapid loss of function and eventually in tissue infarction. Some mechanisms of ischemic brain injury are similar to those in other organs, such as the heart. Brain injury, however, is compounded by excitotoxicity, the phenomenon of overstimulation of glutamate receptors, and by apoptosis, the process of biochemically programmed cell death. Delineating the molecular mechanisms of ischemic brain injury could lay the foundation for the rational development of new stroke treatments.

Global cerebral ischemia, which occurs after complete cessation of blood flow, causes selective neuronal death in certain areas of the brain, including the basal ganglia and hippocampus. This selective neuronal cell death occurs between one and three days after the ischemia. Damage to the basal ganglia produces rigidity of the body musculature, and hippocampal damage produces deficits in recent memory. The mechanism of this delayed selective neuronal death is unknown but may be mediated by apoptosis.

Focal cerebral ischemia is caused by either embolism to or stenosis of cerebral vessels. Sustained reduction in blood flow to a region of brain may cause infarction, resulting in death of neurons and glia. If sustained, death of all cellular elements, including blood vessels, occurs. Surrounding the area of infarction is a region of decreased cerebral blood flow (CBF) termed the *penumbra*.

Definitions of the penumbra vary. Positron emission tomography (PET) scanning has shown that in some patients a decrease in CBF increases the oxygen extraction fraction (OEF) in the same areas. The increase in OEF can be used to define a penumbra. Follow-up studies show that the volume of the penumbra that escapes infarction correlates with neurologic recovery. Evidence of increased OEF can be detected for up to 16 hours, suggesting a time during which the penumbra remains viable. A molecular definition of the penumbra based upon the induction of heat shock proteins is described later in this chapter. The fate of this penumbra is variable, and pharmacological approaches to improving the survival of cells in the penumbra offer hope for improving outcome from stroke.

Advances in understanding the metabolic and molecular responses of brain tissue to ischemia are described in this chapter. Some responses appear to be protective, whereas others act to induce a cascade of injury. Topics discussed include the effects of ischemia on protein synthesis, regulation of metabolic and signaling pathways, excitotoxicity, production of mitochondrial toxins, apoptosis, and induction of stress proteins.

Protein Synthesis Decreases during Cerebral Ischemia

Protein synthesis decreases when CBF falls to 50 percent of normal, long before any detectable changes occur in brain adenosine triphosphate (ATP),[1] and returns to normal when blood flow is restored in areas that do not infarct. Protein synthesis remains depressed in areas that will infarct. Although the mechanism of decreased protein synthesis following ischemia has not been fully elucidated, current evidence points to interactions between transfer RNA (tRNA) and ribosomes as the culprit. Protein synthetic rates are regulated in part by phosphorylation of eukaryotic initiation factor 2 (eIF-2)[2] [see Figure 1]. This critical component of the initiation of protein synthesis is phosphorylated by protein kinase R.[3-5] This pathway appears to

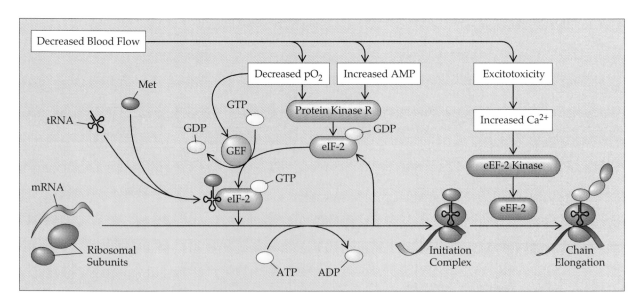

Figure 1 *Hypothetical mechanism of how cerebral ischemia blocks protein translation. Decreased blood flow activates an unknown kinase that phosphorylates protein kinase R. Activation of protein kinase R phosphorylates the α-subunit of eukaryotic initiation factor 2 (eIF-2). This blocks translation of mRNA into protein. Alternatively, decreased blood flow activates guanine exchange factor (GEF), which blocks translation; or decreased blood flow increases glutamate release that alters phosphorylation of elongation factor-2 (eEf-2) and blocks translation.*

regulate protein synthesis during times of stress because mutations of protein kinase R decrease protein synthesis induced by factors such as interferon, and overexpression of protein kinase R produces a block of translation.[6,7] Guanine exchange factor (GEF) also appears to regulate protein synthesis during periods of ischemic cell stress[8].

The factors that activate protein kinase R and GEF are not known. Whatever the process involved, ATP levels are not important for its regulation because brain protein synthesis falls before any decrease in brain ATP. It is possible that either hypoxia, which accompanies a decrease in blood flow, or increases in adenosine monophosphate (AMP), which occur without a significant fall in ATP, may serve as a stimulus for activation of protein kinase R and GEF [*see Figure 1*]. Glutamate-dependent phosphorylation of elongation factor-2 (eEf-2) inhibits protein synthesis in neurons. This is particularly relevant because glutamate is released when CBF decreases and glutamate injures neurons via glutamate receptors, through the phenomenon of excitotoxicity.

Although decreases of protein synthesis are well documented during cerebral ischemia, it is unknown whether such decreases are helpful or harmful. They could be helpful by decreasing energy demand. Alternatively, decreased protein synthesis could be harmful by reducing levels of proteins such as bcl-2, calbindin, and others that either buffer calcium or block proteases, apoptosis, or other adverse cellular responses to ischemia. The kinetics of the response may also be important because inhibition of protein synthesis may be beneficial immediately following ischemia and only become detrimental when ischemia is prolonged. As the dynamics of protein synthesis during and after ischemia are elucidated, it may be possible to manipulate pharmacologically the pertinent factors, such as protein kinase R activity, to minimize injury induced by ischemia.

Hypoxia Induces Gene Responses

Role of Hypoxia in Ischemic Brain

Oxygen and oxidative metabolism are crucial for maintenance of neuronal survival in the brain. During incomplete focal ischemia, CBF decreases, and oxygen becomes the limiting substrate because glucose transport continues. Tissue hypoxia leads to anaerobic glycolysis of glucose and increased production of lactate.[9,10] The resulting lactic acidosis is postulated to be critical for determining whether selective neuronal cell death or infarction occurs during ischemia.[11-13]

Cyanide poisoning in humans and experimental animals that blocks complex IV in the mitochondrial electron transport chain produces widespread neuronal injury, with the basal ganglia being particularly vulnerable. Direct brain injections of cyanide and azide produce neuronal cell death.[14] These data suggest that inhibition of mitochondrial oxidative respiration can kill brain cells and that even when systemic hypoxia is insufficient to damage the brain, local tissue hypoxia may contribute to cell death.

PET scanning reveals that in some patients, OEF increases in areas in which CBF decreases. This increase of OEF, which occurs when CBF ranges between 7 and 17 ml/100 g/min, was used to define the penumbra in a group of 11 patients. The volume of the penumbra that escaped infarction was found to be highly correlated with neurologic recovery.[15] Evidence of increased OEF is detectable for up to 16 hours, suggesting an extended period during which the penumbra is viable. This study confirms the possibility that the penumbra can be defined as the area of hypoxic brain and that functional outcome depends on the fate of the penumbra.

Hypoxia Induces the Hypoxia Inducible Factor–Transcription Factor

The role of hypoxia is complex, as emphasized by findings that it can partially protect the brain against injury. Rat pups exposed for short periods to low levels of oxygen are protected against stroke one day later.[16] Although the mechanism of this hypoxia-induced tolerance is not known, hypoxia inducible factors might play a role in the protection.

Hypoxia inducible factor (HIF-1) is a nuclear protein essential for transcriptional activation of a subset of hypoxia inducible genes that include erythropoietin.[17,18] The mechanism of HIF-1 activation by hypoxia is not known. However, in nitrogen-fixing bacteria, the oxygen-sensing switch is mediated by a protein called FIX-L, a hemeprotein kinase that is active in a low oxygen state. FIX-L phosphorylates and activates an HIF-1–like protein.[18] It is believed, although not yet proved, that there is an oxygen-sensitive hemeprotein kinase in mammalian cells that can activate HIF-1 [see Figure 2].

Glycolytic Enzymes, Glucose Transporters, and iNOS are Hypoxia Inducible Factor Target Genes

HIF-1, once induced by hypoxia and activated by phosphorylation, binds to sequences called hypoxia response elements in the promoters of target genes as well as to enhancer elements in the genes. Enzymes involved in glycolysis

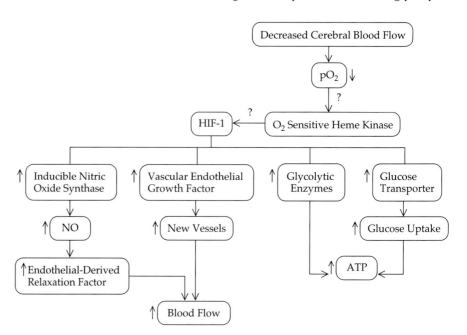

Figure 2   *Ischemia activation of the hypoxia inducible factor (HIF-1) in brain. Hypoxia induction of hypoxia inducible factors leads to induction of specific hypoxia inducible target genes.*

have such elements in their promoter regions, which render them responsive to hypoxia. These genes include phosphofructokinase, aldolase, phosphoglycerate kinase, enolase, pyruvate kinase, and lactate dehydrogenase.[19] Other target genes have been identified, including the glucose transporter 1 and inducible nitric oxide synthase (iNOS).[20,21] Hence, HIF-1 appears to have widespread effects on hypoxic cells, inducing glycolytic enzymes, glucose transporters, vascular growth responses, and other responses to hypoxia [*see Figure 2*]. HIF-1 messenger RNAs (mRNAs) expressed in normal brain are increased following a stroke in an area surrounding the infarction. Understanding the role of HIF-1 and other oxygen-regulated proteins during hypoxia and ischemia may lead to new treatments for ischemia.

Ischemia Turns on Selective Genes

Although ischemia initiates ionic and enzymatic changes that may damage cells, ischemia also initiates new gene expression. In the face of energy deprivation, such genes likely play important roles either in maintaining cellular survival or in promoting programmed cell death. Of these genes, those called immediate early genes, which include *c-fos* and *c-jun*, are among the most thoroughly studied.[22] These genes can be induced by a variety of extracellular and intracellular signals, including calcium, cyclic AMP, neurotrophic factors, and neurotransmitters. Once induced, Fos and Jun proteins join via leucine zippers and bind to specific (AP-1) sites in the promoter regions of target genes to regulate their expression.[23] Genes with AP-1 regulatory sites in their promoters include glial fibrillary acid protein, tyrosine hydroxylase, brain-derived nerve growth factor, fibroblast growth factor, iNOS, heme oxygenase-1 (HO-1), and amyloid precursor protein. Ischemic induction of immediate early genes likely elicits changes in cellular gene expression that affect whether cells live or die [*see Figure 3*].

Focal Ischemia

c-fos, junB, nerve growth factor 1-A (NGFI-A), and many other related immediate early genes are induced in the cerebral cortex and subcortical structures following middle cerebral artery (MCA) occlusion in rats.[22,24-27] This immediate early gene induction occurs within minutes of the MCA ischemia and occurs inside the ischemic MCA territory as well as throughout nonischemic anterior and posterior cerebral artery territories. It is maximal hours after ischemia and decreases in regions of infarction by 24 hours.[24]

These results show the surprising observation that cortex outside the ischemic area is also affected by stroke. An explanation for this widespread response to focal ischemia appears to be spreading depression, a slow-moving wave of cortical depolarization. Sequential waves of depolarization and calcium influx spread throughout a hemisphere following focal ischemia.[28,29] Such spreading depression is prevented by prior administration of *N*-methyl-D-aspartate (NMDA) receptor antagonists such as MK-801. Prior administration of MK-801 also blocks immediate early gene induction after focal ischemia, suggesting that NMDA receptor activation is involved in both responses (spreading depression and activation of immediate early genes).[30,31] This may explain the widespread cortical induction of glial fib-

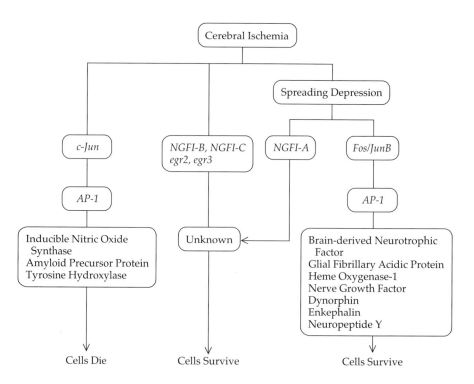

Figure 3 *Ischemia induces a number of transcription factors, including leucine zipper, zinc finger, and other immediate early genes. Each transcription factor induces specific target genes that may mediate cell survival or cell death.*

rillary acid protein in astrocytes and widespread HO-1 induction in microglia even though ischemia was restricted to the MCA territory.[32,33]

What are the significance and consequence of induction of so many immediate early genes after ischemia? First, it is likely that different extracellular and intracellular messengers induce different immediate early genes. Spreading depression induces *c-fos*, hypoxia induces hypoxia inducible factors, and other signals induce the zinc finger immediate early genes. Second, the target genes for the Fos/Jun family members induced by spreading depression are likely different from the target genes for hypoxia inducible factors, which may be different from the target genes for the zinc finger immediate early genes [*see Figure 3*]. Changes in gene expression may be coordinated by relatively few transcription factors.

Global Ischemia

Global ischemia induces region-selective activation of *c-fos*, *c-jun*, *fosB*, *junB*, and *NGFI-A, B,* and *C* in the CA1 and CA3 pyramidal cells and dentate granule cells of the hippocampus.[34-36] Because the genes are induced in cells that live (CA3) as well as in cells that die (CA1), these immediate early gene responses do not appear to play a direct role in cell death. However, the expression of some immediate early gene mRNAs (*c-fos*, *c-jun*, and *NGFI-A*) remains elevated in the CA1 neurons that eventually die.[27,37,38] Interestingly, although the mRNA for these genes is increased, in many cases, the proteins are not expressed,[37] indicating a block of translation. The persistence of untranslated mRNAs appears to correlate with cells destined to die. Failure to synthesize proteins from these mRNAs may result in loss of protein feedback on mRNA synthesis, resulting in persistent mRNA induction.

Glutamate Release during Ischemia Injures the Brain

The excitotoxic theory of brain injury had its beginnings when Rothman and Olney injected large doses of glutamate into young rats and produced brain injury.[39] Glutamate and its nonmetabolizable analogues, such as kainic acid, kill neurons in culture and destroy neurons when injected into the brain. The death of the neurons appears to be mediated by neurotransmitter actions of glutamate at postsynaptic receptors on neurons. Glutamate receptor activation depolarizes neurons and causes calcium to enter the cell. When the neurons become overloaded with calcium, they die.[40] Glutamate acts on at least two types of receptors to kill cells, including the NMDA receptor and the kainic acid/A-amino-3-hydroxy-5-methyl-4-isoxazole proprionic acid (AMPA) receptors.

Glutamate levels increase markedly during cerebral ischemia. This appears to be caused by a combination of ischemic depolarization and release of glutamate from synapses, and energy failure and acidosis, resulting in release of glutamate from glia. This concept of neurotoxicity is supported by experiments showing that drugs that antagonize the NMDA and kainic acid/AMPA receptors decrease stroke size and decrease cell death following focal and global cerebral ischemia in experimental animals. There are several ongoing trials of NMDA antagonists in patients with stroke.

Mitochondrial Toxins Are Produced during Ischemia

A search for genes highly expressed in ischemic versus normal brain disclosed the gene for methylmalonyl CoA mutase (MCM) [*see Figure 4*].[14] MCM mRNA and protein increased markedly in hippocampus following global ischemia.[14] Methylmalonyl CoA is a product of propionate, which is in turn derived from branched-chain amino acids and fatty acids.[41] MCM is a vitamin B_{12}-dependent enzyme that metabolizes methylmalonyl CoA to succinyl CoA[41] [*see Figure 4*].

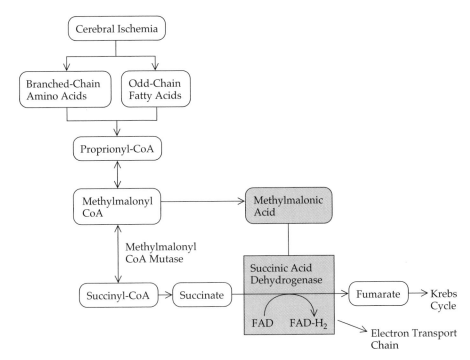

Figure 4 *The role of methylmalonyl CoA mutase (MCM) in the metabolism of odd-chain fatty acids and branched-chain amino acids. The methylmalonic acid that accumulates when MCM function is impaired inhibits succinic acid dehydrogenase function in the electron transport chain as well as in the Krebs cycle.*

Mutations of the *MCM* gene result in accumulation of methylmalonic acid, which is derived from methylmalonyl CoA.[42] The severe metabolic and neurologic abnormalities and rapid death that occur shortly after birth in patients with no MCM protein activity suggest that it may be important for normal brain function and development.[41,42] Methylmalonic acid inhibits succinic acid dehydrogenase, interfering with Krebs cycle and mitochondrial electron transport functions.[43] Structurally related succinic acid dehydrogenase inhibitors, malonate, and 3-nitropropionic acid, kill neurons in experimental animals and in humans,[44] suggesting that methylmalonic acid might also kill neurons. Introduction of methylmalonic acid into rodent brain produced cell death in a pattern similar to the structurally related mitochondrial succinic acid dehydrogenase inhibitors malonate and 3-nitropropionic acid.[14]

These findings suggest that methylmalonic acid may contribute to neuronal injury in human conditions in which it accumulates, including *MCM* gene mutations and vitamin B_{12} deficiency. The induction of methylmalonyl CoA mutase during ischemia might decrease the accumulation of methylmalonic acid. It is not known whether vitamin B_{12} deficiency, which results in elevations of serum methylmalonic acid in five to 15 percent of elderly patients, worsens ischemic injury in these individuals.

Ischemia Produces Oxidative Stress

Oxygen free radicals (reactive, partially reduced forms of oxygen) include superoxide ($O_2^{\cdot-}$), peroxyl radical ($RO_2^{\cdot-}$), nitric oxide ($NO^{\cdot}$), the hydroxyl radical ($OH^{\cdot}$), and singlet oxygen [*see Figure 5*].[45] Free radicals injure tissue following reactions with other cell constituents. The carbon double bonds of polyunsaturated fatty acids, for example, are particularly vulnerable to free radical assault. A number of enzymes, including glutamate synthase, are vulnerable to inactivation by free radicals.[46] Ischemia followed by reperfusion and exposure of tissue to oxygen increases free radical production as evidenced by increases of lipid peroxidation, protein oxidation, and depletion of the antioxidants glutathione and ascorbate.[47,48] Sources of free radical production include mitochondria,[49] endothelial xanthine dehydrogenase, leukocytes,[50] iron stores,[51] and nitric oxide synthase.[52]

Free radicals are produced normally in cells during oxidation reactions. For example, the conversion of oxygen to H_2O and ATP in the electron

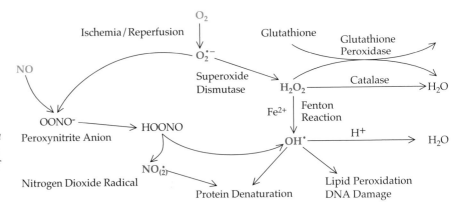

Figure 5 *Free radicals that may be formed during ischemia followed by reoxygenation with reperfusion. Potential targets of some of the free radicals are shown.*

transport chain forms superoxide radical ($O_2^{\bullet-}$), hydrogen peroxide (H_2O_2), and the hydroxyl radical ($OH^{\bullet}$).[45] Although H_2O_2 is not a radical, it reacts with $O_2^{\bullet-}$ and metals, particularly iron, to produce the highly reactive and toxic $OH^{\bullet}$ radical [*see Figure 5*].[45] Oxidation of phospholipids to arachidonic acid and its metabolites also gives rise to free radicals,[53] as does self-oxidation of flavins, catecholamines, and dopamine. Free radicals are also produced by phospholipase A_2, xanthine oxidase, and nitric oxide synthase (NOS).[54] Superoxide radicals produce injury by providing a source of hydroxyl radicals as well as by binding to nitric oxide and forming peroxynitrite and other radicals.

Normal cells have antioxidant enzymes and free radical scavengers to protect against endogenous and exogenous free radicals [*see Figure 5*]. Superoxide dismutase catalyzes the conversion of the superoxide radical to hydrogen peroxide (H_2O_2). Catalase metabolizes H_2O_2 to H_2O and O_2. Glutathione peroxidase converts H_2O_2 to H_2O and O_2 using glutathione as a cofactor. [55] Free radical scavengers normally present within cells include ascorbic acid, vitamin E, β-carotene, and free glutathione.[56] The regulation of free radical production is important for calcium homeostasis because free radicals affect the function of NMDA receptors, voltage-dependent calcium channels, calcium ATPases, calcium exchangers, and calcium-binding proteins.

The roles of the above reactions in cerebral ischemia have not been fully delineated. Because superoxide dismutase administration can decrease stroke size, it is assumed that the superoxide radical promotes ischemic injury.[47] Superoxide dismutase transgenic mice are partially protected from stroke following temporary MCA occlusions but not following permanent MCA occlusions, suggesting that superoxide dismutase is most important during reperfusion with production of free radicals.[48]

Inhibition of cyclooxygenase and lipoxygenase with agents such as piroxicam, flurbiprofen, and indomethacin improves neuronal survival[57,58] after ischemia. Vitamin E[59] decreases ischemic neuronal injury. Free radical scavengers such as dimethylthiourea decrease ischemic neuronal injury in both focal and global models,[60,61] as does the 21-amino steroid tirilazad[62] and related antioxidants.[63] Spin-trapping compounds reduce ischemic brain injury and improve neurologic function.[64] Whether decreasing oxidative stress in patients with stroke will improve outcome is still unknown.

Nitric Oxide Is Produced during Cerebral Ischemia

Nitric oxide synthase metabolizes L-arginine to nitric oxide and citrulline [*see Figure 6*]. Three forms of nitric oxide synthase are named after their cellular origin: neuronal (nNOS), endothelial cell derived (eNOS), and iNOS,[52] which can be induced in a variety of cells, including macrophages and microglia.[65] In the presence of adequate arginine, NOS forms citrulline and nitric oxide. In the absence of arginine, NOS can generate superoxide and hydrogen peroxide.[66]

Nitric oxide gas is a potent vasodilator.[52] The production of nitric oxide in the brain can affect both blood vessels and neighboring cells. Nitric oxide concentrations increase in ischemic brain. Modulation of nitric oxide production can either decrease or worsen ischemic injury, depending upon which cells produce it.[66-68]

Figure 6 *Effects of nitric oxide (NO) in brain. Specific nitric oxide synthases are found in endothelial cells (eNOS), neurons (nNOS), and microglia (iNOS). The NO released from endothelial cells diffuses into muscle cells, stimulates guanosine triphosphate (GTP) and relaxes the cell, and produces dilation of the vessels. The NO produced in neurons and glial cells diffuses to neighboring neurons and promotes cellular injury.*

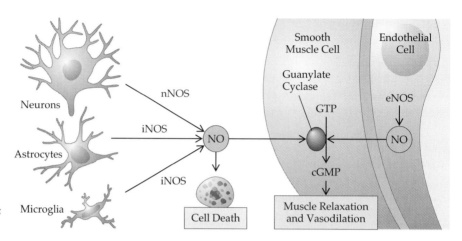

A variety of mediators can activate eNOS, including acetylcholine, substance P, histamine, arginine vasopressin, bradykinin, prostaglandin F, and the endothelin receptor.[65] Activation of eNOS increases production of nitric oxide within endothelial cells, from which it diffuses to adjacent smooth muscle cells to activate cyclic guanosine monophosphate (cGMP). cGMP produces relaxation of smooth muscle cells, vasodilation, and increased CBF [*see Figure 6*]. The nitric oxide released from endothelial cells also decreases aggregation and adherence of platelets and leukocytes. Drugs that inhibit eNOS produce vasoconstriction in vitro and in vivo and worsen ischemic injury.[69] Mice with eNOS knockouts have larger strokes than wild-type littermates. This suggests that eNOS-induced release of nitric oxide from endothelial cells improves ischemic outcome.

The role of NOS-containing cells in the brain is also under investigation. About two percent of brain neurons contain nNOS. Chemical lesions of the neurons containing nNOS improve ischemic injury.[70] Mice with nNOS knockouts have been shown to have smaller strokes than wild-type animals with normal nNOS function.[71] These data suggest that production of nitric oxide by nNOS neurons worsens stroke-related injury.

Both eNOS and nNOS are present at the onset of ischemia. The actions of these enzymes to produce nitric oxide appear to be rapid, occurring within seconds to minutes of the initial ischemia. Therefore, the clinical usefulness of manipulating these enzymes might be limited to prophylaxis against ischemic injury. The role of iNOS in cerebral ischemia has been the least studied. During and immediately after ischemia, iNOS levels are very low. However, iNOS is induced within hours in microglia, astrocytes, and vessels.[72,73] Because nitric oxide diffuses out of cells, it might be predicted that glial nitric oxide synthase would be harmful inasmuch as neuronal production of nitric oxide also appears to be harmful. Indeed, inhibition of iNOS activity several hours after ischemia improves neuronal survival and decreases the size of infarcts.[74,75] Stroke size is decreased in mice with iNOS knockouts.

The data show that nitric oxide acts on vessels (from eNOS) to improve ischemic injury, whereas nitric oxide acts on neurons (from nNOS and iNOS) to worsen ischemic injury. The development of specific drugs that target each form of NOS is currently under way as possible treatments for stroke.

Apoptosis Occurs after Ischemia

Programmed apoptotic cell death requires new mRNA and protein synthesis and is a normal process of cell elimination during normal development. It may be induced abnormally by ischemia,[76] perhaps as a result of neuronal deprivation of trophic factors or because neurons sustain enough cellular injury to trigger a so-called suicide program. The role of apoptosis in cerebral infarction remains uncertain, but induction of apoptotic death programs may account for selective neuronal cell death.

Nuclear damage occurs early during apoptosis, whereas the integrity of plasma and mitochondrial membranes is initially maintained. In contrast, during necrotic cell death, the nuclei, organelles, and plasma membrane swell, with the early loss of plasmalemmal integrity.[77] During apoptosis in some cells, new mRNAs and proteins are synthesized to carry out the suicide program[78]; in other cells, apoptosis occurs by activation of proteins already present in the cell. The cleavage of DNA between nucleosomes by calcium-dependent endonucleases has been demonstrated.[79]

The genes involved in the regulation of apoptosis have been identified in *Caenorhabditis elegans, Drosophila,* and other model systems [*see Figure 7*]. These include (1) genes that prevent cell death, such as *bcl-2* and *bcl-xL*; (2) genes that promote death, such as *bax, bcl-xs, reaper,* the interleukin 1–converting enzyme (ICE) related cysteine proteases (caspases), and endonucleases; and (3) genes that promote the engulfment and destruction of cells.[80-82] The increased production of the protective proteins promotes cell survival, and production of the death proteins produces cell death. Apoptotic cell death does not evoke an inflammatory response. Instead, cells undergoing apoptosis appear to be engulfed quietly by microglia or macrophages without surrounding tissue damage or infiltration by leukocytes or lymphocytes.[83]

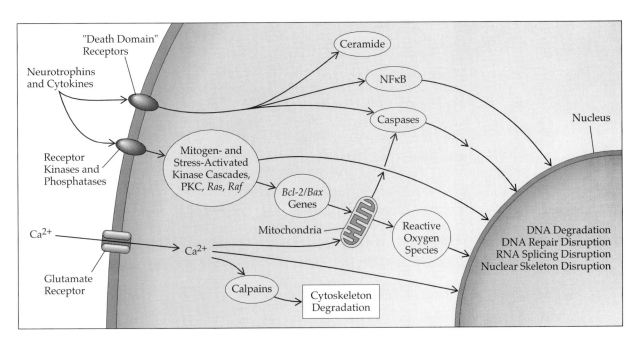

Figure 7 *Different molecules implicated in the control of apoptosis in mammalian cells.*

Once cells are initiated towards the death pathway, specific proteases and endonucleases are activated to carry out the death program. One major family of proteases is known as the caspases, which include ICE (caspase 1), Nedd2 (caspase 2), CPP32 (caspase 3), and others.[84] These proteins have the nearly invariant sequence, . . .QACRG. . ., containing the active site cysteine. Expression of these proteases in cells causes apoptosis that can be prevented by inhibitors of the proteases or introduction of the protective gene, *bcl-2*. [85-87]

Some of the cell death observed in ischemic brain may be related to apoptotic mechanisms. Protein synthesis inhibitors reduce hippocampal cell death following global ischemia in the gerbil[88] and decrease infarct size following focal ischemia.[89] Techniques to demonstrate apoptosis show (DNA laddering and DNA nick end-labeling after global and focal ischemia.[90-92]

Many different pathways appear to promote or block apoptosis [*see Figure 7*]. Reactive oxygen intermediates (e.g., superoxide, OH·) may initiate early phases of apoptosis,[93] although the sustained generation of such compounds may cause necrosis. Transgenic mice that overexpress Bcl-2 have reduced stroke size compared with wild-type mice.[94] Delivery to the brain of Bcl-2 mRNA in defective herpesvirus vectors before focal ischemia decreases injury at the injection sites.[95] The tumor suppressor/transcription factor (p53) may mediate apoptosis in response to DNA damage[96] as it is upregulated in regions of neuronal injury following focal ischemia. Transgenic mice with decreased p53 expression have significantly smaller strokes following MCA occlusions.[97]

Bax, a proapoptotic protein that is a downstream effector of p53, is markedly upregulated in cells that degenerate following ischemia. Other apoptosis-related genes are induced in ischemic cells that will die. Following global ischemia, Bcl-2 and Bcl-x mRNAs are induced in CA1 hippocampal neurons that will die one day later. Bcl-2 induction would protect cells, and the Bcl-x induction may protect or kill cells, depending upon which form of Bcl-x predominates. ICE-like mRNA is also induced in hippocampus after global ischemia.[92] These results are consistent with the hypothesis that CA1 hippocampal neurons die via apoptosis following global ischemia. Ischemia suppression of protein synthesis could prevent the synthesis of Bcl-2, Bcl-xl, and other protective proteins. In the absence of these protective proteins, activation of ICE-like proteases and calcium activation of endonucleases could precipitate apoptosis. The delayed death of CA1 neurons may be explained by the failure to produce sufficient protective proteins followed by activation of death proteins.

Drugs are being developed to facilitate antiapoptotic genes and to block proapoptotic genes. Peptides that inhibit ICE-like proteases decrease infarct size in animals. Animals with knockouts of some but not all of the ICE-like proteases have decreased infarct volumes compared with wild types. The pharmacological manipulation of apoptotic pathways could improve outcome from stroke.

Stress Genes Are Induced in Ischemic Brain

Heat shock proteins (HSPs) are induced by heat stress and other types of injury in response to denatured proteins within cells. Heat shock proteins are induced by heat shock factors, proteins that bind to heat shock elements in HSP gene promoters, to activate transcription of HSP mRNA [see Figure 8], which is translated into HSPs. Heat shock proteins serve as chaperones to (1) renature proteins, (2) keep proteins from folding abnormally, or (3) transfer proteins across membranes of intracellular organelles.[98,99]

One HSP, named HSP70, is expressed constitutively in all cells. This ATPase binds to proteins as they are formed on ribosomes to prevent them from folding abnormally during translation.[100] HSP70 is upregulated by both heat shock and ischemia. The highly inducible *hsp70* gene is expressed at low levels in normal cells.[98,101] HSP70 binds to denatured proteins to prevent further denaturation and to renature partially denatured proteins.

HSP70 Induction after Global Ischemia

Following brief global ischemia, HSP70 protein is expressed selectively in hippocampal CA1 pyramidal neurons in rodent brain.[102,103] This suggests that denatured proteins are present within these neurons and supports the concept that CA1 hippocampal neurons are among the most vulnerable cells to global ischemia in mammalian brain. Injury to CA1 neurons in humans produces a disorder of recent memory, wherein patients are unable to form new lasting memories.

Induction of HSP70 in various regions of the hippocampus may provide insight into the mechanism of selective neuronal vulnerability to ischemia. Within the hippocampus, CA1 neurons are the most vulnerable, followed by dentate hilar neurons, CA3 neurons, dentate granule cell neurons, glia, and capillary endothelial cells.[102] Prolonged global ischemia in the hippocampus induces HSP70 in most regions of the hippocampus, suggesting more extensive protein denaturation.[102] In CA1 pyramidal neurons, pro-

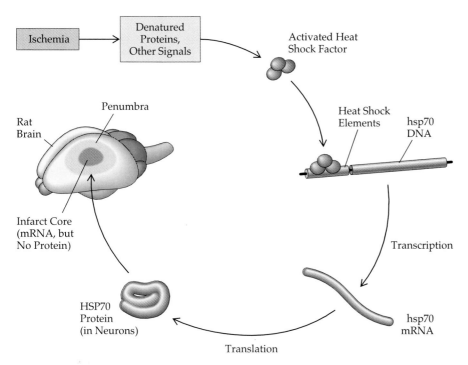

Figure 8 *Ischemia produces denatured proteins that activate heat shock factor proteins. The heat shock factors form a trimer, bind to heat shock elements, and stimulate transcription. The HSP70 protein is translated from the mRNA. HSP70 protein is synthesized in neurons outside the region of infarction. This region of neuronal HSP70 induction is defined as the penumbra.*

longed global ischemia elicits increased expression of HSP mRNA, whereas HSP70 protein is not expressed. This occurs with more severe ischemia because CA1 neuronal damage results in the inability of the cells to translate *hsp70* mRNA into HSP70 protein.[103] Infarction of the hippocampus results in failure to synthesize HSP70 protein in neurons and glial cells, although HSP70 protein is synthesized in endothelial cells. This hierarchy of ischemic vulnerability is typical of all brain regions.

HSP70 Induction in the Penumbra after Focal Ischemia

Following brief MCA ischemia in the rat, *hsp70* mRNA and HSP70 protein are induced in neurons in cortex throughout the MCA distribution, even though there is no evidence of infarction.[104] Following permanent MCA occlusions, HSP70 protein is expressed mainly in capillary endothelial cells within the areas of infarction.[24,105,106] There is little mRNA and protein induction in neurons and glia in the areas of infarction because of the limited availability of ATP.[107] In the cortex outside the areas of infarction, HSP70 protein is expressed mainly in neurons [*see Figure 8*].[106] These HSP70 stained cortical neurons are located in the distribution of the ischemic MCA.[24,105,106]

These data permit formulation of several concepts. First, there is a hierarchy of cellular damage in both global and focal ischemia: neurons are the most vulnerable, glia are next most vulnerable, and endothelial cells the most resistant. Second, an ischemic penumbra can be described in terms of the *hsp70* stress gene response. The ischemic penumbra is defined as the region outside an infarction where neurons express both *hsp70* mRNA and HSP70 protein.[108,109] Third, following mild ischemia, the penumbra includes all cortex in the MCA distribution. As the ischemia is prolonged, however, the ischemic penumbra progressively decreases in size.

Heme Oxygenase-1

Heme oxygenase metabolizes heme-containing proteins to biliverdin, carbon monoxide, and iron.[110] The carbon monoxide released may modulate vascular tone.[111] Heme oxygenase-1 (HO-1) is expressed mainly in glial cells after heat shock and other stresses.[112] The *HO-1* gene has several response elements in its promoter that are activated by oxidative stress and ischemia.

Following MCA infarction, HO-1 is induced in microglial cells throughout the ischemic hemisphere, in the cingulate cortex, and in the occipital cortex.[33] It is possible that ischemia-induced spreading depression induces immediate early genes, including *c-fos* and *c-jun* family members within the microglia, which activate AP-1 sites in the *HO-1* gene and increase HO-1 expression. HO-1 would metabolize heme proteins and increase carbon monoxide release, resulting in dilation of cerebral blood vessels to increase blood flow.

Therapeutic Implications

The remarkable progress in the application of cellular and molecular biology to the neural insult produced by ischemia yields numerous promising avenues for exploration of therapies to minimize brain injury. However, most of the experimental observations remain too rudimentary and imprecise to predict which may truly alleviate stroke-induced injury in humans. Many of these are under investigation, and some have proceeded to advanced clinical trials. Blockade of NMDA channels, and nterventions that protect cells from apoptotic cell death or stimulate protective HSPs, show great promise. Protection from mitochondrial toxins or the adverse effects of nitric oxide provide other examples of potential therapies. It is remarkable how quickly basic science discoveries have already delineated several metabolic pathways that are susceptible to manipulation to minimize the damage caused by cerebral ischemia.

References

1. Hossmann KA: Viability thresholds and the penumbra of focal ischemia. *Ann Neurol* 36:557, 1994

2. Burda J, Martin ME, Garcia A, et al: Phosphorylation of the alpha subunit of initiation factor 2 correlates with the inhibition of translation following transient cerebral ischaemia in the rat. *Biochem J* 302:335, 1994

3. Chen JJ, London IM: Regulation of protein synthesis by heme-regulated eIF-2 alpha kinase. *Trends Biochem Sci* 20:105, 1995

4. Wek RC: eIf-2 kinases: regulators of general and gene-specific translation initiation. *Trends Biochem Sci* 19:491, 1994

5. Clemens MJ, Hershey JW, Hovanessian AC, et al: PKR: proposed nomenclature for the RNA-dependent protein kinase induced by interferon. *J Interferon Res* 13:241, 1993

6. Proud CG: PKR: a new name and new roles. *Trends Biochem Sci* 20:241, 1995

7. Sharp TV, Xiao Q, Jeffrey I, et al: Reversal of the double-stranded-RNA-induced inhibition of protein synthesis by a catalytically inactive mutant of the protein kinase PKR. *Eur J Biochem* 214:945, 1993

8. Hu BR, Wieloch T: Stress-induced inhibition of protein synthesis initiation: modulation of initiation factor 2 and guanine nucleotide exchange factor activities following transient cerebral ischemia in the rat. *J Neurosci* 13:1830, 1993

9. Folbergrova J, Memezawa H, Smith ML, et al: Focal and perifocal changes in tissue energy state during middle cerebral artery occlusion in normo- and hyperglycemic rats. *J Cereb Blood Flow Metab* 12:25, 1992

10. Paschen W, Djuricic B, Mies G, et al: Lactate and pH in the brain: association and dissociation in different pathophysiological states. *J Neurochem* 48:154, 1987

11. Kraig RP, Petito CK, Plum F, et al: Hydrogen ions kill brain at concentrations reached in ischemia. *J Cereb Blood Flow Metab* 7:379, 1987

12. Plum F: What causes infarction in ischemic brain? The Robert Wartenberg Lecture. *Neurology* 33:222, 1983

13. Pulsinelli WA, Levy DE, Sigsbee B, et al: Increased damage after ischemic stroke in patients with hyperglycemia with or without established diabetes mellitus. *Am J Med* 74:540, 1983

14. Narasimhan P, Sklar R, Murrell M, et al: Methyl malonyl coA mutase induction by cerebral ischemia and neurotoxicity of the mitochondrial toxin methylmalonic acid. *J Neurosci* 16:7336, 1996

15. Furlan M, Marchal G, Viader F, et al: Spontaneous neurological recovery after stroke and the fate of the ischemic penumbra. *Ann Neurol* 40:216, 1996

16. Gidday JM, Fitzgibbons JC, Shah AR, et al: Neuroprotection from ischemic brain injury by hypoxic preconditioning in the neonatal rat. *Neurosci Lett* 168:221, 1994

17. Semenza GL, Wang GL: A nuclear factor induced by hypoxia via de novo protein synthesis binds to the human erythropoietin gene enhancer at a site required for transcriptional activation. *Mol Cell Biol* 12:5447, 1992

18. Semenza GL: Regulation of erythropoietin production: new insights into molecular mechanisms of oxygen homeostasis. *Hematol Oncol Clin North Am* 8:863, 1994

19. Semenza GL, Roth PH, Fang HM, et al: Transcriptional regulation of genes encoding glycolytic enzymes by hypoxia-inducible factor 1. *J Biol Chem* 269:23757, 1994

20. Melillo G, Musso T, Sica A, et al: A hypoxia-responsive element mediates a novel pathway of activation of the inducible nitric oxide synthase promoter. *J Exp Med* 182:1683, 1995

21. Ebert BL, Gleadle JM, Or JF, et al: Isoenzyme-specific regulation of genes involved in energy metabolism by hypoxia: similarities with the regulation of erythropoietin. *Biochem J* 313:809, 1996

22. Hsu CY, An G, Liu JS, et al: Expression of immediate early gene and growth factor mRNAs in a focal cerebral ischemia model in the rat. *Stroke* 24:I78, 1993

23. Morgan JI, Curran T: Immediate-early genes: ten years on. *Trends Neurosci* 18:66, 1995

24. Kinouchi H, Sharp FR, Chan PH, et al: Induction of c-fos, junB, c-jun, and hsp70 mRNA in cortex, thalamus, basal ganglia, and hippocampus following middle cerebral artery occlusion. *J Cereb Blood Flow Metab* 14:808, 1994

25. Liu HM: Correlation between proto-oncogene, fibroblast growth factor and adaptive response in brain infarct. *Prog Brain Res* 105:239, 1995

26. An G, Lin TN, Liu JS, et al: Expression of c-fos and c-jun family genes after focal cerebral ischemia [see comments]. *Ann Neurol* 33:457, 1993

27. Honkaniemi J, Sharp FR: Global ischemia induces immediate-early genes encoding zinc finger transcription factors. *J Cereb Blood Flow Metab* 16:557, 1996

28. Nedergaard M, Hansen AJ: Characterization of cortical depolarizations evoked in focal cerebral ischemia. *J Cereb Blood Flow Metab* 13:568, 1993

29. Dietrich WD, Feng ZC, Leistra H, et al: Photothrombotic infarction triggers multiple episodes of cortical spreading depression in distant brain regions. *J Cereb Blood Flow Metab* 14:20, 1994

30. Sharp JW, Sagar SM, Hisanaga K, et al: The NMDA receptor mediates cortical induction of fos and fos-related antigens following cortical injury. *Exp Neurol* 109:323, 1990

31. Kinouchi H, Sharp FR, Chan PH, et al: MK-801 inhibits the induction of immediate early genes in cerebral cortex, thalamus, and hippocampus, but not in substantia nigra following middle cerebral artery occlusion. *Neurosci Lett* 179:111, 1994

32. Kraig RP, Dong LM, Thisted R, et al: Spreading depression increases immunohistochemical staining of glial fibrillary acidic protein. *J Neurosci* 11:2187, 1991

33. Nimura T, Massa SM, Panter S, et al: Heme oxygenase-1 (HO-1) protein induction in rat brain following focal ischemia. *Mol Brain Res* 37:201, 1996

34. Nowak TS Jr: Synthesis of heat shock/stress proteins during cellular injury. *Ann N Y Acad Sci* 679:142, 1993

35. Kiessling M, Stumm G, Xie Y, et al: Differential transcription and translation of immediate early genes in the gerbil hippocampus after transient global ischemia. *J Cereb Blood Flow Metab* 13:914, 1993

36. Dragunow M, Beilharz E, Sirimanne E, et al: Immediate-early gene protein expression in neurons undergoing delayed death, but not necrosis, following hypoxic-ischaemic injury to the young rat brain. *Brain Res Mol Brain Res* 25:19, 1994

37. Nowak TS, Osborne OC, Suga S: Stress protein and proto-oncogene expression as indicators of neuronal pathophysiology after ischemia. *Prog Brain Res* 96:195, 1993

38. Dragunow M, Young D, Hughes P, et al: Is c-Jun involved in nerve cell death following status epilepticus and hypoxic-ischaemic brain injury? *Brain Res Mol Brain Res* 18:347, 1993

39. Rothman SM, Olney JW: Glutamate and the pathophysiology of hypoxic-ischemic brain damage. *Ann Neurol* 19:105, 1986

40. Choi DW: Calcium-mediated neurotoxicity: relationship to specific channel types and role in ischemic damage. *Trends Neurosci* 11:465, 1988

41. Rosenberg LE: The inherited methylmalonic acidemias. *Prog Clin Biol Res* 103:187, 1982

42. Matsui SM, Mahoney MJ, Rosenberg LE: The natural history of the inherited methylmalonic acidemias. *N Engl J Med* 308:857, 1983

43. Dutra JC, Dutra-Filho CS, Cardozo SE, et al: Inhibition of succinate dehydrogenase and beta-hydroxybutyrate dehydrogenase activities by methylmalonate in brain and liver of developing rats. *J Inherit Metab Dis* 16:147, 1993

44. Beal MF, Ferrante RJ, Henshaw R, et al: 3-Nitropropionic acid neurotoxicity is attenuated in copper/zinc superoxide dismutase transgenic mice. *J Neurochem* 65:919, 1995

45. Halliwell B: Reactive oxygen species and the central nervous system. *J Neurochem* 59:1609, 1992

46. Floyd RA, Carney JM: Free radical damage to protein and DNA: mechanisms involved and relevant observations on brain undergoing oxidative stress. *Ann Neurol* 32(suppl):S22, 1992

47. Chan PH: Oxygen radicals in focal cerebral ischemia. *Brain Pathol* 4:59, 1994

48. Chan PH, Epstein CJ, Li Y, et al: Transgenic mice and knockout mutants in the study of oxidative stress in brain injury. *J Neurotrauma* 12:815, 1995

49. Dugan LL, Sensi SL, Canzoniero LM, et al: Mitochondrial production of reactive oxygen species in cortical neurons following exposure to N-methyl-D-aspartate. *J Neurosci* 15:6377, 1995

50. Matsuo Y, Kihara T, Ikeda M, et al: Role of neutrophils in radical production during ischemia and reperfusion of the rat brain: effect of neutrophil depletion on extracellular ascorbyl radical formation. *J Cereb Blood Flow Metab* 15:941, 1995

51. Rehncrona S, Hauge HN, Siesjo BK: Enhancement of iron-catalyzed free radical formation by acidosis in brain homogenates: differences in effect by lactic acid and CO_2. *J Cereb Blood Flow Metab* 9:65, 1989

52. Dawson DA: Nitric oxide and focal cerebral ischemia: multiplicity of actions and diverse outcome. *Cerebrovasc Brain Metab Rev* 6:299, 1994

53. Braughler JM, Hall ED: Involvement of lipid peroxidation in CNS injury. *J Neurotrauma* 9(suppl 1):S1, 1992

54. Ginsberg MD, Watson BD, Busto R, et al: Peroxidative damage to cell membranes following cerebral ischemia: a cause of ischemic brain injury? *Neurochem Pathol* 9:171, 1988

55. Makar TK, Nedergaard M, Preuss A, et al: Vitamin E, ascorbate, glutathione, glutathione disulfide, and enzymes of glutathione metabolism in cultures of chick astrocytes and neurons: evidence that astrocytes play an important role in antioxidative processes in the brain. *J Neurochem* 62:45, 1994

56. Hall ED, Braughler JM, McCall JM: Antioxidant effects in brain and spinal cord injury. *J Neurotrauma* 9(suppl 1):S165, 1992

57. Nakagomi T, Sasaki T, Kirino T, et al: Effect of cyclooxygenase and lipoxygenase inhibitors on delayed neuronal death in the gerbil hippocampus. *Stroke* 20:925, 1989

58. Sasaki T, Nakagomi T, Kirino T, et al: Indomethacin ameliorates ischemic neuronal damage in the gerbil hippocampal CA1 sector. *Stroke* 19:1399, 1988

59. Hara H, Kato H, Kogure K: Protective effect of alpha-tocopherol on ischemic neuronal damage in the gerbil hippocampus. *Brain Res* 510:335, 1990

60. Kiyota Y, Pahlmark K, Memezawa H, et al: Free radicals and brain damage due to transient middle cerebral artery occlusion: the effect of dimethylthiourea. *Exp Brain Res* 95:388, 1993

61. Pahlmark K, Folbergrova J, Smith ML, et al: Effects of dimethylthiourea on selective neuronal vulnerability in forebrain ischemia in rats. *Stroke* 24:731, 1993

62. Hall ED: Neuroprotective actions of glucocorticoid and nonglucocorticoid steroids in acute neuronal injury. *Cell Mol Neurobiol* 13:415, 1993

63. Hara H, Kogure K: Prevention of hippocampus neuronal damage in ischemic gerbils by a novel lipid peroxidation inhibitor (quinazoline derivative). *J Pharmacol Exp Ther* 255:906, 1990

64. Carney JM, Floyd RA: Brain antioxidant activity of spin traps in Mongolian gerbils. *Methods Enzymol* 234:523, 1994

65. Zhang ZG, Chopp M, Zaloga C, et al: Cerebral endothelial nitric oxide synthase expression after focal cerebral ischemia in rats. *Stroke* 24:2016, 1993

66. Faraci FM, Brian JE Jr: Nitric oxide and the cerebral circulation. *Stroke* 25:692, 1994

67. Dalkara T, Moskowitz MA: The complex role of nitric oxide in the pathophysiology of focal cerebral ischemia. *Brain Pathol* 4:49, 1994

68. Iadecola C, Pelligrino DA, Moskowitz MA, et al: Nitric oxide synthase inhibition and cerebrovascular regulation. *J Cereb Blood Flow Metab* 14:175, 1994

69. Furfine ES, Harmon MF, Paith JE, et al: Potent and selective inhibition of human nitric oxide synthases: selective inhibition of neuronal nitric oxide synthase by S-methyl-L-thiocitrulline and S-ethyl-L-thiocitrulline. *J Biol Chem* 269:26677, 1994

70. Ferriero DM, Sheldon RA, Black SM, et al: Selective destruction of nitric oxide synthase neurons with quisqualate reduces damage after hypoxia-ischemia in the neonatal rat. *Pediatr Res* 38:912, 1995

71. Huang Z, Huang PL, Panahian N, et al: Effects of cerebral ischemia in mice deficient in neuronal nitric oxide synthase. *Science* 265:1883, 1994

72. Galea E, Reis DJ, Xu H, et al: Transient expression of calcium-independent nitric oxide synthase in blood vessels during brain development. *FASEB J* 9:1632, 1995

73. Iadecola C, Zhang F, Xu S, et al: Inducible nitric oxide synthase gene expression in brain following cerebral ischemia. *J Cereb Blood Flow Metab* 15:378, 1995

74. Zhang F, Casey RM, Ross ME, et al: Aminoguanidine ameliorates and L-arginine worsens brain damage from intraluminal middle cerebral artery occlusion. *Stroke* 27:317, 1996

75. Iadecola C, Zhang F, Xu X: Inhibition of inducible nitric oxide synthase ameliorates cerebral ischemic damage. *Am J Physiol* 268:R286, 1995

76. Johnson EJ, Greenlund LJ, Akins PT, et al: Neuronal apoptosis: current understanding of molecular mechanisms and potential role in ischemic brain injury. *J Neurotrauma* 2:843, 1995

77. Kerr JF, Wyllie AH, Currie AR: Apoptosis: a basic biological phenomenon with wide-ranging implications in tissue kinetics. *Br J Cancer* 26:239, 1972

78. Schwartz LM, Kosz L, Kay BK: Gene activation is required for developmentally programmed cell death. *Proc Natl Acad Sci USA* 87:6594, 1990

79. Grasl KB, Ruttkay NB, Koudelka H, et al: In situ detection of fragmented DNA (TUNEL assay) fails to discriminate among apoptosis, necrosis, and autolytic cell death: a cautionary note. *Hepatology* 21:1465, 1995

80. Vaux DL, Cory S, Adams JM: *Bcl-2* gene promotes haemopoietic cell survival and cooperates with *c-myc* to immortalize pre-B cells. *Nature* 335:440, 1988

81. Yuan J, Shaham S, Ledoux S, et al: The *C. elegans* cell death gene *ced-3* encodes a protein similar to mammalian interleukin-1 beta-converting enzyme. *Cell* 75:641, 1993

82. Steller H, Abrams JM, Grether ME, et al: Programmed cell death in *Drosophila*. *Philos Trans R Soc Lond B Biol Sci* 345:247, 1994

83. Nitatori T, Sato N, Waguri S, et al: Delayed neuronal death in the CA1 pyramidal cell layer of the gerbil hippocampus following transient ischemia is apoptosis. *J Neurosci* 15:1001, 1995

84. Thornberry NA, Bull HG, Calaycay JR, et al: A novel heterodimeric cysteine protease is required for interleukin-1 beta processing in monocytes. *Nature* 356:768, 1992

85. Gagliardini V, Fernandez PA, Lee RK, et al: Prevention of vertebrate neuronal death by the crmA gene. *Science* 263:826, 1994

86. Milligan CE, Prevette D, Yaginuma H, et al: Peptide inhibitors of the ICE protease family arrest programmed cell death of motoneurons in vivo and in vitro. *Neuron* 15:385, 1995

87. Kumar S: Inhibition of apoptosis by the expression of antisense Nedd2. *FEBS Lett* 368:69, 1995

88. Shigeno T, Yamasaki Y, Kato G, et al: Reduction of delayed neuronal death by inhibition of protein synthesis. *Neurosci Lett* 120:117, 1990

89. Linnik MD, Zobrist RH, Hatfield MD: Evidence supporting a role for programmed cell death in focal cerebral ischemia in rats. *Stroke* 24:2002, 1993

90. MacManus JP, Hill IE, Preston E, et al: Differences in DNA fragmentation following transient cerebral or decapitation ischemia in rats. *J Cereb Blood Flow Metab* 15:728, 1995

91. Du C, Hu R, Csernansky CA, et al: Very delayed infarction after mild focal cerebral ischemia: a role for apoptosis? *J Cereb Blood Flow Metab* 16:195, 1996

92. Honkaniemi J, Massa SM, Breckinridge M: Global ischemia induces apoptosis-associated genes in hippocampus. *Brain Res Mol Brain Res* 42:79, 1996

93. Greenlund LJ, Deckwerth TL, Johnson EJ: Superoxide dismutase delays neuronal apoptosis: a role for reactive oxygen species in programmed neuronal death. *Neuron* 14:303, 1995

94. Martinou JC, Dubois-Dauphin M, Staple JK, et al: Overexpression of Bcl-2 in transgenic mice protects neurons from naturally occurring cell death and experimental ischemia. *Neuron* 13:1017, 1994

95. Linnik MD, Zahos P, Geschwind MD, et al: Expression of Bcl-2 from a defective herpes simplex virus-1 vector limits neuronal death in focal cerebral ischemia. *Stroke* 26:1670, 1995

96. Hainaut P: The tumor suppressor protein p53: a receptor to genotoxic stress that controls cell growth and survival. *Curr Opin Oncol* 7:76, 1995

97. Crumrine RC, Thomas AL, Morgan PF: Attenuation of p53 expression protects against focal ischemic damage in transgenic mice. *J Cereb Blood Flow Metab* 14:887, 1994

98. Massa SM, Swanson RA, Sharp FR: The stress gene response in brain. *Cerebrovasc Brain Metab Rev* 8:95, 1996

99. Nowak TS Jr, Jacewicz M: The heat shock/stress response in focal cerebral ischemia. *Brain Pathol* 4:67, 1994

100. Beckmann RP, Mizzen LE, Welch WJ: Interaction of Hsp 70 with newly synthesized proteins: implications for protein folding and assembly. *Science* 248:850, 1990

101. Longo FM, Wang S, Narasimhan P, et al: cDNA cloning and expression of stress-inducible rat *hsp70* in normal and injured rat brain. *J Neurosci Res* 36:325, 1993

102. Gonzalez MF, Lowenstein D, Fernyak S, et al: Induction of heat shock protein 72-like immunoreactivity in the hippocampal formation following transient global ischemia. *Brain Res Bull* 26:241, 1991

103. Nowak TS: Localization of 70 kDa stress protein mRNA induction in gerbil brain after ischemia. *J Cereb Blood Flow Metab* 11:432, 1991

104. Sharp FR, Kinouchi H, Koistinaho J, et al: HSP70 heat shock gene regulation during ischemia. *Stroke* 24:172, 1993

105. Kinouchi H, Sharp FR, Koistinaho J, et al: Induction of heat shock *hsp70* mRNA and HSP70 kDa protein in neurons in the "penumbra" following focal cerebral ischemia in the rat. *Brain Res* 619:334, 1993

106. Kinouchi H, Sharp FR, Hill MP, et al: Induction of 70-kDa heat shock protein and *hsp70* mRNA following transient focal cerebral ischemia in the rat. *J Cereb Blood Flow Metab* 13:105, 1993

107. Kobayashi S, Welsh FA: Regional alterations of ATP and heat-shock protein-72 mRNA following hypoxia-ischemia in neonatal rat brain. *J Cereb Blood Flow Metab* 15:1047, 1995

108. Sharp FR: Stress proteins are sensitive indicators of injury in the brain produced by ischemia and toxins. *J Toxicol Sci* 20:450, 1995

109. Sharp FR, Sagar SM: Alterations in gene expression as an index of neuronal injury: heat shock and the immediate early gene response. *Neurotoxicology* 15:51, 1994

110. Ewing JF, Maines MD: Glutathione depletion induces heme oxygenase-1 (HSP32) mRNA and protein in rat brain. *J Neurochem* 60:1512, 1993

111. Ewing JF, Haber SN, Maines MD: Normal and heat-induced patterns of expression of heme oxygenase-1 (HSP32) in rat brain: hyperthermia causes rapid induction of mRNA and protein. *J Neurochem* 58:1140, 1992

112. Maines MD, Eke BC, Weber CM, et al: Corticosterone has a permissive effect on expression of heme oxygenase-1 in CA1-CA3 neurons of hippocampus in thermal-stressed rats. *J Neurochem* 64:1769, 1995

Acknowledgments

Figures 1 and 6 through 8   Seward Hung.

Figures 2 and 3 through 5   Marcia Kammerer.

CHAPTER 8

Molecular Basis of HIV-Associated Neurologic Disease

Lennart Mucke, M.D., Manuel Buttini, Ph.D.

As a consequence of the acquired immunodeficiency syndrome (AIDS) pandemic, human immunodeficiency virus (HIV) infection may by now be the most prevalent viral infection of the nervous system as well as the most frequent cause of dementia in young Americans. Roughly 30 million adults and children have already been infected by HIV type 1 (HIV-1) worldwide, according to estimates of the World Health Organization, and this number will likely continue to rise. A substantial number of HIV-infected people develop HIV-1-associated neurologic diseases causing severe pain, weakness, or dementia.

HIV-Associated Neurologic Diseases

HIV-1 infection can be associated with neurologic diseases at nearly every stage, with diverse effects on the nervous system [*see Figure 1; Table 1*]. Detailed reviews of the epidemiology and differential diagnosis of these diseases[1-5] and of their management[6,7] are available. Neurologic diseases associated with HIV-1 infections include dementia, vacuolar myelopathy, aseptic meningitis, peripheral neuropathies, and transient neurologic deficits. Although their exact etiologies have not yet been determined, HIV-1 is presumed to play a key role in their pathogenesis. In addition to these primary HIV-1–associated neurologic diseases, neoplasms and a variety of opportunistic diseases afflict the nervous system of AIDS patients as a consequence of immunosuppression, including primary central nervous system (CNS) lymphoma, progressive multifocal leukoencephalopathy, cerebral toxoplasmosis, cryptococcal meningitis, and diverse neurologic manifestations of cytomegalovirus infection.

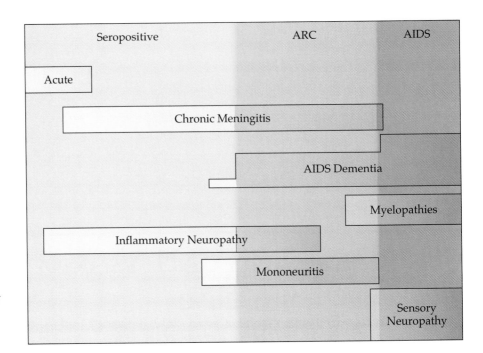

Figure 1 Development of HIV-1–associated neurologic diseases in relation to stage of viral infection. ARC is AIDS-related complex.

There is currently no cure or effective symptomatic treatment for any of the primary HIV-1–associated neurologic diseases. An improvement in this situation likely will require a better understanding of the molecular mechanisms underlying these conditions. This chapter focuses on HIV-1–associated dementia (subsequently referred to as HIV dementia), which has been studied extensively at the molecular level. Space restrictions on the number of references permitted here precluded citation of key sources, including many articles reporting original observations. Furthermore, HIV-1–associated vacuolar myelopathy, which causes progressive spastic paraparesis and ataxia, and peripheral neuropathies, which can result in severe chronic pain, also contribute significantly to morbidity in AIDS patients and, hence, clearly merit further investigation. However, because relatively little is known about their molecular basis, these entities are not discussed here.

HIV Dementia

HIV dementia [*see Table 2*] is characterized by impairments in attention and memory and in the speed and agility of dealing with complex or unrelated issues; it can culminate in severe, global loss of mental functions, resulting in a nearly vegetative state. HIV dementia is often associated with blunting of animation, interest, and volition and may be accompanied by motor impairments that are characterized initially by slowing of rapid or fine movements and later by progression to incoordination and weakness. The nomenclature, staging, and differential diagnosis of HIV dementia have been described.[1-3,4,8]

Whereas earlier studies indicated that eight to 66 percent of persons with AIDS might develop HIV dementia, the introduction of more stringent definitions of this entity has resulted in more consistent and conservative estimates of five to 15 percent. Notably, the diagnosis now requires the absence of other causes of cognitive, motor, or behavioral symptoms or signs, such as opportunistic infections, malignancies, or substance abuse. Because these conditions are common in persons with AIDS, this strict definition likely results in an underestimation of the prevalence of HIV dementia, particularly in moribund patients with advanced AIDS. This issue is exemplified by the relatively frequent coexistence of HIV-1 and cytomegalovirus infections in the same patient. Cytomegalovirus can cause retinitis, mononeuritis multiplex, myelitis/polyradiculopathy, ventriculoencephalitis, and encephalitis with dementia. Consequently, cytomegalovirus-induced CNS impairments may obscure coexisting CNS alterations induced directly by HIV-1.[3] Not surprisingly, the prevalence of HIV-1–associated minor cognitive or motor deficits among asymptomatic HIV-1–infected or AIDS patients is even more difficult to establish; it may be as high as 50 percent in AIDS patients.[8] In

Table 1   Summary of HIV-1–Associated Neurologic Diseases by Neuroanatomical Localization

Central Nervous System	Meninges and Other Structures around CNS
	  Early and late-middle course
	    Aseptic meningitis and asymptomatic HIV infection
	  Late course
	    Cryptococcal meningitis
	    Tuberculous meningitis
	    HIV headache
	
	Brain
	  Focal—late course
	    Cerebral toxoplasmosis
	    Primary CNS lymphoma
	    Progressive multifocal leukoencephalopathy
	  Diffuse—early course
	    Postinfectious encephalomyelitis
	  Diffuse—late course
	    AIDS dementia complex
	    Cytomegalovirus encephalitis
	
	Spinal cord
	  Late course
	    Vacuolar myelopathy (AIDS dementia complex)

Peripheral Nervous System and Muscle	Nerve and Root
	  Early course
	    Brachial plexitis, focal neuropathy, and polyneuropathy
	  Middle course
	    Subacute and chronic demyelinating polyneuropathy
	    Benign mononeuritis multiplex
	  Late course
	    Distal predominantly sensory polyneuropathy
	    Cytomegalovirus polyradiculopathy
	    Mononeuritis multiplex: late cytomegalovirus
	
	Muscle
	  Late course
	    Inflammatory myopathy
	    Noninflammatory myopathies
	    Zidovudine myopathy

Table 2 Clinical Features Suggestive of HIV Dementia

HIV-1 seropositivity and CD4 level ≤ 200 cells/mm³

History of progressive subacute mental decline: apathy, memory loss, and slowed cognition

Physical findings include slowed limb and eye movements, hyperreflexia, hypertonia, release signs, and an HIV dementia score ≤ 10

Neuropsychological testing demonstrates worsening performance in at least two areas, including frontal lobe tasks, motor speed, and nonverbal memory

CSF studies show elevated p24, IgG, and protein levels and the absence of other infections (e.g., cryptococcal infections and syphilis)

Cranial imaging reveals cerebral atrophy and white matter abnormalities, including hyperintensities on MRI and hypodensities on CT; exclusion of opportunistic processes

Exclusion of major psychiatric and metabolic disorders and substance abuse

---

fact, because of the dynamic interplay between viral adaptations and the development of more effective treatments against opportunistic infections and HIV-1 itself, it may currently be impossible to determine the actual incidence of structural and functional CNS impairments in HIV-1–infected persons with or without AIDS. Given the rapid development of antiretroviral therapies, such estimates will likely be obsolete by the time they reach the practitioner. Nonetheless, it is of interest to determine the likelihood of neuropsychological impairments in asymptomatic HIV-1–positive persons. The number of such persons is likely to rise with the development of more effective AIDS treatments, and such impairments could have negative effects on work performance and quality of life.

Although the association of specific neurologic diseases with HIV-1 infection indicates that the virus plays a significant role in their pathogenesis, it has been difficult to determine whether HIV-1 functions as a primary etiologic factor or acts more indirectly. For HIV-1 to affect the nervous system directly, it must be able to gain access to neural tissues.

Neuroinvasiveness of HIV-1

Early after systemic infection, HIV-1 invades the cerebrospinal fluid (CSF) space and possibly also the brain parenchyma, presumably via infected monocytes/macrophages and/or T cells that can traverse the blood-CSF and blood-brain barriers. Adhesion molecules, cytokines, chemokines, and leukotrienes may all play important roles in this process[9] and, hence, constitute potentially important targets for prophylactic therapeutic interventions.

For unknown reasons, HIV-1–infected cells concentrate in the cerebral white matter, basal ganglia, thalamus, and brain stem,[10] a distribution that may well relate to the subcortical nature of HIV dementia. Although disturbances or loss of neuronal function is the likely substrate of HIV-1–associated neurologic disease, HIV-1 does not appear to infect neurons or oligodendrocytes in vivo. As emphasized early in the epidemic,[11,12] this suggests indirect mechanisms of HIV-1–associated neuronal impairment. Evidence for replication of HIV-1 virus in the CNS has been found primarily in blood-derived macrophages and resident microglia[2]; that is, cells of the mononuclear phagocyte lineage. Mononuclear phagocytes have important functions in wound repair and in defense against infections. HIV-1 infection may impair such microglial/macrophage functions and thereby contribute to CNS damage. In vitro microbial products and host-derived injury response factors can also induce microglia/macrophages to produce neurotoxins, and it is this latter mechanism that has been the focus of most recent molecular research on HIV dementia.

Although the HIV-1 regulatory protein Nef has been identified by in situ hybridization and by immunostaining in reactive astrocytes in brains of children and adults with AIDS,[13-15] there is little in vivo evidence for replication of HIV-1 in astrocytes. However, in vitro studies indicate that persistent astroglial infections could be transiently converted into productive infections by exposure of astrocytes to proinflammatory cytokines.[16]

These studies indicate that HIV-1 is clearly neuroinvasive. What, then, are the structural and molecular consequences of HIV-1 infection in the CNS, and how do they relate to the clinical impairments in people with HIV disease?

Correlation Studies: Guilt by Association

Correlations between clinical and neuropathologic findings may help determine whether HIV dementia reflects the loss or structural damage of CNS cells and, if so, which CNS components are most affected. In addition, correlations between clinical/neuropathologic findings and immunochemical/biochemical measurements may help identify potential molecular mediators of HIV-1–induced CNS impairment. Both types of studies have been carried out and have provided a number of potentially important, albeit frequently inconsistent, lead findings.

Does HIV dementia result from the loss of neurons or glia or from structural damage to these cells? Up to 90 percent of AIDS patients may show neuropathologic alterations affecting diverse components of the nervous system. Typical changes in AIDS, even in the absence of opportunistic infections, include brain atrophy, white matter pallor, astrocytosis, microglial nodules, multinucleated giant cells, and evidence for increased permeability of the blood-brain barrier.[17,18] In addition, there is frequently evidence for subtle structural neuronal damage, including decreased density of presynaptic terminals and vacuolar degeneration of neuronal dendrites [*see Figure 2*], as well as selective loss of neuronal subpopulations, particularly in the frontal cortex and hippocampus [*see Table 3*].

CNS alterations in either HIV dementia or HIV encephalitis have been studied [*see Table 3*]. These entities are related but not identical. HIV dementia is a clinical entity, as defined, whereas HIV encephalitis is a neuropathologic entity characterized by multinucleated giant cells, microglial nodules, microgliosis, astrocytosis, and expression of HIV antigens.[17] However, not all of these features are required for the diagnosis of HIV encephalitis, and different studies have used different combinations of these characteristics to define HIV encephalitis. Unfortunately, inconsistencies in the definitions of HIV dementia and HIV encephalitis and insufficient emphasis on a clear differentiation between the two entities have stalled progress in this complex field.

Because structural neuronal damage and glial responses to injury can appear within 24 hours of an insult and AIDS-related deaths are frequently as-

Figure 2 Vacuolar degeneration of large pyramidal neuron and loss of presynaptic terminals in a patient with HIV encephalitis. Neocortical sections from persons with (a) and without (b) HIV encephalitis were double-immunolabeled with antibodies against MAP-2 (marker of neuronal somata and dendrites [green]) and synaptophysin (marker of presynaptic terminals [red]). Note the vacuolization of the large apical dendrite and the overall rarefaction of presynaptic terminals in the patient with HIV encephalitis (a). Images were generated by laser scanning confocal microscopy.

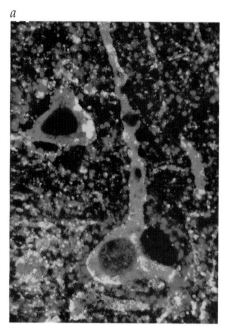 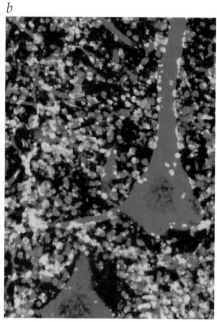

sociated with dramatic premortem changes in multiple organ systems, including the CNS, in any given AIDS patient the state of the CNS at autopsy may be histopathologically and biochemically quite different from that at the time of the last neuropsychologic evaluation. This makes it difficult to correlate pathological and clinical findings. To identify reliably neuropathologic changes that underlie HIV dementia, it is necessary to subtract rigorously in such patients all alterations found in carefully matched HIV-infected nondemented controls who died under otherwise similar circumstances. However, it is not easy to find large numbers of well-matched cases, and, hence, this approach has not been used universally.

Although different interpretations are possible, we conclude from the studies summarized [*see Table 3*] that overall loss of neurons does not account for HIV dementia. Nonetheless, the correlation of HIV encephalitis with loss of neuronal subpopulations and dropout of neuronal dendrites and synapses raises the question of whether these structural alterations may disrupt important neuronal circuits. Studies in which these changes are correlated directly with HIV dementia are in progress.

Although increased numbers of glial fibrillary acidic protein (GFAP)–positive astrocytes are seen frequently in CNS HIV-1 infections, astrocytosis does not appear to correlate with HIV dementia. Although myelin pallor demonstrated by decreased white matter staining with Luxol fast blue is frequently associated with HIV dementia, it does not appear to represent classic demyelination[18]; its precise ultrastructural substrate and potential functional correlates remain to be determined.

Taken as a whole, these studies suggest that HIV dementia is characterized by a discrepancy between relatively subtle structural disruptions of CNS integrity and profound functional neuronal impairments. These findings are consistent with the view that HIV dementia results from the action of soluble factors that directly or indirectly derange neuronal function.[11,12] Notably, the neuropathologic observations also suggest that, at least in vivo, these factors are unlikely to be potent inducers of widespread neuronal death.

The search for soluble inducers of neuronal dysfunction has focused on viral products [*see Table 4*] and host-derived factors [*see Table 5*]. Although this issue is not without controversy, the majority of studies have found that HIV dementia correlates, at least to some extent, with viral burden or expression of viral products. Because some HIV-1 products can be detected more readily than others, it is important to remember that ease of detectability does not necessarily imply high neuropathogenic potential, and vice versa. HIV-1 gp120 may be a good case in point. It has been much easier to detect HIV-1 gp41 than gp120 in the CNS or CSF of patients with HIV encephalitis.[19,20] However, because gp120 and gp41 are derived from the same gp160 precursor molecule,[21] it seems unlikely that in the HIV-1–infected CNS gp41 is generated without concomitant production of gp120. Furthermore, in vitro and in vivo studies indicate that gp120 has significant neuropathogenic potential.[22] Undetectable levels of a particular factor may simply reflect the inadequacy of single end point assessments in a chronic disease or the low sensitivity of the detection system, and single measurements of global tissue concentrations could easily miss temporally and focally elevated factors that may contribute significantly to CNS impairment over prolonged periods of time.[23]

Levels of numerous host-derived factors can be altered in the CSF or CNS of AIDS patients [*see Table 5*]. These alterations could (1) be a byproduct of cellular injury and merely indicate disease severity, (2) reflect compensatory ben-

Table 3 CNS Integrity—HIV Dementia/HIV Encephalitis Relation

Neuropathological Alteration	Correlation with HIV Dementia	Correlation with HIV Encephalitis
Cerebral/cortical atrophy	No (85-*BNWSe, 86-BNWe)	
Decreased dendritic and synaptic density		Yes (81-NMb)
Increased clustering of neurons	Yes (78-*NOa)	
Increased number of GFAP-positive astrocytes	No (18-NWIe)	No (82-NIOe, 81-BNWIb)
Loss of calbindin-positive neurons		Yes (79-NIb)
Loss of parvalbumin-positive neurons		Yes (80-HIb)
Loss of somatostatin-positive neurons		Yes (99-NIb)
Myelin pallor	Yes (18-WLc, 83-Wla); No (84-*WLc)	
Neuronal loss	No (75-*NSa, 76-NSa)	No (76-NSa, 77-NSe)

*Prospective design.

B—basal ganglia/deep grey matter; H—hippocampus; N—neocortex; W—white matter.

I—immunohistochemistry; L—Luxol fast blue staining; M—computer-aided analysis of confocal microscopic images of immunolabeled sections; O—other; S—stereology.

Most relevant control group assessed: a—seropositive nondemented persons; b—seropositive persons without HIV encephalitis; c—seronegative and seropositive nondemented persons; d—seronegative persons with other neurological diseases; e—seronegative nondemented persons.

Numbers refer to references listed at the end of the chapter.

Table 4 Virus—HIV Dementia/HIV Encephalitis Relation

HIV-1	*Correlation with HIV Dementia*	*Correlation with HIV Encephalitis*
Antigen		
p24	Yes (87-*a, 88-*a, 89-*Ca, 90-*Ca, 91-*Ca); No (43-c)	Yes (87-*b)
gp41	Yes (43-c); No (92-*a)	
Nef	Yes (15-a)	No (15-b)
mRNA		
gag, tat/rev	Yes (93-*c)	
DNA		
env	No (94-*)	
gag		Yes (10-b)
pol	Yes (87-*a, 95-a)	Yes (87-*b)

*Prospective design.

Antigens were detected by antigen capture assay, immunoblot, immunohistochemistry, or solid phase immunoassay; RNA by reverse transcriptase-polymerase chain reaction (RT-PCR); and DNA by polymerase chain reaction (PCR). Most studies examined brain tissues. Those that analyzed cerebrospinal fluid are specifically indicated (C).

Most relevant control group assessed: a—seropositive nondemented persons; b—seropositive persons without HIV encephalitis; c—seronegative and seropositive nondemented persons.

Numbers refer to references listed at the end of the chapter.

eficial responses of the body to infection and CNS impairment, or (3) causally contribute to neurologic disease manifestations. Experimental manipulations are required to differentiate among these possibilities. Notably, elevations of tumor necrosis factor–α (TNF-α) and quinolinic acid have been found to correlate with HIV dementia in more than one study, making these molecules particularly suspect. So far, the analysis of cytokine and excitotoxin levels has only rarely been combined with an examination of corresponding antagonists or soluble and membrane-bound receptors, any of which could significantly influence the biologic impact of elevated cytokine and excitotoxin levels on the CNS. Therefore, the correlation of cytokines or other factors with disease manifestations does not necessarily reflect pathogenic potential.[23]

Multifactoriality and Its Conceptual Traps

HIV dementia is frequently associated with high viral burden and elevated cytokine levels [*see Tables 4 and 5*] but can also occur in the absence of these changes, suggesting that HIV dementia is multifactorial in etiology. For a number of reasons, multifactorial diseases are frequently associated with apparent inconsistencies in correlation studies. Because the CNS can respond to diverse triggers of neuropathogenesis with a limited number of alterations, similar structural and functional CNS changes can be caused by distinct factors. This is well illustrated by Alzheimer's disease, which can be caused by a variety of different point mutations in any of three genes but occurs more frequently in the absence of these mutations [*see Chapter 3*]. Thus, inconsistent associations between suspected etiologic factors and clinical findings do not disprove that these factors can contribute significantly to the disease process. Inconsistent findings among studies can also relate to the need for different etiologic factors to synergize to induce disease. For example, a viral product or cytokine may, by itself, be relatively innocuous until combined with another insult. Confusion often arises at branchpoints of complex pathogenetic cascades. For example, viral products could directly injure nerve cells while triggering cytokine production by other cells. As a result, cytokine levels may correlate well with neuronal injury, independent of whether these molecules are actually neurotoxic. When reading the literature on multifactorial diseases such as HIV dementia and Alzheimer's disease, it is important to bear these caveats in mind so as not to mistake complexity for incongruity.

Cellular Mediators

A variety of experimental approaches have been explored to identify the cellular and molecular mediators of HIV-associated CNS impairments. Here we discuss primarily results obtained in cell culture or rodent models. Other useful HIV-1–related models, such as simian and feline immunodeficiency virus infections, have been reviewed.[24,25]

Whereas neurons appear to be the primary cellular victims in HIV dementia, mononuclear phagocytes, including CNS microglia, have emerged as the likeliest pathogenetic culprits. Mononuclear phagocytes appear to be

Table 5 Host Factor—HIV Dementia/HIV Encephalitis Relation

Molecule	Correlation with HIV Dementia	Correlation with HIV Encephalitis
Cytokines and receptors		
IL-1α	Increase (96-*CMd)	
IL-1β	Decrease (97-*BRc)	No (62-BMb, 23-CB)
IL-1 receptors		Increase (98-BMb)
IL-4	Decrease (97-*BRc)	
IL-6	Increase (96-*CMd); No (97-*BRc)	No (62-BPb, 23-CM)
TGF-β	No (97-*BRc)	
TGF-β receptors		Increase (98-BMb)
TNF-α	Increase (92-*BRa, 106-*BRa, 97-*BRc, 96-*Cmd); No (95-BMa, 100-*CMa)	Increase (23-CM); No (62-BMb)
TNF-α receptors		No (62-BMb)
Macrophage activation markers and others		
β$_2$-microglobulin	Increase (101-*CMa)	
HAM-56	Increase (92-*BMa)	
iNOS	Increase (43-*BMRc)	
Neopterin	Increase (101-*Cma); No (102-BMa)	Increase (102-BMa)
Platelet-activating factor	Increase (103-CMa)	
Prostaglandin E$_2$	Increase (104-*CMa)	
Quinolinic acid	Increase (95-BPa, 105-*CMa)	No (23-CBM)

*Prospective design.

B—brain tissue; C—cerebrospinal fluid; iNOS—inducible nitric oxide synthase.

Most relevant control group assessed: a—seropositive nondemented persons; b—seropositive persons without HIV encephalitis; c—seronegative and seropositive nondemented persons; d—seronegative persons with other neurologic diseases; e—seronegative nondemented persons.

Molecules/antigens (M) were detected by enzyme-linked immunoabsorbent assay, gas chromatography/mass spectometry, immunohistochemistry, binding of labeled cytokines, or radioimmunoassay. RNAs (R) were detected by reverse transcriptase-polymerase chain reaction.

No—no correlation found.

Numbers refer to references listed at the end of the chapter.

the main cellular mediators of inflammatory reactions in the CNS. They produce an array of potentially neurotoxic factors, such as cytokines, free oxygen radicals, arachidonic acid metabolites, excitatory amino acids, and amines, in response to a variety of pathological stimuli; HIV-1 infection appears to further augment this capacity.[26-29]

Although evidence that astrocytes are also involved in the pathogenesis of HIV-1–associated neurologic disease is accumulating rapidly, their role in this process is likely quite complex. In vitro, astrocytes appear to exert regulatory effects on production of neurotoxic mediators by both HIV-1–infected and –uninfected monocytes/macrophages. Depending on the activation status of the macrophages, or perhaps because of other in vitro parameters, the astroglial influences were either stimulatory[30] or inhibitory.[26,31] Astrocytes appear to fulfill a range of functions that support CNS homeostasis and protect neurons as well as other CNS cells from injury.[32] It is therefore interesting that HIV-1 products and cytokines that are elevated in the CNS of AIDS patients can alter astroglial properties in vitro[33-36] and in vivo.[20,37,38] Conceivably, HIV-1–induced or cytokine-induced disturbances of astroglial support functions may contribute indirectly to neuronal injury or dysfunction.[33,39] Although this hypothesis is attractive, the pathogenetic importance of astroglial impairments in HIV dementia, or any other neurologic disease for that matter, remains to be demonstrated.

Molecular Mechanisms

Because widespread neuronal impairment in HIV dementia can be associated with relatively low numbers of infected cells, it was suggested early on that CNS cells may be affected by extracellular exposure to detrimental factors (released from infected cells or from cell-free virus) that can diffuse widely through the CNS parenchyma.[11,12] A number of viral and host-derived molecules have been identified that could act as such mediators of HIV-induced CNS impairment.

HIV Proteins, Excitotoxicity, and Nitric Oxide

Numerous in vitro and in vivo studies indicate that HIV-1 gp120 can induce neuronal dysfunction, damage, or death, depending on its concentration and the presence of several cellular and molecular cofactors.[22,36] Neuropathogenic gp120 effects can be inhibited by antagonists of the *N*-methyl-D-aspartate (NMDA) subtype of glutamate receptors in cell culture,[22] brain slices,[40] and transgenic mice.[41] Thus, the aberrant stimulation of NMDA receptors appears to be a critical component of gp120-induced neurotoxicity. In vitro studies indicate that the effects of gp120 on neuronal NMDA receptors may be indirect, involving an essential microglial/macrophage loop.[22] It is likely that factors other than gp120 can also induce the production of microglia/macrophage-derived neurotoxins.

Excitotoxicity appears to be a common pathogenetic pathway in diverse neurodegenerative disorders [*see Chapters 4 and 7*], which makes this process a particularly attractive target for therapeutic interventions. In fact, the NMDA receptor antagonist memantine is being tested for efficacy in HIV dementia as concurrent treatment with antiretroviral therapy in a phase II, randomized, double-blind, placebo-controlled, multicenter clinical trial (NIH AIDS Clinical Trials Group Protocol 301).

Exposure of monocytes, glia, or mixed neuronal/glial cultures to recombinant HIV-1 envelope proteins (gp160, gp120, or gp41) increases nitric oxide synthase (NOS) expression or activity,[42,43] and neuropathogenic effects induced by gp120 or gp41 can be prevented by NOS inhibitors.[40,43,44] The product of NOS, nitric oxide (NO), is a labile, free radical gas that appears to play important roles in physiologic as well as pathophysiological processes; it is generated by different forms of NOS that are encoded by separate genes located on different chromosomes [*see Chapter 7*]. Although numerous mechanisms have been proposed by which NO might damage cells,[45,46] the exact processes underlying NO-mediated neurotoxicity are unknown. Overstimulation of glutamate receptors results in elevated intracellular calcium levels which, in turn, can result in NOS activation and NO production. Therefore, NO may play an important role in excitotoxic neuronal damage.[47,48]

Modification of NOS activity is a potentially interesting therapeutic target. Inhibitors that are relatively selective for specific NOS isoforms are being developed and have already been shown to be neuroprotective in rodent models of stroke, Huntington's disease, and Parkinson's disease.[49,50] If these compounds also show promise in clinical trials, they may well provide new therapies for HIV dementia.

In addition to its effects on microglia and neurons, gp120 affects astrocytes both in vitro and in vivo.[34-36] Some of these gp120 effects could contribute directly to the astrocytosis that is so prominent in AIDS and can be detected even in the brains of asymptomatic HIV-1–infected individuals.[17,35,51] Other effects of gp120 on astrocytes might promote the invasion of the CNS by infected monocytes.[36]

The nonstructural regulatory HIV-1 protein Tat, which transactivates the viral long terminal repeat, has also been implicated in HIV dementia. Like gp120, Tat can be released from infected cells and taken up by uninfected cells. Tat can accumulate in the nucleus, where it may affect a variety of host functions, and has been shown to exert neurotoxic effects both in vitro and in vivo.[52-54] Exposure of slices of human cortex or rat hippocampus to recombinant Tat in vitro results in neuronal depolarizations that appear to be mediated primarily by non–NMDA-type glutamate receptors and resemble responses induced by kainate.[54] However, Tat-induced neurotoxicity in cultures of human fetal neurons is blocked not only by antagonists of non-NMDA receptors but also by antagonists of NMDA receptors.[54]

In summary, these studies indicate that several HIV-1 proteins can induce molecular changes in different CNS cell types. The most detrimental of these effects appear to be mediated by microglia/macrophages and to involve overstimulation of neuronal glutamate receptors as well as consequent destabilization of neuronal calcium homeostasis and NO production. Consequently, microglial activity, excitotoxicity, and NOS regulation constitute potentially important targets for the treatment of HIV dementia.

HIV-1 Strains

Because HIV-1 strains found in the CNS are primarily macrophage-tropic, it is widely assumed that macrophage tropism is required for HIV-1 neuroinvasiveness. A related but largely unresolved issue is whether neuroinvasive HIV-1 strains differ in their neuropathogenic potential. This is an important question, particularly because such differences could explain, at least in part, why viral load does not always correlate with degree of neurologic impairment [*see Table 4*].

Comparisons of limited sequence elements of brain-derived HIV-1 clones from small groups of demented versus nondemented HIV-1–infected patients have revealed significant differences in specific amino acid positions within gp120 between the two clinical groups.[55] It will be important to determine whether strategic mutations of the identified residues change the neurobiological effects of gp120 in vitro and in vivo and to examine wider portions of the HIV-1 genome in larger cohorts of HIV-1–infected persons with and without dementia.

Cytokines

Several of the correlation studies listed in Table 4 have identified patients with HIV dementia in whom cerebral viral load is minimal. These cases emphasize the potential importance of host-derived factors [*see Table 5*]. Cytokines and lipid-derived biomediators may trigger and critically influence the extent of neurotoxin production by monocytic phagocytes within the CNS.[9] Cytokines such as TNF-α, interleukin-1 (IL-1), and IL-6 are key mediators of the inflammatory cascade, and dysregulation of their synthesis, secretion, or action is suspected to be responsible for inflammation-induced pathology, including CNS alterations associated with AIDS.[9,56,57] We focus here on TNF-α, which seems to be the most intensely studied suspect in this group.

TNF-α increases the synthesis of virus in HIV-1–infected monocytes.[58] However, in human mixed brain cell cultures, it actually suppresses HIV-1 expression, presumably via the induction of chemokines.[59] The neuropathogenic potential of TNF-α appears to be equally controversial. Although TNF-α is neuroprotective in cultures of rat hippocampal neurons,[60] it is neurotoxic in cultures of human fetal neurons.[61]

Increased TNF-α receptor immunoreactivity in brains of AIDS patients is found primarily on activated microglia and macrophages,[62] consistent with an indirect mode of action. However, as these investigators note, lack of immunostaining does not exclude TNF-α receptor expression by other CNS cell types. In vitro studies of primary human fetal astrocytes have recently revealed an intriguing indirect mechanism by which TNF-α could induce excitotoxic neuronal damage: TNF-α–inhibited high-affinity astroglial glutamate uptake,[63] which is essential for the efficient removal of potentially excitotoxic amino acid transmitters from the extracellular milieu of the CNS.[64] Moreover, exposure of astrocytes to HIV-1 gp120 triggered an increase in astroglial glutamate efflux.[34] Taken together, these observations suggest that viral and host-derived factors could collaborate in HIV dementia to dysregulate astroglial transport of excitatory amino acid transmitters, making this process a potentially interesting target for therapeutic interventions.

High-level expression of TNF-α in the CNS of transgenic mice[65] results in a severe neurologic disease that in many respects resembles chronic progressive multiple sclerosis but differs from HIV encephalitis in that it is characterized pathologically by extensive infiltration of the CNS parenchyma by T lymphocytes. Remarkably, intraperitoneal injections of a chimeric hamster/mouse monoclonal antibody to murine TNF-α prevented the clinical disease in TNF-α transgenic mice and ameliorated the histopathologic CNS alterations.[65] Because of the prominent inflammatory infiltration induced by TNF-α in the CNS of transgenic mice, these studies have not yet resolved whether overexpression of TNF-α per se is neurotoxic in vivo. Experiments are in progress to determine whether TNF-α has any detrimental CNS effects in immunodeficient mice (Iain Campbell, personal communication, 1997.) These studies should provide further insights into the neurotoxic potential of TNF-α and may generate a model that more closely simulates the situation encountered in the brains of immunosuppressed AIDS patients.

In vivo, phosphodiesterase-IV inhibitors, which suppress TNF-α gene expression by enhancing intracellular levels of cyclic adenosine monophosphate, reduce disease manifestations in several models of peripheral inflammation[66] and prevent autoimmune demyelination in rats and nonhuman primates.[67,68]

Interestingly, TNF-α receptor knockout mice in which TNF-α receptor expression was ablated by targeted gene disruption showed increased susceptibility to neuronal injury induced by ischemia or excitotoxicity, and a markedly attenuated microglial response to brain injury.[69] These provocative findings challenge two widely held concepts in HIV dementia research: that TNF-α is neurotoxic and that neuronal injury depends on microglial activation. Because they were obtained in knockout mice, these data cannot rule out an immunopathologic role of TNF-α in human AIDS. However, they do emphasize that upmodulation of cytokines and other factors by infection or injury may fulfill important protective functions. Consequently, the pharmacological suppression of such responses could have detrimental side effects. Various strategies targeting TNF-α are being assessed in clinical trials aimed primarily at peripheral AIDS manifestations. In view of the above considerations, it will be important to monitor carefully the effects, beneficial or detrimental, of TNF-α inhibitors on HIV-1–associated neurologic disease.

Apoptosis

A concept that is gaining momentum rapidly in molecular neurology is that neuronal apoptosis may contribute to neurodegenerative disorders [*see Chapters 4 and 7*]. Apoptosis is an active cell suicide process involving a sequence of regulated cellular events that culminate in chromatin condensation, DNA fragmentation, membrane blebbing, and cell death.[70] The notion that apoptotic neuronal death contributes to neurodegeneration is attractive because a variety of factors and drugs can interfere effectively with apoptotic pathways even in the continued presence of the apoptosis-inducing environmental stimulus.

Evidence for apoptotic cell death is found in the brains of patients with AIDS.[71] Whereas control patients with Alzheimer's disease show apoptosis of neurons and astrocytes, AIDS patients show apoptosis of neurons, astrocytes, and endothelial cells.[71] It is speculated that endothelial apoptosis might contribute to the disruption of the blood-brain barrier found so frequently in AIDS. Although the number of apoptotic CNS cells may be higher in people with HIV dementia or HIV encephalitis than in AIDS patients without these conditions, apoptotic cells are also found in the brains of AIDS patients without dementia or encephalitis.[71] Therefore, and because widespread loss of neurons does not appear to be a characteristic feature of HIV dementia [see Table 3], it remains unclear whether HIV-1–associated apoptosis of CNS cells contributes to neurologic impairments in AIDS or merely reflects parallel pathogenetic processes.

Therapeutic Targets and Future Perspectives

The studies discussed identify a number of potentially important mediators of HIV dementia [see Figure 3]. From a therapeutic perspective, it is important to determine which of these mediators are most critical to development of neurologic impairment. Convenient small animal models could significantly facilitate and expedite this process. Some aspects of HIV-1 neuropathogenesis are difficult to study in such models because they relate to species-specific components of HIV-1 infection or cognitive deficits. However, many aspects of neuropathogenesis do not appear to be species specific and, hence, can be studied effectively in rodent models. This is perhaps best illustrated by the recent advances made with transgenic models simulating specific aspects of two other complex dementing illnesses: Alzheimer's disease [see Chapter 3] and prion-induced spongiform encephalopathies [see Chapter 10].

The ultimate test of the relative neuropathogenic importance of specific factors is their inhibition in human subjects. Unfortunately, results from CNS-targeted therapeutic trials can be quite difficult to interpret because of limitations in the bioavailability and specificity of pharmacological compounds. For example, many antiretroviral drugs do not penetrate the blood-brain barrier effectively, and some may have unexpected direct effects on CNS functions. The latter issue may be particularly difficult to control given the potential for drug interactions in patients who almost routinely require treatment with multiple drugs.

Nonetheless, potentially therapeutic levels of zidovudine (AZT) can be achieved in the CSF of AIDS patients over a range of daily doses,[72] and numerous retrospective and observational studies indicate that zidovudine may ameliorate and prevent HIV-1–associated neurologic disease.[7] Zidovudine may prevent the spread of HIV in the brain by inhibiting reverse transcriptase or interfering with the production, release, or action of viral neurotoxins by some other mechanism. However, it is also possible that some neurologic benefits originally ascribed to zidovudine therapy reflect, at least in part, better general medical management, earlier aggressive treatment of opportunistic infections, or genetic shifts of HIV away from potentially neuropathic strains.[8] Urgently needed definitive answers from controlled prospective studies are pending at this time.

The development of better antiretroviral therapies has clearly given rise to hope among patients and physicians alike. However, it also raises some concerns. It is not completely clear whether the suppression of HIV-1 in the

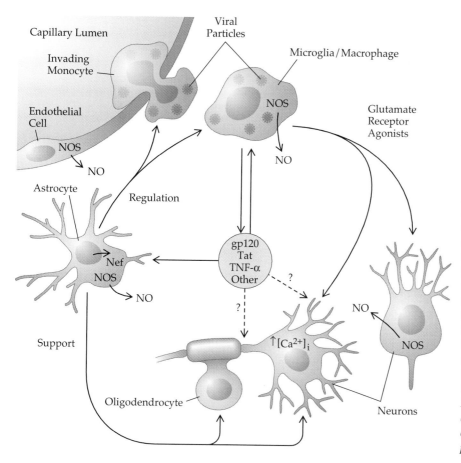

Figure 3 *Invading macrophages and resident microglia permit HIV-1 replication and secrete viral products as well as host-derived injury response factors. Some of these molecules may stimulate microglia/macrophages to produce glutamate receptor agonists that can induce dysfunction or excitotoxic injury in susceptible neurons. In addition, these factors alter astroglial functions that may also be affected more directly by persistent infection and astroglial production of regulatory HIV-1 proteins. Several different CNS cell types can produce nitric oxide (NO), which, depending on a number of variables, can have either beneficial or detrimental effects. Neurons expressing nitric oxide synthase (NOS) may be relatively resistant to HIV-1–induced excitotoxicity and could contribute to HIV dementia pathogenesis via NO production*

blood and the consequent preservation of the immune system and prolonged survival will result in an increase or decrease in the number of persons with HIV-1–related CNS impairments. This is because many of the drugs used to treat AIDS do not readily penetrate the blood-brain barrier,[7] and because it is not known if the persistence of HIV-1 in the CNS requires continual reinfection of the CNS from the periphery. In the worst-case scenario, the CNS might be turned into a sanctuary for HIV, providing a source of reinfection elsewhere in the body. Furthermore, although available AIDS treatments may eradicate replicating virus, they cannot eliminate persistent infections that may result in the production of viral products such as Tat or Nef that have the potential to disturb CNS cell functions in the absence of viral replication.

The increasing number of people infected with HIV-1 makes the development of better treatments for HIV-1–associated neurologic diseases an important and urgent issue. Although suppression of potentially detrimental CNS inflammation may turn out to be a rewarding strategy, we believe that the most effective therapeutic targets may be located at the extreme ends of the pathogenetic cascade. Specifically, we anticipate that future studies will reveal the exact mechanisms by which neuronal function becomes deranged in patients with AIDS and that this knowledge will lead to the design of effective symptomatic treatments. Most important, however, is the prospect that eradication of HIV by public education, vaccine development, and improved antiviral treatments will make HIV-associated neurologic diseases a matter of historical interest. Unfortunately, for now, HIV remains a mounting force with which neurologists worldwide will increasingly have to reckon.

References

1. Working Group of the American Academy of Neurology AIDS Task Force: Nomenclature and research case definitions for neurologic manifestations of human immunodeficiency virus-type 1 (HIV-1) infection. *Neurology* 41:778, 1991

2. Price RW, Perry SW: *HIV, AIDS, and the Brain.* Raven Press, New York, 1994

3. McCutchan JA: Clinical impact of cytomegalovirus infections of the nervous system in patients with AIDS. *Clin Infect Dis* 21:S196, 1995

4. Price RW: Neurological complications of HIV infection. *Lancet* 348:445, 1996

5. Dal Pan GJ, McArthur JC: Neuroepidemiology of HIV infection. *Neurol Clin* 14:359, 1996

6. Simpson DM, Berger JR: Neurologic manifestations of HIV infection. *Med Clin North Am* 80:1363, 1996

7. Melton ST, Kirkwood CK, Ghaemi SN: Pharmacotherapy of HIV dementia. *Ann Pharmacother* 31:457, 1997

8. Grant I, Heaton RK, Atkinson JH, et al: Neurocognitive disorders in HIV-1 infection. *Curr Top Microbiol Immunol* 202:11, 1995

9. Nottet HSLM, Gendelman HE: Unraveling the neuroimmune mechanisms for the HIV-1-associated cognitive/motor complex. *Immunol Today* 16:441, 1995

10. Achim CL, Wang R, Miners DK, et al: Brain viral burden in HIV infection. *J Neuropathol Exp Neurol* 53:284, 1994

11. Price RW, Brew B, Sidtis J, et al: The brain in AIDS: central nervous system HIV-1 infection and AIDS dementia complex. *Science* 239:586, 1988

12. Johnson RT, McArthur JL, Narayan O: The neurobiology of human immunodeficiency virus infections. *FASEB J* 2:2970, 1988

13. Saito Y, Sharer LR, Epstein LG, et al: Overexpression of *nef* as a marker for restricted HIV-1 infection of astrocytes in postmortem pediatric central nervous tissues. *Neurology* 44:474, 1994

14. Tornatore C, Chandra R, Berger JR, et al: HIV-1 infection of subcortical astrocytes in the pediatric central nervous system. *Neurology* 44:481, 1994

15. Ranki A, Myberg M, Ovod V, et al: Abundant expression of HIV Nef and Rev proteins in brain astrocytes in vivo is associated with dementia. *AIDS* 9:1001, 1995

16. Tornatore C, Nath A, Amemiya K, et al: Persistent human immunodeficiency virus type 1 infection in human fetal glial cells reactivated by T-cell factor(s) or by the cytokines tumor necrosis factor alpha and interleukin-1 beta. *J Virol* 65:6094, 1991

17. Budka H: Cerebral pathology in AIDS: a new nomenclature and pathogenetic concepts. *Curr Opin Neurol Neurosurg* 5:917, 1992

18. Power C, Kong P-A, Crawford TO, et al: Cerebral white matter changes in acquired immunodeficiency syndrome dementia: alterations of the blood-brain barrier. *Ann Neurol* 34:339, 1993

19. Hill JM, Mervis RF, Avidor R, et al: HIV envelope protein-induced neuronal damage and retardation of behavioral development in rat neonates. *Brain Res* 603:222, 1993

20. Toggas SM, Masliah E, Rockenstein EM, et al: Central nervous system damage produced by expression of the HIV-1 coat protein gp120 in transgenic mice. *Nature* 367:188, 1994

21. McCune JM, Rabin LB, Feinberg MB, et al: Endoproteolytic cleavage of gp160 is required for the activation of human immunodeficiency virus. *Cell* 53:55, 1988

22. Lipton SA: HIV-related neuronal injury: potential therapeutic intervention with calcium channel antagonists and NMDA antagonists. *Mol Neurobiol* 8:181, 1994

23. Achim CL, Heyes MP, Wiley CA: Quantitation of human immunodeficiency virus, immune activation factors, and quinolinic acid in AIDS brains. *J Clin Invest* 91:2769, 1993

24. Narayan O, Joag SV, Stephens EB: Selected models of HIV-induced neurological disease. *Curr Top Microbiol Immunol* 202:152, 1995

25. Henriksen SJ, Prospero-Garcia O, Phillips TR, et al: Feline immunodeficiency virus as a model for study of lentivirus infection of the central nervous system. *Curr Top Microbiol Immunol* 202:167, 1995

26. Giulian D, Vaca K, Corpuz M: Brain glia release factors with opposing actions upon neuronal survival. *J Neurosci* 13:29, 1993

27. Gendelman HE, Tardieu M: Macrophages/microglia and the pathophysiology of CNS injuries in AIDS. *J Leukoc Biol* 56:387, 1994

28. Giulian D, Yu J, Li X, et al: Study of receptor-mediated neurotoxins released by HIV-1-infected mononuclear phagocytes found in human brain. *J Neurosci* 16:3139, 1996

29. Pulliam L, Gascon R, Stubblebine M, et al: Unique monocyte subset in patients with AIDS dementia. *Lancet* 349:692, 1997

30. Genis P, Jett M, Bernton EW, et al: Cytokines and arachidonic acid metabolites produced during human immunodeficiency virus (HIV)-infected macrophage-astroglia interactions: implications for the neuropathogenesis of HIV disease. *J Exp Med* 176:1703, 1992

31. Nottet HSLM, Jett M, Flanagan CR, et al: A regulatory role for astrocytes in HIV-1 encephalitis: an overexpression of eicosanoids, platelet-activating factor, and tumor necrosis factor-α by activated HIV-1-infected monocytes is attenuated by primary human astrocytes. *J Immunol* 154:3567, 1995

32. Eddleston MP, Mucke L: Molecular profile of reactive astrocytes: implications for their role in neurologic disease. *Neuroscience* 54:15, 1993

33. Pulliam L, West D, Haigwood N, et al: HIV-1 envelope gp120 alters astrocytes in human brain cultures. *AIDS Res Hum Retroviruses* 9:439, 1993

34. Benos DJ, Hahn BH, Bubien JK, et al: Envelope glycoprotein gp120 of human immunodeficiency virus type 1 alters ion transport in astrocytes: implications for AIDS dementia complex. *Proc Natl Acad Sci USA* 91:494, 1994

35. Wyss-Coray T, Masliah E, Toggas SM, et al: Dysregulation of signal transduction pathways as a potential mechanism of nervous system alterations in HIV-1 gp120 transgenic mice and humans with HIV-1 encephalitis. *J Clin Invest* 97:789, 1996

36. Shrikant P, Benos DJ, Tang LP, et al: HIV glycoprotein 120 enhances intercellular adhesion molecule-1 gene expression in glial cells. *J Immunol* 156:1307, 1996

37. Campbell IL, Abraham CR, Masliah E, et al: Neurologic disease induced in transgenic mice by cerebral overexpression of interleukin 6. *Proc Natl Acad Sci USA* 90:10061, 1993

38. Wyss-Coray T, Feng L, Masliah E, et al: Increased central nervous system production of extracellular matrix components and development of hydrocephalus in transgenic mice overexpressing transforming growth factor-β1. *Am J Pathol* 147:53, 1995

39. Mucke L, Eddleston M: Astrocytes in infectious and immune-mediated diseases of the central nervous system. *FASEB J* 7:1226, 1993

40. Raber J, Toggas SM, Lee S, et al: Central nervous system expression of HIV-1 gp120 activates the hypothalamic-pituitary-adrenal axis: evidence for involvement of NMDA receptors and nitric oxide synthase. *Virology* 226:362, 1996

41. Toggas SM, Masliah E, Mucke L: Prevention of HIV-1 gp120-induced neuronal damage in the central nervous system of transgenic mice by the NMDA receptor antagonist memantine. *Brain Res* 706:303, 1996

42. Koka P, He K, Zack JA, et al: Human immunodeficiency virus 1 envelope proteins induce interleukin 1, tumor necrosis factor α, and nitric oxide in glial cultures derived from fetal, neonatal, and adult human brain. *J Exp Med* 182:941, 1995

43. Adamson DC, Wildemann B, Sasaki M, et al: Immunologic NO synthase: elevation in severe AIDS dementia and induction by HIV-1 gp41. *Science* 274:1917, 1996

44. Dawson VL, Dawson TM, Uhl GR, et al: Human immunodeficiency virus type 1 coat protein neurotoxicity mediated by nitric oxide in primary cortical cultures. *Proc Natl Acad Sci USA* 90:3256, 1993

45. Lipton SA, Choi Y-B, Pan Z-H, et al: A redox-based mechanism for the neuroprotective and neurodestructive effects of nitric oxide and related nitroso-compounds. *Nature* 364:626, 1993

46. Zhang J, Pieper A, Snyder SH: Poly(ADP-ribose) synthetase activation: an early indicator of neurotoxic DNA damage. *J Neurochem* 65:1411, 1995

47. Dawson VL, Kizushi VM, Huang PL, et al: Resistance to neurotoxicity in cortical cultures from neuronal nitric oxide synthase-deficient mice. *J Neurosci* 16:2479, 1996

48. Strijbos PJLM, Leach MJ, Garthwaite J: Vicious cycle involving Na+ channels, glutamate release, and NMDA receptors mediates delayed neurodegeneration through nitric oxide formation. *J Neurosci* 16:5004, 1996

49. Iadecola C: Bright and dark sides of nitric oxide in ischemic brain injury. *Trends Neurosci* 20:132, 1997

50. Matthews RT, Yang L, Beal MF: S-Methylthiocitrulline, a neuronal nitric oxide synthase inhibitor, protects against malonate and MPTP neurotoxicity. *Exp Neurol* 143:282, 1997

51. Lenhardt TM, Super MA, Wiley CA: Neuropathological changes in an asymptomatic HIV seropositive man. *Ann Neurol* 23:209, 1988

52. Sabatier J-M, Vives E, Mabrouk K, et al: Evidence for neurotoxic activity of tat from human immunodeficiency virus type 1. *J Virol* 65:961, 1991

53. Weeks BS, Lieberman DM, Johnson B, et al: Neurotoxicity of the human immunodeficiency virus type 1 Tat transactivator to PC12 cells requires the Tat amino acid 49-58 basic domain. *J Neurosci* 42:34, 1995

54. Magnuson DSK, Knudsen BE, Geiger JD, et al: Human immunodeficiency virus type 1 Tat activates non-*N*-methyl-D-aspartate excitatory amino acid receptors and causes neurotoxicity. *Ann Neurol* 37:373, 1995

55. Power C, McArthur JC, Johnson RT, et al: Demented and nondemented patients with AIDS differ in brain-derived human immunodeficiency virus type 1 envelope sequences. *J Virol* 68:4643, 1994

56. Tyor WR, Wesselingh SL, Griffin JW, et al: Unifying hypothesis for the pathogenesis of HIV-associated dementia complex, vacuolar myelopathy, and sensory neuropathy. *J Acquir Immune Defic Syndr Hum Retrovirol* 9:379, 1995

57. Heyser CJ, Masliah E, Samimi A, et al: Progressive decline in avoidance learning paralleled by inflammatory neurodegeneration in transgenic mice expressing interleukin 6 in the brain. *Proc Natl Acad Sci USA* 94:1500, 1997

58. Locardi C, Petrini C, Boccoli G, et al: Increased human immunodeficiency virus (HIV) expression in chronically infected U937 cells upon in vitro differentiation by hydroxy-vitamin D3: roles of interferon and tumor necrosis factor in regulation of HIV production. *J Virol* 64:5874, 1990

59. Lokensgard JR, Gekker G, Ehrlich LC, et al: Proinflammatory cytokines inhibit HIV-1SF162 expression in acutely infected human brain cell cultures. *J Immunol* 158:2449, 1997

60. Cheng B, Christakos S, Mattson MP: Tumor necrosis factors protect neurons against metabolic-excitotoxic insults and promote maintenance of calcium homeostasis. *Neuron* 12:139, 1994

61. Talley AK, Dewhurst S, Perry SW, et al: Tumor necrosis factor alpha-induced apoptosis in human neuronal cells: protection by the antioxidant *N*-acetylcysteine and the genes *bcl-2* and *crmA*. *Mol Cell Biol* 15:2359, 1995

62. Sippy BD, Hofman FM, Wallach D, et al: Increased expression of tumor necrosis factor-α receptors in the brains of patients with AIDS. *J Acquir Immune Defic Syndr Hum Retrovirol* 10:511, 1995

63. Fine SM, Angel RA, Perry SW, et al: Tumor necrosis factor–α inhibits glutamate uptake by primary human astrocytes. *J Biol Chem* 271:15303, 1996

64. Rothstein JD, Dykes-Hoberg M, Pardo CA, et al: Knockout of glutamate transporters reveals a major role for astroglial transport in excitotoxicity and clearance of glutamate. *Neuron* 16:675, 1996

65. Probert L, Akassoglou K, Pasparakis M, et al: Spontaneous inflammatory demyelinating disease in transgenic mice showing central nervous system-specific expression of tumor necrosis factor α. *Proc Natl Acad Sci USA* 92:11294, 1995

66. Sekut L, Yarnall D, Stimpson SA, et al: Anti-inflammatory activity of phosphodiesterase (PDE)-IV inhibitors in acute and chronic models of inflammation. *Clin Exp Immunol* 100:126, 1995

67. Sommer N, Löschmann P-A, Northoff GH, et al: The antidepressant rolipram suppresses cytokine production and prevents autoimmune encephalomyelitis. *Nat Med* 1:244, 1995

68. Genain CP, Roberts T, Davis RL, et al: Prevention of autoimmune demyelination in non-human primates by a cAMP-specific phosphodiesterase inhibitor. *Proc Natl Acad Sci USA* 92:3601, 1995

69. Bruce AJ, Boling W, Kindy MS, et al: Altered neuronal and microglial responses to excitotoxic and ischemic brain injury in mice lacking TNF receptors. *Nat Med* 2:788, 1996

70. Park DS, Stefanis L, Yan CYI, et al: Ordering the cell death pathway. *J Biol Chem* 271:21898, 1996

71. Shi B, De Girolami U, He J, et al: Apoptosis induced by HIV-1 infection of the central nervous system. *J Clin Invest* 98:1979, 1996

72. Burger DM, Kraaijeveld CL, Meenhorst PL, et al: Penetration of zidovudine into the cerebrospinal fluid of patients infected with HIV. *AIDS* 7:1581, 1993

73. Power C, Johnson RT: HIV-1 associated dementia: clinical features and pathogenesis. *Can J Neurol Sci* 22:92, 1995

74. Power C, Selnes OA, Grim JA, et al: HIV dementia scale: a rapid screening test. *J Acquir Immune Defic Syndr Hum Retroviruses* 8:273, 1995

75. Everall IP, Glass JD, McArthur J, et al: Neuronal density in the superior frontal and temporal gyri does not correlate with the degree of human immunodeficiency virus-associated dementia. *Acta Neuropathol* 88:538, 1994

76. Weis S, Haug H, Budka H: Neuronal damage in the cerebral cortex of AIDS brains: a morphometric study. *Acta Neuropathol* 85:185, 1993

77. Everall IP, Luthert PJ, Lantos PL: Neuronal number and volume alterations in the neocortex of HIV infected individuals. *J Neurol Neurosurg Psychiatry* 56:481, 1993

78. Asare E, Dunn G, Glass J, et al: Neuronal pattern correlates with the severity of human immunodeficiency virus-associated dementia complex. *Am J Pathol* 148:31, 1996

79. Masliah E, Ge N, Achim CL, et al: Differential vulnerability of calbindin-immunoreactive neurons in HIV encephalitis. *J Neuropathol Exp Neurol* 54:350, 1995

80. Masliah E, Ge N, Achim CL, et al: Selective neuronal vulnerability in HIV encephalitis. *J Neuropathol Exp Neurol* 51:585, 1992

81. Masliah E, Achim CL, Ge N, et al: Spectrum of human immunodeficiency virus-associated neocortical damage. *Ann Neurol* 32:321, 1992

82. Weis S, Haug H, Budka H: Astroglial changes in the cerebral cortex of AIDS brains: a morphometric and immunohistochemical investigation. *Neuropathol Appl Neurobiol* 19:329, 1993

83. Schmidbauer M, Huemer M, Cristina S, et al: Morphological spectrum, distribution and clinical correlation of white matter lesions in AIDS brains. *Neuropathol Appl Neurobiol* 18:489, 1992

84. Glass JD, Wesselingh SL, Selnes OA, et al: Clinical-neuropathologic correlation in HIV-associated dementia. *Neurology* 43:2230, 1993

85. Subbiah P, Mounton P, Fedor H, et al: Stereological analysis of cerebral atrophy in human immunodeficiency virus-associated dementia. *J Neuropathol Exp Neurol* 55:1032, 1996

86. Oster S, Christoffersen P, Gundersen HJG, et al: Cerebral atrophy in AIDS: a stereological study. *Acta Neuropathol* 85:617, 1993

87. Bell JE, Donaldson YK, Lowrie S, et al: Influence of risk group and zidovudine therapy on the development of HIV encephalitis and cognitive impairment in AIDS patients. *AIDS* 10:493, 1996

88. Brew BJ, Rosenblum M, Cronin K, et al: AIDS dementia complex and HIV-1 brain infection: clinical-virological correlations. *Ann Neurol* 38:563, 1995

89. Brew BJ, Paul MO, Nakajima G, et al: Cerebrospinal fluid HIV-1 p24 antigen and culture: sensitivity and specificity for AIDS-dementia complex. *J Neurol Neurosurg Psychiatry* 57:784, 1994

90. Singer EJ, Syndulko K, Fahy-Chandon BN, et al: Cerebrospinal fluid p24 antigen levels and intrathecal immunoglobulin G synthesis are associated with cognitive disease severity in HIV-1. *AIDS* 8:197, 1994

91. Royal W, Selnes OA, Concha M, et al: Cerebrospinal fluid human immunodeficiency virus type 1 (HIV-1) p24 antigen levels in HIV-1-related dementia. *Ann Neurol* 36:32, 1994

92. Glass JD, Fedor H, Wesselingh SL, et al: Immunocytochemical quantitation of human immunodeficiency virus in the brain: correlations with dementia. *Ann Neurol* 38:755, 1995

93. Wesselingh SL, Glass J, McArthur JC, et al: Cytokine dysregulation in HIV-associated neurological disease. *Adv Neuroimmunol* 4:199, 1994

94. Johnson RT, Glass JD, McArthur JC, et al: Quantitation of human immunodeficiency virus in brains of demented and nondemented patients with acquired immunodeficiency syndrome. *Ann Neurol* 39:392, 1996

95. Sei S, Saito K, Stewart SK, et al: Increased human immunodeficiency virus (HIV) type 1 DNA content and quinolinic acid concentration in brain tissues from patients with HIV encephalopathy. *J Infect Dis* 172:638, 1995

96. Perrella O, Carrieri PB, Guarnaccia D, et al: Cerebrospinal fluid cytokines in AIDS dementia complex. *J Neurol* 239:387, 1992

97. Wesselingh SL, Power C, Glass JD, et al: Intracerebral cytokine messenger RNA expression in acquired immunodeficiency syndrome dementia. *Ann Neurol* 33:576, 1993

98. Masliah E, Ge N, Achim CL, et al: Cytokine receptor alterations during HIV infection in the human central nervous system. *Brain Res* 663:1, 1994

99. Fox L, Alford M, Achim C, et al: Neurodegeneration of somatostatin-immunoreactive neruons in HIV encephalitis. *J Neuropathol Exp Neurol* 56:360, 1997

100. Grimaldi LM, Martino GV, Franciotta DM, et al: Elevated alpha-tumor necrosis factor

levels in spinal fluid from HIV-1-infected patients with central nervous system involvement. *Ann Neurol* 29:21, 1991

101. Brew BJ, Dunbar N, Pemberton L, et al: Predictive markers of AIDS dementia complex: CD4 cell count and cerebrospinal fluid concentrations of β2-microglobulin and neopterin. *J Infect Dis* 174:294, 1996

102. Shaskan EG, Brew BJ, Rosenblum M, et al: Increased neopterin levels in brains of patients with human immunodeficiency virus type 1 infection. *J Neurochem* 59:1541, 1992

103. Gelbard HA, Nottet HSLM, Swindells S, et al: Platelet-activating factor: a candidate human immunodeficiency virus type 1-induced neurotoxin. *J Virol* 68:4628, 1994

104. Griffin DE, Wesselingh SL, McArthur JC: Elevated central nervous system prostaglandins in human immunodeficiency virus-associated dementia. *Ann Neurol* 35:592, 1994

105. Heyes MP, Brew BJ, Saito K, et al: Inter-relationships between quinolinic acid, neuroactive kynurenines, neopterin and J_2-microglobulin in cerebrospinal fluid and serum of HIV-1-infected patients. *J Neuroimmunol* 40:71, 1992

106. Glass JD, Wesselingh SL, Selnes OA, et al: Clinical-neuropathologic correlation in HIV-associated dementia. *Neurology* 43:2230, 1993

Acknowledgments

This project was supported by National Institutes of Health grants NS33056 (L.M.) and NS34602 (L.M.). We thank B. Navia and R. Price for information on the pharmacotherapy of HIV dementia, S. Ordway for editorial assistance, and R. Haines for help in preparing the manuscript.

Figure 1   Marcia Kammerer.

Figure 2   Lennart Mucke.

Figure 3   Seward Hung.

CHAPTER 9

Molecular Analysis of Parkinson's Disease

Robert H. Edwards, M.D.

Parkinson's disease (PD) is second in incidence only to Alzheimer's disease as a cause of neural degeneration. Its precise pathogenesis remains a mystery despite the recognition decades ago of selective neuronal loss in the substantia nigra. Only rarely has a genetic component been found; most patients appear to be affected randomly. After onset, the disease progresses inexorably over years to extreme disability. Because the pathogenesis of PD is unknown, current therapy addresses the symptoms rather than the cause. The underlying disease continues to progress despite treatment and eventually becomes refractory to it, resulting in severe incapacitation. In this chapter, evidence for a possible pathogenetic process in PD is examined, arising from several different lines of investigation.

Parkinson's disease presents clinically with a classic triad of symptoms: bradykinesia, rigidity, and tremor. Perhaps the most crucial for diagnosis is bradykinesia, or paucity of movement. Patients with PD characteristically move less than normal individuals, with effects on spontaneous posture readjustment, blinking, and purposeful voluntary movements. Rigidity almost always accompanies the bradykinesia and involves an increase in tone independent of the rate of passive movement. In addition, PD patients at rest often have a slow, pill-rolling tremor. Association with an underlying tremor produces the cogwheel rigidity characteristic of PD. Parkinson's disease may also cause a characteristic abnormality of eye movement, with deficits in saccadic pursuit and reduced vertical gaze. Abnormalities of posture and gait are common. Patients with PD have trouble starting to walk and make repeated small movements referred to as *festination*. They also tend to freeze under certain circumstances and to fall backwards. A substantial proportion (as many as 20 to 30 percent) of PD patients also suffer from either depression or dementia.

Parkinson's disease results from degeneration of a subset of dopaminergic neurons in the brain stem, located in the substantia nigra pars compacta (SNc). In addition to expressing the transmitter dopamine, these cells also contain large amounts of melanin and are hence heavily pigmented, leading to the speculation that melanin may in some way contribute to their degeneration. However, careful morphometric analysis indicates preferential involvement of the ventral tier of substantia nigra cells, and this subpopulation contains less melanin than the dorsal tier of neurons. Interestingly, normal aging affects principally the dorsal tier, supporting a role for melanin in this process rather than in PD[1] [*see Figure 1*]. In addition, PD preferentially affects the lateral parts of the substantia nigra. Taken together, the observations suggest that PD differs in some fundamental way from the normal aging process. Further, a distinct group of midbrain dopamine neurons in the ventral tegmental area (VTA) adjacent to the substantia nigra resemble those of the substantia nigra and also contain dopamine, but they show much less involvement in PD, consistent with a preferential involvement of more lateral cell groups. To a lesser extent, PD also affects other monoamine cell populations, including the noradrenergic locus coeruleus and serotoninergic raphe nuclei.

These observations of selective vulnerability presumably indicate important features in the pathogenesis of PD. In addition to cell loss, PD is associated with characteristic intracytoplasmic inclusions, known as Lewy bodies, that contain intermediate filaments and ubiquitin. Although these have been described in other locations and disorders, the appearance of Lewy bodies in brain-stem monoamine nuclei is virtually pathognomonic of PD. The presence of Lewy bodies in the substantia nigra at postmortem examination

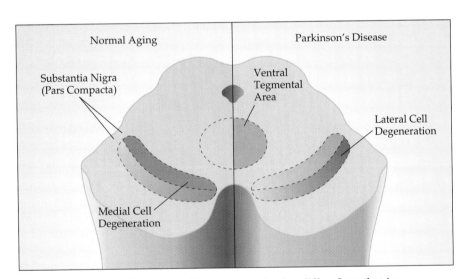

Figure 1 *Midbrain dopamine cell loss in normal aging differs from that in Parkinson's disease. Normal aging involves preferential loss of dopamine cells in the dorsal tier (dark blue) of the substantia nigra, pars compacta (SNc) relative to the ventral tier (gray to light blue). Within the ventral tier, medial cells degenerate to a greater extent than lateral cells. In contrast, Parkinson's disease involves preferential degeneration of the ventral tier (dark red), in a gradient from lateral to medial. The more medial ventral tegmental area (VTA) also contains dopamine neurons that show some involvement in Parkinson's disease but less than that of the substantia nigra.*

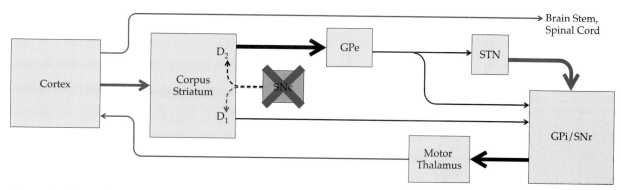

Figure 2 *Dopamine neurons in the substantia nigra pars compacta (SNc) project rostrally to GABAergic neurons in the corpus striatum. Stimulation of D1-like receptors on these GABAergic cells can be considered excitatory (blue arrows) and of D2-like receptors, inhibitory (black arrows). Cells with predominantly D1-like input then project through the direct pathway to the substantia nigra pars reticulata (SNr), where they provide inhibitory input. Striatal cells expressing primarily D2 receptors project through the indirect pathway to the globus pallidus pars externa (GPe), where they also provide inhibitory input. GABAergic GPe neurons then inhibit cells in the subthalamic nucleus (STN) that send excitatory connections to the globus pallidus pars interna (GPi), a cell group functionally analogous to the SNr. Thus, the striatal neurons inhibit cells in the SNr through the direct pathway and excite cells in the GPi through the indirect pathway. Both GPi and SNr cells then project to the motor thalamus, where they inhibit the generation of movement. Because dopamine excites the direct pathway and inhibits the indirect pathway, its loss in Parkinson's disease leads to reduced activity in the direct pathway and increased activity in the indirect pathway, with the net effect of increased activity in the SNr and GPi and reduced movement. Originally developed from this circuit, surgical lesion of the indirect pathway in the GPi effectively reduces symptoms in Parkinson's disease.*

of clinically unaffected individuals even suggests the existence of presymptomatic PD. Thus, Lewy bodies appear to develop early in PD and, like the selective cell loss, presumably reflect the basic pathogenetic mechanism.

Pathophysiology of Parkinson's Disease

Dopaminergic projections to the striatum form synapses on the major cell type found in this brain region, medium spiny neurons that use the inhibitory transmitter γ-aminobutyric acid (GABA). These GABAergic cells in turn project back to the midbrain through two apparently distinct pathways[2] [*see Figure 2*]. Neurons in the direct pathway express the neural peptides dynorphin and substance P and form inhibitory synapses directly onto cells in the pars reticulata of the substantia nigra (SNr). In contrast, striatal neurons in the indirect pathway contain enkephalin and project to cells in the pars externa of the globus pallidus (GPe). Neurons in the GPe also use the transmitter GABA and form inhibitory synapses onto cells in the subthalamic nucleus. Subthalamic nucleus cells use glutamate as a transmitter and hence make excitatory connections onto neurons in the pars interna of the globus pallidus (GPi), cells that are functionally similar to neurons in the SNr. Hence, the indirect pathway involves one excitatory and two inhibitory connections and so has an effect opposite that of the direct pathway, which involves only one inhibitory synapse. In addition, medium spiny neurons giving rise to the two striatal outflow pathways appear to contain different proportions of the two types of dopamine receptor. D1-like dopamine receptors (D1 and D5) act principally to increase cyclic adenosine monophosphate (cAMP) levels, whereas D2-like receptors (D2, D3, and D4) reduce cAMP levels and neural activity by activating K+ channels and inhibiting Ca2+ channels.[3] Striatal neurons projecting in the direct pathway express predominantly D1-like receptors that respond to dopamine by excitation. Medium spiny neurons in the indirect pathway express predominantly D2-like receptors that respond to dopamine by inhibition. Dopamine normally activates the direct pathway and inhibits the indirect pathway. As a result, activity in the direct pathway inhibits SNr, and inhibition of the indirect pathway also induced by dopamine has the same net effect on the GPi through one excitatory and two inhibitory synapses. SNr/GPi in turn inhibit nuclei of the thalamus involved in the generation of movement. Movement therefore requires disinhibition of SNr/GPi, and dopamine promotes this effect through both direct and indirect pathways. The deficiency in dopamine that occurs in PD results in insufficient activation of the direct pathway and excessive activation (through a loss of inhibition) of the indirect pathway. In this case, there is inadequate inhibition of the SNr/GPi, and these nuclei continue to inhibit the generation of movement in the thalamus. Although this scheme may undergo considerable revision, it suggests a pathophysiological mechanism responsible for hypokinetic disorders such as PD as well as one avenue for symptomatic intervention.

Treatment Approaches in Parkinson's Disease

In the early stages of PD, replacement of the lost dopamine ameliorates many of the symptoms of PD. Indeed, clinical response to the precursor levodopa (L-Dopa) helps to establish the diagnosis because a similar syndrome may result from loss of cells that receive the dopaminergic input, and dopamine replacement cannot activate this pathway in the absence of the postsynaptic cell. In addition, the treatment of early PD is so effective that it can be difficult to distinguish treated patients from unaffected individuals. Dopaminergic agonists such as bromocriptine can also be used to treat PD, but these agents are generally not as effective as L-Dopa, possibly because the dopamine derived from L-Dopa is stored within the remaining dopamine neurons and so undergoes release in a physiologically appropriate manner, whereas the direct stimulation of dopamine receptors by pharmacological agonists cannot be regulated. However, certain symptoms of PD may not respond to L-Dopa as dramatically as the bradykinesia does. In particular, gait and balance difficulties tend not to respond to L-Dopa, possibly because these symptoms reflect the loss of nondopaminergic cells such as the noradrenergic neurons of the locus coeruleus or the serotoninergic neurons of the raphe nuclei. Other drugs alleviate particular symptoms of PD; for example, anticholinergic agents are particularly effective in reducing rest tremor. Interference with dopamine metabolism by, for example, inhibition of catechol O-methyltransferase may also prove clinically useful. However, the symptoms of PD become increasingly difficult to control as the disease progresses. L-Dopa begins to wear off more quickly, possibly because it can no longer be stored in the diminishing number of residual dopaminergic terminals. This results in so-called off periods that require more closely spaced dosing. In addition, peak dopamine concentrations begin to produce excessive movements or dyskinesia, and patients can cycle rapidly between these two states, possibly because the exogenous dopamine can no longer be released in a regulated way by the remaining terminals.

A variety of approaches to manage end-stage PD are being developed. Dopamine agonists can ameliorate symptoms late in PD but lack the efficacy of L-Dopa and are generally used to reduce the amount of L-Dopa required at earlier stages in the disease. Surgical lesion of the hyperactive indirect striatal outflow pathway originally proposed by Mahlon deLong on the basis of the striatonigral pathways can also reduce symptoms quite effectively,[2] as can transplantation of fetal dopamine neurons[4] and, to a lesser extent, transplantation of adult adrenal medulla.[5] In addition, growth factors such as brain-derived neurotrophic factor (BDNF) and glial-derived neurotrophic factor (GDNF) promote cell survival and dopamine release.[6,7] Other, uncharacterized factors are even required for expression of the dopaminergic phenotype.[8] However, the underlying disorder continues to progress despite the symptomatic treatment used to date, indicating the need for a better understanding of the basic pathogenetic mechanism.

The nature of disease progression in PD provides important clues to the pathogenesis. Patients do not become symptomatic until they have lost most of the cells in the SNc, indicating the existence of a significant preclinical phase. Although normal aging also involves a loss of midbrain dopamine neurons, this appears to occur at a slower rate than in PD. Interestingly, extrapolation backward of the cell counts from patients with early PD suggests that the disease involves exponential cell loss.[1] The presence of microglia in the degenerating SNc at postmortem examination also supports an active process.[9] These observations suggest that PD involves a final common mechanism of progressive degeneration that may be triggered when the SNc cell number drops below a critical threshold; however, the initial events that commit to this pathway remain unknown [*see Figure 3*]. A diminished number of cells at birth, accelerated normal aging, or other acute events may predispose to PD, and we will eventually have to account for the triggering events as well as the final mechanism of degeneration. The existence of a preclinical phase also indicates the possibility for early intervention to prevent the development of symptoms. However, we will need to understand the pathogenesis of PD to identify those at risk in order to take advantage of this window of opportunity.

The MPTP Model of Parkinson's Disease

One exogenous toxin produces a syndrome indistinguishable in many respects from PD. Initially identified by William Langston in a group of young narcotics abusers with an acute neurologic syndrome, exposure to 1-methyl-4-phenyl-1,2,3,6-tetrahydropyridine (MPTP) produces bradykinesia and rigidity that, like idiopathic PD, responds well to treatment with L-Dopa.[10] Pathologically, MPTP toxicity produces massive cell loss with the same predilection for dopamine neurons, in particular SNc rather than VTA. It also affects other monoamine populations, such as the locus coeruleus and raphe nuclei, to approximately the same extent as PD. Further, MPTP toxicity can produce Lewy bodies. Other toxins can produce a parkinsonian syndrome, but these either affect dopaminergic targets in the

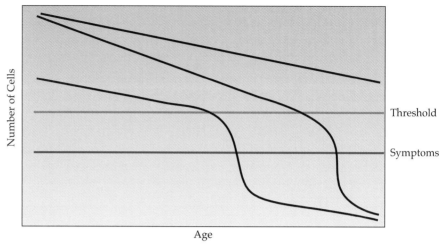

Figure 3   *The upper black line represents normal aging, with a gradual loss of midbrain dopamine cells throughout life that does not result in symptoms. Symptomatic Parkinson's disease appears to develop only when less than 20 to 50 percent of dopamine neurons in the substantia nigra remain (indicated by the blue line). In addition, extrapolation backward of cell counts from patients dying at different intervals after disease onset suggests rapid progression in the final stages (lower black line), possibly triggered by a decline in the number of cells below a critical threshold (gray line). However, the nature of progression in early stages remains unknown. Parkinson's disease may result from either an acceleration of the normal aging process, with a normal number of cells at birth, or a reduced number of cells at birth followed by a normal rate of cell loss. In either case, the individuals would reach the critical threshold at which the disease is triggered.*

striatum (such as carbon monoxide or manganese toxicity) or have a much less selective action on SNc neurons (such as 6-hydroxydopamine). In light of the unique similarities between MPTP toxicity and PD, the mechanism of MPTP damage has dramatically influenced thinking about the pathogenesis of PD, has helped in making numerous predictions about the idiopathic disorder, and even led to changes in clinical management.

Over the past 15 years, work by many investigators has elucidated particular features of MPTP toxicity [*see Figure 4*]. MPTP appears to penetrate the blood-brain barrier readily as a moderately lipophilic compound. Inside the brain but outside the vulnerable neurons, monoamine oxidase B (MAO-B) converts MPTP to the active metabolite *N*-methyl-4-phenylpyridinium (MPP+). A block of this conversion by inhibition of MAO-B with drugs such as pargyline and deprenyl completely prevents MPTP toxicity.[11,12] Plasma membrane monoamine transporters that normally terminate synaptic transmission by removing transmitter from the extracellular space then recognize and accumulate MPP+ as a result of its structural similarity to the normal monoamine substrates.[13] Like deprenyl, pharmacological inhibition of these reuptake systems also blocks MPTP toxicity. Glia and other neurons lack these transport activities, apparently accounting for the selective vulnerability of monoamine cells to MPTP toxicity. Inside the cell, MPP+ enters mitochondria and appears to produce injury by inhibiting complex I of the respiratory chain.[14,15]

The mechanism of MPTP toxicity leads to several predictions about the pathogenesis of PD. One is that the idiopathic disorder may also involve a disorder of mitochondrial function. Defects in complex I activity have been observed in brain tissue from affected patients[16,17] [*see Chapter 15*]. Surprisingly, similar defects have also been described in nonneural tissue from these patients, and defects in different respiratory chain activities have been reported in other neurodegenerative disorders, such as Alzheimer's disease.[18] Some data even suggest that PD may be transmitted maternally through mitochondrial inheritance.[19,20] However, mitochondrial disturbances accumulate with age, making it crucial to establish the specificity of these findings for PD. In addition, defects in respiratory function may simply accumulate in the course of PD rather than participate in pathogenesis as they do in MPTP toxicity. Nonetheless, even if it does not directly predispose to the disease through maternal inheritance, an acquired abnormality in mitochondrial function might help to propagate the original insult through reduced adenosine triphosphate (ATP) production and energy failure. In any case, the defects may at least provide a marker for the early diagnosis of PD and have certainly stimulated a very interesting new line of investigation.

Despite the striking similarities, the clinical syndrome of MPTP toxicity also differs in certain crucial respects from PD. First, MPTP toxicity evolves

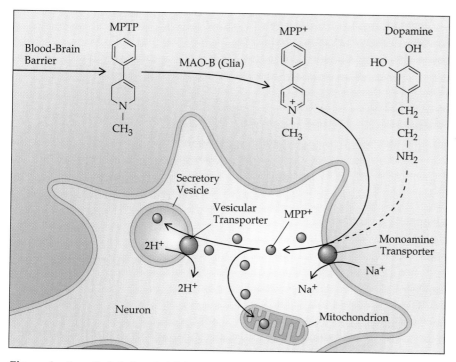

Figure 4 *1-methyl-4-phenyl-1,2,3,6-tetrahydropyridine (MPTP) readily penetrates the blood-brain barrier because of its lipophilic properties. Converted into the active metabolite N-methyl-4-phenylpyridinium (MPP+) by monoamine oxidase B (MAO-B), the toxin is accumulated by plasma membrane monoamine transporters in dopamine neurons. These transporters use the Na+ gradient across the plasma membrane to drive uptake and normally terminate the synaptic action of monoamines such as dopamine. Inside the cell, MPP+ enters mitochondria, where it appears to act by inhibiting complex I in the respiratory chain, resulting in energy failure and cell death. Alternatively, vesicular monoamine transport can sequester the toxin in secretory vesicles and so protect the cell. In contrast to the plasma membrane transporters, this vesicular transporter uses the proton electrochemical gradient across the vesicle membrane, exchanging two lumenal protons for one cytoplasmic molecule of monoamine or toxin.*

over an extremely short time, days to weeks rather than many years. Although low-grade exposure to an exogenous toxin might produce disease with this protracted time course, the inexorable, gradual progression of typical PD seems incompatible with this possibility. Alternatively, an acute toxic insult might reduce dopaminergic cell number below the threshold required to trigger the self-propagating degeneration yet remain above the number required to produce symptoms [*see Figure 3*]. Supporting this possibility, several individuals and animals exposed briefly to MPTP acquired a slowly progressive illness at a discrete interval after exposure.[21] Second, the epidemiology does not show a clustering of PD cases suggestive of toxic exposure. Third, the MPTP model also predicts that inhibition of MAO-B may protect against progression in idiopathic PD. An unknown exogenous toxin may, for example, require activation by MAO-B. Because this has important clinical implications, extensive drug trials have been under way for several years to assess a role for MAO-B in PD. Initial studies suggested mild slowing of disease progression by the selective MAO-B inhibitor deprenyl.[22] However, this drug also ameliorates the symptoms of PD by reducing the metabolism of dopamine, and this effect, rather than an effect on disease progression, appears to have accounted for most of the benefit observed.[23,24] These observations indicate that despite the importance of MPTP toxicity as a model for PD, it differs in crucial ways from the idiopathic disorder. Recent investigation has begun to suggest the nature of these differences, in one case by focusing on a poorly understood feature of MPTP toxicity itself.

Monoamine Transport in Brain

Plasma membrane uptake of MPP+ appears to account for the selective vulnerability of monoamine cells to MPTP, but discrepancies suggest an important role for other factors. Plasma membrane transport normally functions to remove transmitter from the synapse and uses the Na+ gradient across the plasma membrane to drive the cotransport of transmitter. In monoamines, it is the site of action, for example, of antidepressants and of cocaine. Studies have identified distinct transport mechanisms for dopamine, norepinephrine, and serotonin, and molecular cloning has shown that these three transporters belong to a large family of closely related neurotransmitter transport proteins.[25] Although the three proteins have similar substrate preference, they differ in sensitivity to drugs and in localization, with dopamine neurons expressing the dopamine transporter, locus coeruleus neurons the norepinephrine transporter, and raphe neurons the serotonin transporter. It is significant that all of these transporters recognize and transport MPP+, apparently accounting for the selective vulnerability of monoamine cells to MPTP toxicity.[26,27] However, only the SNc shows substantial cell loss after exposure to MPTP. Variation in the number or activity of plasma membrane monoamine transporters could account for the relative sparing of VTA, locus coeruleus, and raphe nuclei. Indeed, the dopamine transporter has a higher turnover number and may occur at higher levels on SNc than VTA cells. However, the norepinephrine transporter has a much higher affinity for MPP+ than the dopamine transporter. In addition, chromaffin cells in the adrenal medulla accumulate larger amounts of MPP+ than any other tissue and show little cell loss.[28,29] These observations indicate that although required for MPTP toxicity, plasma membrane uptake does not suffice to make a cell vulnerable to MPTP; therefore, other factors must have a role.

Molecular Models of Dopaminergic Cell Degeneration

In an effort to identify additional factors that influence sensitivity to MPTP, our group used a genetic strategy. Rat pheochromocytoma PC12 cells express a plasma membrane monoamine transporter and show susceptibility to MPP+. However, PC12 cells show toxicity only after exposure to relatively large concentrations of MPP+, consistent with their derivation from the adrenal medulla, a tissue resistant to MPTP toxicity. Surprisingly, a fibroblast cell line that lacks plasma membrane monoamine transport activity shows more sensitivity to MPP+ than do PC12 cells, suggesting that PC12 cells express a mechanism for resistance to the toxin. To identify this mechanism, we transferred genes from the MPP+-resistant PC12 cells into the sensitive fibroblasts and subjected the derived cells to selection with the toxin. A single fibroblast cell clone became resistant to MPP+.

The resistance of these cells disappeared after treatment with the drug reserpine,[30] which inhibits the transport of monoamine transmitters into synaptic vesicles. Like other classical transmitters, monoamines are synthesized in the cytoplasm and so require transport into secretory vesicles for release by regulated exocytosis. In contrast to plasma membrane neurotransmitter transporters, the vesicular transport of all classical neurotransmitters uses the proton electrochemical gradient generated across the vesicle membrane by a vacuolar H+-adenosine triphosphatase that pumps protons into the lumen of the vesicle.[31,32] In monoamines, the activity involves the exchange of two lumenal protons for one cytoplasmic monoamine. Reserpine inhibits vesicular monoamine transport and lowers blood pressure effectively, but the side effect of psychological depression has limited its clinical use. Indeed, the behavioral effects of reserpine originally gave rise to the now generally accepted hypothesis that monoamine level influences mood. However, in contrast to plasma membrane neurotransmitter transport, the proteins responsible for vesicular transport remained unknown. The ability of reserpine to reverse the resistance to MPP+ suggested that vesicular monoamine transport can package toxins as well as transmitters into secretory vesicles. Vesicular transport presumably lowers the cytoplasmic concentration of toxin, thereby limiting its access to the primary site of action in mitochondria [*see Figure 4*]. Although previous work had suggested that vesicular monoamine transporters (VMATs) could recognize MPP+ as a substrate, the genetic strategy described here provided the first evidence that the activity protected against the toxin. It also provided a way to identify the proteins responsible for this activity.

Through repeated selection in MPP+, we identified a single gene that confers resistance to the toxin.[33,34] The DNA sequence predicted a protein with 12 transmembrane domains, a structure typical for a transport protein. However, this sequence showed little similarity to other known transporters, including the plasma membrane neurotransmitter transporters. Biochemical analysis showed that it also encoded vesicular monoamine transport with all the functional characteristics described in classic studies. Thus, the sequence defined a novel family of proteins that mediate the packaging of neurotransmitters into synaptic vesicles. This family now includes two VMATs and a transporter for acetylcholine. VMAT1 occurs in the adrenal medulla and in scattered nonneural cells elsewhere, whereas VMAT2 occurs in monoamine neurons in the brain and peripheral nervous system. Although differences in activity may contribute to the selective vulnerability of midbrain dopamine neurons, both VMATs can protect against MPP+ toxicity. A detailed functional analysis also indicates that VMAT2, although expressed in the more vulnerable neurons, has a higher affinity for MPP+ and other monoamine substrates than VMAT1, the isoform expressed in the toxin-resistant adrenal medulla. Interestingly, the identification of these two isoforms also accounts for a clinical pharmacological observation: Unlike reserpine, which inhibits both VMATs, tetrabenazine inhibits only VMAT2 and thus has a predominantly central action that has made it more useful than reserpine in the treatment of movement disorders.[35]

Although the sequence of the VMATs has little similarity to other known transporters, it has weak but definite similarity to a class of bacterial proteins. These proteins confer resistance to a variety of antibiotics and appear to act by extruding them from bacteria, a process topologically equivalent to transport into secretory vesicles. In addition, the bacterial transport activities involve proton exchange and so resemble vesicular neurotransmitter transport in terms of bioenergetics as well. Further, reserpine inhibits one of the bacterial transporters in this family, providing pharmacological evidence of similarity. VMAT thus appears to have evolved from ancient detoxification systems. Along with the ability to protect against MPP+, this suggests that VMAT has two biologic roles, one in signaling and the other in neural protection. Indeed, VMAT has an extremely high affinity for its monoamine substrates, indicating the ability to maintain very low cytosolic concentrations. However, there is little epidemiological evidence for an exogenous toxin in PD.

Dopamine Toxicity in Parkinson's Disease

Role of Free Radicals

A substantial body of evidence implicates dopamine as an endogenous toxin in PD. Dopamine shows potent toxicity when applied to cultured neural cells.[36,37] Indeed, dopamine is more toxic to many cells than the neurotoxin MPP+. In contrast to glutamate toxicity, which occurs through the activation of specific glutamate receptors, dopamine toxicity is not receptor mediated. Rather, dopamine oxidizes to produce hydrogen peroxide. Catalase, which degrades hydrogen peroxide, protects against dopamine toxicity, indicating a primary role for hydrogen peroxide in the process. Hydrogen peroxide presumably acts as a free radical to damage cellular constituents, including lipid as well as protein. The question remains, however, whether this in vitro evidence of dopamine toxicity has relevance for idiopathic PD. Clinicians have been concerned about whether therapy with L-Dopa, the precursor of dopamine, may accelerate progression of the disease.

Considerable evidence now indicates that PD involves oxidative stress caused by an increase in free radicals[38-40] [*see also Chapter 15*]. Like hydrogen peroxide, the more highly reactive superoxide and hydroxyl radicals modify lipid and protein, damaging the cell. Indeed, postmortem brain tissue from patients with PD shows changes in lipid composition consistent with excess free radicals.[41] In addition, the antioxidant glutathione is reduced, and heavy metals, which promote the formation of free radicals, accumulate in PD.[42] Because these changes may simply reflect the cell loss that occurs in end-stage PD, it is particularly interesting that incidental Lewy body disease diagnosed strictly postmortem in individuals with no symptoms of PD also showed a decrease in the active form of glutathione, providing evidence of oxidative stress early in the course of the disease.[43]

Oxidative stress may derive from several sources. First, exogenous toxins can increase free radicals. In MPTP toxicity, energy failure caused by inhibition of respiration presumably interferes with the ability to protect against oxidative stress, leading to the accumulation of free radicals. Indeed, overexpression in transgenic mice of superoxide dismutase, the enzyme that metabolizes superoxide radicals to hydrogen peroxide, protects against MPTP toxicity.[44] In addition, the toxin 6-hydroxydopamine oxidizes very easily to produce free radicals. Second, oxidative stress may derive from an endogenous toxin. This may also involve either mitochondrial dysfunction, resulting in energy failure, or direct chemical oxidation. Dopamine oxidizes almost as readily as its 6-hydroxy derivative, indicating the potential for this normal transmitter to produce oxidative stress. Hence, the presence of oxidative stress in idiopathic PD is consistent with the role of dopamine as an endogenous neurotoxin.

Toxicity of Endogenous Amines

The action of dopamine as an endogenous toxin would help to account for the selective vulnerability of dopamine neurons. Norepinephrine and serotonin oxidize less readily than dopamine does, presumably accounting for their relative sparing in PD. Preferential involvement of large dopamine cells may reflect the increased dopamine synthesis required to support release at extensive projections. Differences in turnover may also contribute to the greater susceptibility of SNc cells than VTA cells. Paradoxically, the intrinsic toxicity of dopamine may also help to account for the selective vulnerability of dopamine neurons to exogenous toxins such as MPTP. A reduction in neural activity produced by glutamate receptor antagonists reduces MPTP toxicity, possibly by reducing energy expenditure in the face of respiratory inhibition. However, reduced activity would also decrease dopamine turnover and hence the exposure to its toxic breakdown products.

The intrinsic toxicity of monoamine transmitters suggests the existence of systems that normally protect against their toxicity. This focuses attention on more experimentally tractable factors, such as proteins, rather than the more elusive free radicals themselves. Mitochondria maintain energy reserves, enzymes such as superoxide dismutase metabolize free radicals, and a variety of enzymes maintain the level of the principal cellular antioxidant, reduced glutathione. Growth factors such as BDNF and GDNF reduce the toxicity of MPTP and may act by enhancing antioxidant defenses.[6,45] In addition, because dopamine is a normal substrate of the VMATs, these proteins may protect against dopamine as well as MPP+ toxicity. If so, this predicts that dopamine exerts its toxic action in the cytoplasm and that storage in secretory vesicles either protects against its oxidation or sequesters the transmitter in such a way that oxidation does not damage vital cell functions.

Effects of Amphetamine

The action of amphetamine supports a role for cytoplasmic dopamine in neural degeneration. Amphetamines increase the synaptic level of monoamines by inducing flux reversal through the plasma membrane monoamine transporters from the cytoplasm into the extracellular space.[46] However, the normal amounts of monoamine in the cytoplasm are low (0.1 to 1 μm in the case of dopamine), requiring efflux from the vesicular pool for subsequent efflux into the synapse. Vesicular efflux may involve exchange of luminal monoamine for cytoplasmic amphetamine, followed by the nonspecific diffusion of lipophilic amphetamine back to the cytoplasm for another round of exchange. Alternatively, amphetamines may simply act as weak bases to disrupt the proton electrochemical gradient ($\Delta\mu_{H+}$) across the vesicle membrane that normally drives transmitter packaging.[47] Previous studies have shown that in addition to driving forward transport, $\Delta\mu_{H+}$ also prevents efflux, presumably by blocking reverse flux through the VMATs. Amphetamines dissipate $\Delta\mu_{H+}$, allowing flux reversal and massively elevating cytoplasmic concentrations of monoamine that then enter the synapse.

Amphetamines also produce considerable toxicity to monoamine neurons.[48] Although this may result from dissipation of the plasma membrane Na^+ gradient by flux reversal of the Na^+-coupled plasma membrane transporters, it may also derive from increased cytoplasmic monoamine and oxidative stress. Indeed, dopamine neurons in culture show hot spots of oxidative stress after exposure to amphetamine that presumably reflect a local increase in transmitter concentration adjacent to the sites of vesicular storage and its subsequent oxidation.[49] Amphetamines may produce degeneration by removing one of the mechanisms that protect against the intrinsic toxicity of dopamine, vesicular monoamine transport. Further, the intrinsic toxicity of dopamine may account for the apparent evolution of vesicular neurotransmitter transport from bacterial detoxification systems indicated by the sequence of the proteins.

Although vesicular monoamine transport may protect against endogenous as well as exogenous toxins, a primary role for dopamine toxicity in PD predicts several important discrepancies from the MPTP model. First, the susceptible cells synthesize dopamine. For this reason, the toxicity of dopamine does not depend to the same extent on plasma membrane uptake as an exogenous toxin such as MPP^+. Inhibition of plasma membrane monoamine transport may thus have less benefit for idiopathic PD than for MPTP toxicity. Second, dopamine oxidation would be expected to injure more than simply the mitochondria, and incidental Lewy body disease shows little evidence of respiratory dysfunction, suggesting that this may occur only late in the disorder.[43] Third, dopamine neurons contain the A isoform of monoamine oxidase (MAO-A) rather than MAO-B,[50] the enzyme implicated in the activation of MPTP. If dopamine acts as an endogenous toxin, inhibition of MAO-B would be expected to have little effect on disease progression, consistent with the results of clinical trials.[23,24] Further, the metabolism of dopamine may have an important role in PD.

Oxidation of Dopamine

Dopamine can oxidize through two distinct mechanisms. Chemical auto-oxidation occurs rapidly in the absence of any protein catalyst to generate hydrogen peroxide and quinones. Using cultured neural cells to study the toxicity of added dopamine, catalase strongly protects, indicating an important role for the released hydrogen peroxide in this model of degeneration. However, the quinone product of auto-oxidized dopamine is itself highly reactive, generating additional oxidative stress [*see Figure 5*]. On the other hand, enzymatic oxidation of dopamine by MAO eventually yields the inert metabolite 3,4-dihydroxyphenylacetaldehyde as well as hydrogen peroxide. Although hydrogen peroxide is therefore produced by both chemical and enzymatic degradation of dopamine, the enzymatic route yields an inert metabolite, whereas chemical oxidation produces highly reactive quinones. Thus, enzymatic oxidation in dopamine neurons by MAO-A may have a protective role. In contrast to MPTP toxicity, in which inhibition of MAO-B protects, inhibition of the A isoform possibly more relevant to PD may accelerate disease progression, prompting great caution in the use of such agents in the idiopathic disorder.

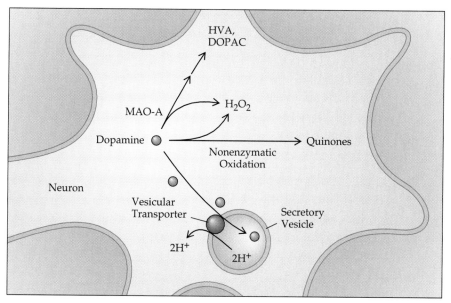

Figure 5 *The cell can dispose of dopamine in at least three basic ways. First, enzymatic oxidation by monoamine oxidase A (MAO-A) produces hydrogen peroxide (H_2O_2) and, together with a variety of other metabolic enzymes, inert metabolites such as 3,4-dihydroxyphenylacetaldehyde (DOPAC) and homovanillic acid (HVA). Nonenzymatic oxidation of dopamine occurs in a variety of aqueous solutions and also produces hydrogen peroxide. However, it also produces quinones that are highly reactive and may well contribute to oxidative stress. Third, vesicular storage sequesters dopamine into secretory vesicles for regulated release. The extremely high affinity of vesicular transport proteins suggests that they regulate the cytoplasmic concentration of transmitter and hence its availability for the first two oxidative reactions.*

To understand the role of oxidative stress and dopamine toxicity in idiopathic PD, we need to develop improved animal models that more closely resemble idiopathic PD. However, the hypothesis itself suggests a number of ways in which to reproduce the human disorder. In particular, disruption of the mechanisms that normally protect against dopamine toxicity may unleash the intrinsic toxicity of this transmitter and so produce the characteristic cell loss within a time frame more consistent with a slowly progressive neurodegenerative disease. The ability to create a model of PD by disrupting these mechanisms would strongly support a role for dopamine toxicity in the human disorder. It would also provide a way to identify clues for presymptomatic diagnosis in humans. In addition, it would enable direct testing of a wide range of drugs that might slow or even arrest the usually inexorable progression of PD.

Inherited and Acquired Contributions to Parkinson's Disease

Why does PD affect certain individuals and not others? A primary role for dopamine toxicity suggests several explanations. In particular, it suggests an imbalance between the production of dopamine and the mechanisms that normally protect against its toxicity. Increased production may result from increased neural activity or a defect in the negative feedback system that reduces tyrosine hydroxylase activity and hence biosynthesis when cytoplasmic levels rise. Reduced protection may result from a decrease in vesicular storage, in enzymatic oxidation of dopamine and its metabolites, in the levels of reduced glutathione, or in the diverse cellular processes that maintain these mechanisms (such as through the production of adenosine triphosphate by mitochondrial oxidative phosphorylation). These two basic disease mechanisms may also have either an inherited or an acquired cause.

Epidemiological studies have suggested a small role for inheritance in PD. Despite the original reports of high familial incidence apparently attributable to the inclusion of cases with essential tremor rather than true PD, studies indicate that classic PD has a low familial incidence. Although it remains controversial, some investigators report an incidence not significantly higher than the incidence of PD in the general population.[51] Parkinsonian syndromes with earlier onset may have a higher familial incidence but do not always represent classic PD. The initial twin studies also did not support a large inherited contribution to PD, with similar levels of concordance in both monozygotic and dizygotic twins.[52-54] However, the level of concordance was extremely low in both cases, making a comparison between them unreliable as an indication of the hereditary component. More recently, imaging with fluorine-18-fluorodopa (^{18}F-Dopa) has enabled a more careful analysis of the unaffected twins.[55,56] Because PD has a long presymptomatic phase, a clinically unaffected twin may have the disorder but simply not yet have developed symptoms. Interestingly, reexamination of the twins by imaging with ^{18}F-Dopa showed a high incidence of abnormality in the clinically unaffected twin. However, the incidence of imaging abnormality was very high in both the monozygotic and dizygotic pairs, making a genetic predisposition again somewhat less likely. On the other hand, imaging of families with PD has also shown abnormalities in clinically un-affected individuals, suggesting that low penetrance may also account for the apparently low familial incidence of PD.

A small number of families exhibit PD with mendelian inheritance.[57,58] Despite initial questions about the relationship of these clinical syndromes to classic PD, pathological as well as clinical investigation has indicated a strong similarity, if not identity, to the usual idiopathic disorder. Because the apparently low penetrance of PD observed so far in twins and in families further suggests that sporadic PD may also have a major genetic component, identification of the genes responsible for familial PD may have relevance for sporadic PD as well. A report of linkage to the long arm of chromosome 4 in one large kindred led to a search for candidate genes.[59] A mutation responsible for disease in this family occurs in the α-synuclein gene,[60] which encodes a protein found in nerve terminals but whose function remains unknown.[61] Interestingly, a cleavage product of α-synuclein accumulates in the plaques of Alzheimer's disease, suggesting that it may also occur in the Lewy bodies of Parkinson's disease.[62] Unfortunately, rodents normally contain the pathogenic residue at this site,[60] suggesting that the creation of a mouse model may be rather difficult. Speculations on the relationship of α-synuclein to oxidative stress are tantalizing targets for further investigation; perhaps an oxidizing environment promotes the abnormal processing or deposition of the α-synuclein cleavage product.

Additional genes may predispose to PD. Indeed, debrisoquine hydroxylase, an enzyme implicated in the metabolism of MPTP, has two functionally distinct alleles. The low-metabolizing allele has shown some association with idiopathic PD in certain populations,[63-65] and polymorphisms in other genes may similarly predispose to PD. In addition to providing information about pathogenesis, these genetic loci will help to identify individuals at risk for the disease in whom the appropriate measures may prevent its development. Genetics can also implicate mechanisms related to experience rather than inheritance that may operate in sporadic PD.

Neural Activity in Degeneration

Neural activity appears to have an important role in neural degeneration. Activity induces the release of glutamate at excitatory synapses and, like dopamine, the normal transmitter glutamate can produce significant toxicity. Unlike dopamine, however, glutamate acts through specific receptors that mediate Ca^{2+} entry.[66,67] Interestingly, glutamate receptor antagonists appear to reduce the toxicity of MPTP, implicating activity in the action of even this exogenous toxin.[68] Activity also appears to have a role in the toxicity of amphetamines.[69] Thus, sporadic PD may also involve an increase in neural activity that could trigger degeneration by increasing the turnover of dopamine. Indeed, differences in activity could account for the apparently sporadic nature of the disease and its low penetrance even in familial forms. A particular mouse mutant further supports a role for neural activity in PD.

The mouse mutation *weaver* involves a developmental defect in the migration of cerebellar neurons. Although the mice have impaired survival, careful anatomic analysis has indicated that midbrain dopamine neurons

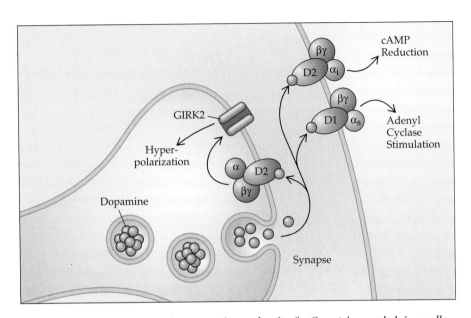

Figure 6 *The weaver mutation occurs in a subunit of a G protein-coupled, inwardly rectifying K+ (GIRK2) channel. Although the exact receptors physiologically responsible for activating GIRK2 remain unclear, GIRK2 is expressed in dopamine neurons and could mediate the effects of D2-like autoreceptors on the cell body, dendrites, and terminals of these neurons. In this situation, dopamine released by the cell would activate D2-like dopamine receptors on the same cell, initiating the dissociation of a heterotrimeric G protein, activation of GIRK2, and hence hyperpolarization. This negative feedback system provides a brake on dopamine release. However, the mutation does not appear to involve a simple loss of function but rather a toxic gain of function in which reduced ionic selectivity results in Na+ influx and depolarization, as well as constitutive activation independent of signaling through G proteins. Postsynaptic D1 and D2 receptors couple through α_s to stimulate adenyl cyclase and α_i to reduce cyclic adenosine monophosphate.*

also degenerate relatively selectively within weeks after birth. The pattern of degeneration strongly resembles that observed in PD, with preferential involvement of SNc cells and relative sparing of VTA and other monoamine neurons.[70] Interestingly, the mutation in *weaver* occurs in one subunit of a G protein–coupled, inwardly rectifying K+ channel (GIRK2).[71] Subunits of a heterotrimeric G protein activate these channels in response to stimulation of particular cell surface receptors [*see Figure 6*]. The cell then hyperpolarizes, reducing its activity. As anticipated, midbrain dopamine neurons express GIRK2. Although the receptors to which GIRK2 couples in these cells remain unclear, they may mediate the response of D2-like dopamine autoreceptors expressed on the same cells that provide negative feedback for released dopamine. The *weaver* mutation reduces the cation selectivity of GIRK2, allowing Na+ as well as K+ to flow through the channel.[72,73] In addition, it reduces the coupling to G proteins, indicating constitutive activity. Together with the increased Na+ conductance, this suggests that the mutant channel acts to depolarize the cell and increase firing. Similar to glutamate toxicity, increased neural activity may thus lead to cell death. However, additional data suggest that the mutation may result in either a loss or a gain of function in this particular K+ channel,[74] raising the possibility that a defect in the dopamine feedback pathway might increase dopamine production and oxidative stress and so be responsible for the degeneration. *Weaver* thus provides a model for dopamine cell degeneration that appears to involve excessive activation. Although GIRK2 occurs in several neuronal populations, *weaver* produces degeneration only in midbrain dopamine cells,[75] again supporting a role for dopamine toxicity in their selective vulnerability.

Conclusion

Previous work has emphasized the role of small molecules such as MPTP and free radicals in the pathogenesis of PD. Indeed, MPTP produces a syndrome with remarkable similarity to idiopathic PD. However, there is little evidence for an exogenous toxin in PD, and the idiopathic disorder shows several important differences from MPTP toxicity. In addition to the marked difference in time course, the progression of PD does not clearly respond to inhibition of monoamine oxidase. The analysis of tissue from patients with PD suggests that increased oxidative stress has an important role in this mysterious degeneration. However, free radicals are extremely difficult to study directly. In contrast, the proteins that influence MPTP toxicity and oxidative stress are much more experimentally tractable.

Identified as a protein that confers resistance to MPP+ toxicity, the vesicular monoamine transporter normally functions to package monoamine transmitters into secretory vesicles. However, its ability to protect against MPP+ and its sequence similarity to bacterial antibiotic resistance proteins suggest that it also has a primary function in neural protection. In the absence of evidence for an exogenous toxin in PD, and in light of the considerable intrinsic toxicity of dopamine, vesicular monoamine transport may thus protect against the toxicity of this normal neurotransmitter. Indeed, the action of dopamine as an endogenous toxin would help to account for the oxidative stress observed in PD as well as the selective vulnerability of dopaminergic neurons and the toxicity of amphetamines.

Multiple mechanisms may normally protect dopamine cells from degeneration in PD. In addition to vesicular monoamine transport, enzymes involved in dopamine metabolism, such as MAO-A, and proteins that regulate neuronal excitability, such as the GIRK2 subunit of a K+ channel, appear to influence susceptibility to degeneration. Proteins involved in free-radical metabolism and the production of energy that drives many of these processes may also have a role. Genetic studies in familial and perhaps even sporadic PD may suggest additional mechanisms and identify the proteins involved. However, unlike the concept of oxidative stress, the identification of these proteins could suggest ways to test directly the primary role of dopamine toxicity.

Interference with the genes that protect against dopamine toxicity should unleash the intrinsic toxicity of this normal transmitter and produce a model of PD. This would create an animal model relevant for idiopathic human PD and help to establish the role of dopamine as an endogenous toxin. It would also help to characterize the pathogenetic mechanism of PD using a combination of anatomic, biochemical, pharmacological, and genetic approaches. In addition, the model would greatly accelerate drug testing and focus attention on a limited set of therapeutic agents that can be tested more effectively in human trials.

References

1. Fearnley JM, Lees AJ: Aging and Parkinson's disease: substantia nigra regional selectivity. *Brain* 114:2283, 1991

2. DeLong MR: Primate models of movement disorders of basal ganglia origin. *Trends Neurosci* 13:281, 1990

3. Gingrich JA, Caron MG: Recent advances in the molecular biology of dopamine receptors. *Annu Rev Neurosci* 16:299, 1993

4. Olanow CW, Kordower JH, Freeman TB: Fetal nigral transplantation as a therapy for Parkinson's disease. *Trends Neurosci* 19:102, 1996

5. Diamond SG, Markham CH, Rand RW, et al: Four-year follow-up of adrenal-to-brain transplants in Parkinson's disease. *Arch Neurol* 51:559, 1994

6. Hyman C, Hofer M, Barde Y-A, et al: BDNF is a neurotrophic factor for dopaminergic neurons of the substantia nigra. *Nature* 350:230, 1991

7. Gash DM, Zhang Z, Ovadia A, et al: Functional recovery in parkinsonian monkeys treated with GDNF. *Nature* 380:252, 1996

8. Zetterstrom RH, Solomin L, Jansson L, et al: Dopamine neuron agenesis in Nurr1-deficient mice. *Science* 276:248, 1997

9. McGeer PL, Itagaki S, Akiyama H, et al: Rate of cell death in parkinsonism indicates active neuropathological process. *Ann Neurol* 24:574, 1988

10. Langston JW, Ballard P, Tetrud JW, et al: Chronic parkinsonism in humans due to a product of meperidine analog synthesis. *Science* 219:979, 1983

11. Langston JW, Irwin I, Langston EG, et al: Pargyline prevents MPTP-induced parkinsonism in primates. *Science* 225:1480, 1984

12. Markey S, Johannessen J, Chiueh C, et al: Intraneuronal generation of a pyridinium metabolite may cause drug-induced parkinsonism. *Nature* 311:464, 1984

13. Snyder S, D'Amato R, Nye J, et al: Selective uptake of MPP+ by dopamine neurons is required for MPTP toxicity: studies in brain synaptosomes and PC12 cells. *MPTP: A Neurotoxin Producing a Parkinsonian Syndrome.* Markey SP, Castagnoli N, Trevor AJ, Eds. Academic Press, New York, 1986, p 191

14. Ramsay RR, Singer TP: Energy-dependent uptake of n-methyl-4-phenylpyridinium, the neurotoxic metabolite of 1-methyl-4-phenyl-1,2,3,6-tetrahydropyridine, by mitochondria. *J Biol Chem* 261:7585, 1986

15. Krueger MJ, Singer TP, Casida JE, et al: Evidence that the blockade of mitochondrial respiration by the neurotoxin 1-methyl-4-phenylpyridinium (MPP+) involves binding at the same site as the respiratory inhibitor, rotenone. *Biochem Biophys Res Commun* 169:123, 1990

16. Schapira AH: Nuclear and mitochondrial genetics in Parkinson's disease. *J Med Genet* 32:411, 1995

17. Shoffner JM: Mitochondrial defects in basal ganglia diseases. *Curr Opin Neurol* 8:474, 1995

18. Parker WDJ, Mahr NJ, Filley CM, et al: Reduced platelet cytochrome c oxidase activity in Alzheimer's disease. *Neurology* 44:1086, 1994

19. Swerdlow RH, Parks JK, Miller SW, et al: Origin and functional consequences of the complex I defect in Parkinson's disease. *Ann Neurol* 40:663, 1996

20. Wooten GF, Currie LJ, Bennett JP, et al: Maternal inheritance in Parkinson's disease. *Ann Neurol* 41:265, 1997

21. Vingerhoets FJ, Snow BJ, Tetrud JW, et al: Positron emission tomographic evidence for progression of human MPTP-induced dopaminergic lesions. *Ann Neurol* 36:765, 1994

22. Parkinson Study Group: Effect of deprenyl on the progression of disability in early Parkinson's disease. *N Engl J Med* 321:1364, 1989

23. Parkinson Study Group: Effects of tocopherol and deprenyl on the progression of disability in early Parkinson's disease. *N Engl J Med* 328:176, 1993

24. Parkinson Study Group: Impact of deprenyl and tocopherol treatment on Parkinson's disease in DATATOP subjects not requiring levodopa. *Ann Neurol* 39:29, 1996

25. Amara SG, Kuhar MJ: Neurotransmitter transporters: recent progress. *Annu Rev Neurosci* 16:73, 1993

26. Buck KJ, Amara SG: Chimeric dopamine-norepinephrine transporters delineate structural domains influencing selectivity for catecholamines and 1-methyl-4-phenylpyridinium. *Proc Natl Acad Sci USA* 91:12584, 1994

27. Wall SC, Gu H, Rudnick G: Biogenic amine flux mediated by cloned transporters stably expressed in cultured cell lines: amphetamine specificity for inhibition and efflux. *Mol Pharm* 47:544, 1995

28. Johannessen JN, Chiueh CC, Burns RS, et al: Differences in the metabolism of MPTP in the rodent and primate parallel differences in sensitivity to its neurotoxic effects. *Life Sci* 36:219, 1985

29. Reinhard JF Jr, Carmichael SW, Daniels AJ: Mechanisms of toxicity and cellular resistance to 1-methyl-4-phenyl-1,2,3,6-tetrahydropyridine and 1-methyl-4-phenylpyridinium in adrenomedullary chromaffin cell cultures. *J Neurochem* 55:311, 1990

30. Liu Y, Roghani A, Edwards RH: Gene transfer of a reserpine-sensitive mechanism of resistance to MPP+. *Proc Natl Acad Sci USA* 89:9074, 1992

31. Schuldiner S, Shirvan A, Linial M: Vesicular neurotransmitter transporters: from bacteria to humans. *Physiol Rev* 75:369, 1995

32. Liu Y, Edwards RH: The role of vesicular transport proteins in synaptic transmission and neural degeneration. *Ann Rev Neurosci* 20:125, 1997

33. Liu Y, Peter D, Roghani A, et al: A cDNA that supresses MPP+ toxicity encodes a vesicular amine transporter. *Cell* 70:539, 1992

34. Erickson JD, Eiden LE, Hoffman BJ: Expression cloning of a reserpine-sensitive vesicular monoamine transporter. *Proc Natl Acad Sci USA* 89:10993, 1992

35. Peter D, Jimenez J, Liu Y, et al: The chromaffin granule and synaptic vesicle amine transporters differ in substrate recognition and sensitivity to inhibitors. *J Biol Chem* 269:7231, 1994

36. Rosenberg PA: Catecholamine toxicity in cerebral cortex in dissociated cell culture. *J Neurosci* 8:2887, 1988

37. Michel PP, Hefti F: Toxicity of 6-hydroxydopamine and dopamine for dopaminergic neurons in culture. *J Neurosci Res* 26:428, 1990

38. Fahn S, Cohen G: The oxidant stress hypothesis in Parkinson's disease: evidence supporting it. *Ann Neurol* 32:804, 1992

39. Beal MF: Aging, energy and oxidative stress in neurodegenerative diseases. *Ann Neurol* 38:357, 1995

40. Simonian NA, Coyle JT: Oxidative stress in neurodegenerative diseases. *Annu Rev Pharm Toxicol* 36:83, 1996

41. Dexter DT, Carter CJ, Wells FR, et al: Basal lipid peroxidation in substantia nigra is increased in Parkinson's disease. *J Neurochem* 52:1830, 1989

42. Riederer P, Sofic E, Rausch W-D, et al: Transition metals, ferritin, glutathione, and ascorbic acid in Parkinson brains. *J Neurochem* 52:515, 1989

43. Jenner P, Dexter DT, Sian J, et al: Oxidative stress as a cause of nigral cell death in Parkinson's disease and incidental Lewy body disease. *Ann Neurol* 32(suppl):82, 1992

44. Przedborski S, Kostic V, Jackson-Lewis V, et al: Transgenic mice with increased Cu/Zn-superoxide dismutase activity are resistant to N-methyl-4-phenyl-1,2,3,6-tetrahydropyridine-induced neurotoxicity. *J Neurosci* 12:1658, 1992

45. Tomac A, Lindqvist E, Lin LF, et al: Protection and repair of the nigrostriatal dopaminergic system by GDNF in vivo. *Nature* 373:335, 1995

46. Giros B, Jaber M, Jones SR, et al: Hyperlocomotion and indifference to cocaine and amphetamine in mice lacking the dopamine transporter. *Nature* 379:606, 1996

47. Sulzer D, Rayport S: Amphetamine and other psychostimulants reduce pH gradients in midbrain dopaminergic neurons and chromaffin granules: a mechanism of action. *Neuron* 5:797, 1990

48. Steele TD, McCann UD, Ricaurte GA: 3,4-Methylenedioxymethamphetamine (MDMA, ecstasy): pharmacology and toxicology in animals and humans. *Addiction* 89:539, 1994

49. Cubells JF, Rayport S, Rajendran G, et al: Methamphetamine neurotoxicity involves vacuolation of endocytic organelles and dopamine-dependent intracellular oxidative stress. *J Neurosci* 14:2260, 1994

50. Westlund KN, Denney RM, Rose RM, et al: Localization of distinct monoamine oxidase A and monoamine oxidase B cell populations in human brainstem. *Neuroscience* 25:439, 1988

51. Duvoisin RC, Gearing FR, Schweitzer MD, et al: A family study of parkinsonism. *Progress in Neurogenetics*. Barbeau A, Brunette JR, Eds. Excerpta Medica, Amsterdam. p 492, 1969

52. Duvoisin RC, Eldridge R, Williams A, et al: Twin study of Parkinson's disease. *Neurology* 31:77, 1981

53. Marsden CD: Parkinson's disease in twins. *J Neurol Neurosurg Psychiatry* 50:105, 1987

54. Vieregge P, Shiffke KA, Friedrich HJ, et al: Parkinson's disease in twins. *Neurology* 42:1453, 1992

55. Sawle GV, Wroe SJ, Lees AJ, et al: The identification of presymptomatic parkinsonism: clinical and [18F]dopa positron emission tomography studies in an Irish kindred. *Ann Neurol* 32:609, 1992

56. Holthoff VA, Vieregge P, Kessler J, et al: Discordant twins with Parkinson's disease: positron emission tomography and early signs of impaired cognitive circuits. *Ann Neurol* 36:176, 1994

57. Waters CH, Miller CA: Autosomal dominant Lewy body parkinsonism in a four-generation family. *Ann Neurol* 35:59, 1994

58. Golbe LI, DiIorio G, Sanges G, et al: Clinical genetic analysis of Parkinson's disease in the Contursi kindred. *Ann Neurol* 40:767, 1996

59. Polymeropoulos MH, Higgins JJ, Golbe LI, et al: Mapping of a gene for Parkinson's disease to chromosome 4q21-q23. *Science* 274:1197, 1996

60. Polymeropoulos MH, Lavedan C, Leroy E, et al: Mutation in the α-synuclein gene identified in families with Parkinson's disease. *Science* 276:2045, 1997

61. Maroteaux L, Campanelli JT, Scheller RH: Synuclein: a neuron-specific protein localized to the nucleus and presynaptic nerve terminal. *J Neurosci* 8:2804, 1988

62. Iwai A, Masliah E, Yoshimoto M, et al: The precursor protein of non-Aβ component of Alzheimer's disease amyloid is a presynaptic protein of the central nervous system. *Neuron* 14:467, 1995

63. Plante-Bordeneuve V, Davis MB, Maraganore DM, et al: Debrisoquine hydroxylase gene polymorphism in familial Parkinson's disease. *J Neurol Neurosurg Psychiatry* 57:911, 1994

64. Sandy MS, Armstrong M, Tanner CM, et al: CYP2D6 allelic frequencies in young-onset Parkinson's disease. *Neurology* 47:225, 1996

65. Gasser T, Muller-Myhsok B, Supala A, et al: The CYP2D6B allele is not overrepresented in a population of German patients with idiopathic Parkinson's disease. *J Neurol Neurosurg Psychiatry* 61:518, 1996

66. Choi DW: Glutamate neurotoxicity and diseases of the nervous system. *Neuron* 1:623, 1988

67. Dugan LL, Choi DW: Excitotoxicity, free radicals and cell membrane changes. *Ann Neurol* 35:S17, 1994

68. Turski L, Bressler K, Rettig K-J, et al: Protection of substantia nigra from MPP+ neurotoxicity by N-methyl-d-aspartate antagonists. *Nature* 349:414, 1991

69. Sonsalla PK, Nicklas WJ, Heikkila RE: Role for excitatory amino acids in methamphetamine-induced nigrostriatal dopaminergic toxicity. *Science* 243:398, 1989

70. Roffler-Tarlov S, Pugatch D, Graybiel AM: Patterns of cell and fiber vulnerability in the mesostriatal system of the mutant mouse weaver. II. High affinity uptake sites for dopamine. *J Neurosci* 10:1990

71. Patil N, Cox DR, Bhat D, et al: A potassium channel mutation in weaver mice implicates membrane excitability in granule cell differentiation. *Nat Genet* 11:126, 1995

72. Slesinger PA, Patil N, Liao YJ, et al: Functional effects of the mouse weaver mutation on G protein-gated inwardly rectifying K+ channels. *Neuron* 16:321, 1996

73. Navarro B, Kennedy ME, Velimirovic B, et al: Nonselective and G betagamma-insensitive weaver K+ channels. *Science* 272:1950, 1996

74. Surmeier DJ, Mermelstein PG, Goldowitz D: The weaver mutation of GIRK2 results in a loss of inwardly rectifying K+ current in cerebellar granule cells. *Proc Natl Acad Sci USA* 93:11191, 1996

75. Liao YJ, Jan YN, Jan LY: Heteromultimerization of G protein-gated inwardly rectifying K+ channel proteins GIRK1 and GIRK2 and their altered expression in weaver brain. *J Neurosci* 16:7137, 1996

Acknowledgments

Figures 1, 2, and 4 through 6   Seward Hung.

Figure 3   Marcia Kammerer.

CHAPTER 10

Molecular Neurology of Prion Diseases

Stanley B. Prusiner, M.D.

The story of prions is a fascinating saga in the annals of biomedical science. My own studies began with a patient dying of Creutzfeldt-Jakob disease (CJD) in the fall of 1972. Only a few months earlier, I had begun a residency in neurology and was most impressed by a disease process that could kill my patient in two months by destroying her brain while her body remained unaware of this process. No febrile response, no leucocytosis or pleocytosis, no humoral immune response, yet I was told that she was infected with a so-called *slow virus*. The term slow virus had been coined by Bjorn Sigurdsson in 1954 when he was working in Iceland on scrapie and visna of sheep.[1] Five years later, William Hadlow suggested that kuru, a disease of New Guinea highlanders, was similar to scrapie and thus was also caused by a slow virus.[2] Seven more years passed before the transmissibility of kuru was established when Carleton Gajdusek and colleagues passaged the disease to chimpanzees inoculated intracerebrally.[3] Just as Hadlow had made the intellectual leap between scrapie and kuru, Igor Klatzo and colleagues made a similar connection between kuru and CJD.[4] In both instances, these neuropathologists were struck by the similarities in light microscopic pathology of the central nervous system that kuru exhibited with scrapie or CJD. In 1968, the transmission of CJD to chimpanzees after intracerebral inoculation was reported.[5]

My own fascination with CJD quickly shifted to scrapie once I learned of the fascinating radiobiological data that Tikvah Alper and colleagues had collected on the scrapie agent as measured by bioassays in mice.[6-8] This shift from scrapie to CJD would eventually prove to be artificial because the same disease process causes both of these disorders. The scrapie agent had been found to be extremely resistant to inactivation by ultraviolet and ionizing radiation, as was later shown for the CJD agent.[9] It seemed to me that the most intriguing question was the chemical nature of the scrapie agent. Alper's data had evoked a torrent of hypotheses concerning the composition of the scrapie agent. Suggestions ranged from small DNA viruses to membrane fragments to polysaccharides to proteins, the last of which eventually proved to be correct.[10-15]

My own work focused on the development of improved bioassays that would allow purification of the infectious pathogen causing scrapie and CJD. After some initial studies on the sedimentation properties of scrapie infectivity in mouse spleens and brains, I concluded that hydrophobic interactions were responsible for the nonideal physical behavior of the scrapie particle.[16,17] Indeed, the scrapie agent presented a biochemical nightmare: infectivity was often spread through many fractions generated by attempts at purification. These findings demanded new approaches and better assays.

In the pages that follow, our current understanding of the prion particles that cause scrapie and CJD is described. Indeed, prions are unprecedented infectious pathogens that cause (by an entirely novel mechanism) a group of invariably fatal, neurodegenerative diseases. Prion diseases may present as genetic, infectious, or sporadic disorders, all of which involve modification of the prion protein (PrP). CJD generally presents as a progressive dementia, whereas scrapie of sheep and bovine spongiform encephalopathy (BSE) are generally manifested as ataxic illnesses.[18]

Prions are devoid of nucleic acid and seem to be composed exclusively of a modified isoform of PrP designated PrP^{Sc}. The normal, cellular PrP, denoted PrP^{C}, is converted into PrP^{Sc} through a process whereby some of its α-helical structure is partially converted into β-sheet.[19] This structural transition is ac-

---

Table 1 Evidence That Prions Are Composed of PrP^{Sc} Molecules and Devoid of Nucleic Acid

PrP^{Sc} and scrapie infectivity copurify by biochemical and immunologic procedures.

The unusual properties of PrP^{Sc} mimic those of prions. Many different procedures that modify or hydrolyze PrP^{Sc} inactivate prions.

Levels of PrP^{Sc} are directly proportional to prion titers. Nondenatured PrP^{Sc} has not been separated from scrapie infectivity.

No evidence for a virus has been found either in terms of a viruslike particle or a nucleic acid genome.

Accumulation of PrP^{Sc} causes the pathology of prion diseases, including PrP amyloid plaques that are pathognomonic.

PrP gene mutations are genetically linked to inherited prion disease and cause formation of PrP^{Sc}.

Overexpression of PrP^{C} increases the rate of PrP^{Sc} formation, which shortens the incubation time. Knockout of the PrP gene eliminates the substrate necessary for PrP^{Sc} formation and prevents both prion disease and replication.

Species variations in the PrP sequence are responsible, at least in part, for the species barrier that is found when prions are passaged from one host to another as manifested by prolonged incubation times.

PrP^{Sc} preferentially binds to homologous PrP^{C}, resulting in formation of the same nascent PrP^{Sc} and prion infectivity.

Chimeric PrP genes change susceptibility to prions from different species and support production of artificial prions with novel properties that are not found in nature.

Prion diversity is enciphered within the conformation of PrP^{Sc}. Strains can be generated by passage through hosts with different PrP genes. Prion strains are maintained by PrP^{C}/PrP^{Sc} interactions.

Human prions from fCJD(E200K) and fatal familial insomnia patients impart different properties to chimeric MHu2M PrP in transgenic mice, which provides a mechanism for strain propagation.

companied by profound changes in the physicochemical properties of the PrP. Whereas PrP^C is soluble in nondenaturing detergents, PrP^{Sc} is not.[20] PrP^C is readily digested by proteases, whereas PrP^{Sc} is partially resistant.[21] The species of a particular prion is encoded by the sequence of the chromosomal PrP gene of the mammals in which it last replicated. In contrast to pathogens with a nucleic acid genome, prions encipher strain-specific properties in the tertiary structure of PrP^{Sc}. Transgenetic studies argue that PrP^{Sc} acts as a template upon which PrP^C is refolded into a nascent PrP^{Sc} molecule through a process facilitated by another protein (provisionally designated protein X).

Studies of prions have wide implications ranging from basic principles of protein conformation to the development of effective therapies for prion diseases. In this chapter, the structural biology of prion proteins as well as the genetics and molecular neurology of prion diseases are reviewed. How information is enciphered within the infectious prion particle is described. Whereas both the primary structure and posttranslational chemical modifications determine the tertiary structure of PrP^C, the conformation of PrP^C is modified by PrP^{Sc} as it is refolded into a nascent molecule of PrP^{Sc}.[22] The hallmark common to all of the prion diseases, whether sporadic, dominantly inherited, or acquired by infection, is that they involve the aberrant metabolism of the prion protein.[23]

The Prion Concept

The prion diseases were long thought to be caused by slow-acting viruses. These diseases were often referred to as slow virus diseases, transmissible spongiform encephalopathies, or unconventional viral diseases.[1,24] Considerable effort was expended searching for the scrapie virus, yet no discovery was made either of a viruslike particle or a genome composed of RNA or DNA.

A greatly improved bioassay made possible the development of an effective purification protocol to enrich fractions for scrapie infectivity.[25] The titers of scrapie agent in partially purified fractions from infected Syrian hamster (SHa) brain were found to decrease when exposed to procedures that modify or hydrolyze proteins.[26] Conversely, scrapie infectivity was unaltered by procedures that selectively modify or hydrolyze nucleic acids. The inability to demonstrate any dependence of infectivity on a polynucleotide[26] was in agreement with earlier studies reporting the extreme resistance of infectivity to UV irradiation at 254 nm.[7] The best current working definition of a prion is a *pro*teinaceous *in*fectious particle that lacks nucleic acid distinguishing it from both viroids and viruses.[26]

Whereas no single finding has proved the existence of prions, the remarkable convergence of experimental data accumulated over the past two decades has built a convincing edifice [*see Table 1*]. In contrast, many of these experimental results demand that if the scrapie virus exists, then PrP^{Sc} should be an integral component of this virus [*see Table 2*].

Discovery of the Prion Protein

A protein discovered by the enrichment of fractions from SHa brain for scrapie infectivity[27] is the protease-resistant core of PrP^{Sc}, denoted PrP 27-

---

Table 2   Characteristics of the Scrapie Virus

The virus is tightly bound to PrP^{Sc} or possesses a coat protein that shares antigenic sites and physical properties with those of PrP^{Sc}, which explains copurification.

Structural properties of the virus are the same as PrP^{Sc}; procedures that modify PrP^{Sc} inactivate the virus.

The viral genome is protected from procedures modifying nucleic acid by PrP^{Sc}.

No nucleic acid > 50 nucleotides is found in purified preparations; it has unusual properties and might be small and thus difficult to detect.

The virus is hidden by PrP^{Sc} and thus not found.

PrP amyloid plaques and spongiform degeneration result from induction of PrP^{Sc} induced by the virus.

The virus uses PrP^C as a receptor; the virus has higher affinity for mutant than wt receptor.

At least two different ubiquitous viruses must exist.

Species specificity and artificial prions with new host range demand that PrP^C or PrP^{Sc} be tightly bound to the virus.

PrP^{Sc} is a cofactor for the virus, which is necessary for infectivity.

30, with an apparent molecular weight (M_r) of 27 to 30 kd.[28,29] Copurification of PrP 27-30 and scrapie infectivity show that the physicochemical properties as well as the antigenicity of these two entities are similar[30] [*see Table 1*]. In terms of the virus hypothesis, it is necessary to postulate that the virus has a coat protein highly homologous with PrP or that the virus binds tightly to PrP^{Sc}. In either case, the PrP-like coat proteins or the PrP-^{Sc}/virus complexes must display properties indistinguishable from PrP^{Sc} alone [*see Table 2*].

Determination of a single, unique sequence for the NH$_2$-terminus of PrP 27-30 permitted the synthesis of isocoding mixtures of oligonucleotides that were subsequently used to identify incomplete PrP complimentary DNA (cDNA) clones from hamster[21] and mouse (Mo).[31] cDNA clones encoding the entire open reading frames (ORF) of SHa and Mo PrP were eventually recovered.[29,32]

Prion Protein Isoforms

The finding that PrP messenger RNA (mRNA) levels are similar in normal, uninfected, and scrapie-infected tissues caused skepticism as to whether PrP 27-30 was related to the infectious prion particle.[31] Further biochemical analysis of the prion protein revealed a protease-sensitive protein that is soluble in nondenaturing detergents, which was designated PrP^C. [20,21]

PrP^C and PrP^{Sc} have the same covalent structure, and each consists of 209 amino acids [*see Figure 1*]. The N-terminal sequencing, the deduced amino acid sequences from PrP cDNA, and immunoblotting studies argue that PrP 27-30 is a truncated protein of about 142 residues that is derived from PrP^{Sc} by limited proteolysis of the N-terminus.[28,29,32]

In general, approximately 10^5 PrP^{Sc} molecules correspond to one median infectious dose (ID$_{50}$) unit of prions. We believe that PrP^{Sc} is best defined as the abnormal isoform of PrP that stimulates conversion of PrP^C into nascent PrP^{Sc}, which then accumulates to cause disease. Although resistance to limited proteolysis has proved to be a convenient tool for detecting PrP^{Sc}, it has not been found to be useful in all studies.[33,34] Some investigators equate protease resistance with PrPSc and use the term PrP-res.[35]

Although insolubility as well as protease resistance were used in our initial studies to differentiate PrPSc from PrPC,[20] subsequent investigations have shown that these properties are only surrogate markers, as are high β-sheet content and polymerization into amyloid.[36-42] Only when these surrogate markers are present are they useful; certainly, their absence does not establish sterility for prion infectivity. PrPSc is usually not detected by Western immunoblotting if less than 10^5 ID$_{50}$ units/ml of prions are present in a sample [*see Figure 1a*][43]; furthermore, PrPSc from different species may exhibit different degrees of protease resistance.

In our experience, the method of sample preparation from scrapie-infected brain also influences the sensitivity of PrPSc immunodetection, in part because PrPSc is not uniformly distributed in the brain.[44,45] Some experiments in which PrPSc detection proved difficult in partially purified preparations[46,47] were repeated with crude homogenates in which PrPSc was readily measured.[48,49]

Cell Biology of PrP^Sc Formation

In scrapie-infected cells, PrPC molecules destined to become PrPSc exit to the cell surface prior to conversion into PrPSc.[50-52] Like other glycosyl-phosphatidyl inositol (GPI)–anchored proteins, PrPC appears to reenter the cell through a subcellular compartment bounded by cholesterol-rich, detergent-

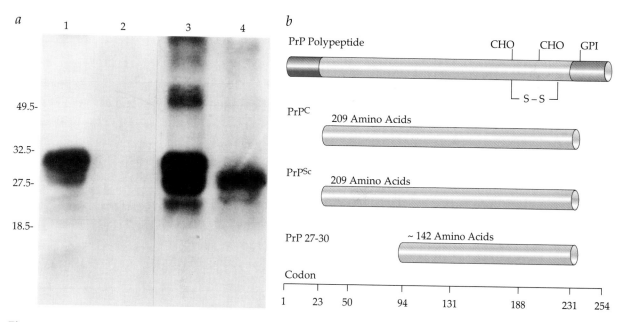

Figure 1   *Prion protein isoforms. (a) Western immunoblot of brain homogenates from uninfected (lanes 1 and 2) and prion-infected (lanes 3 and 4) Syrian hamsters (SHa). Samples in lanes 2 and 4 were digested with 50 μg/ml of proteinase K for 30 minutes at 37° C. PrPC in lanes 2 and 4 was completely hydrolyzed under these conditions, whereas approximately 67 amino acids were digested from the NH$_2$-terminus of PrPSc to generate PrP 27-30. After polyacrylamide gel electrophoresis and electrotransfer, the blot was developed with anti-PrP R073 polyclonal rabbit antiserum. Molecular size markers are in kilodaltons. (b) Bar diagram of SHaPrP, which consists of 254 amino acids. After processing of the NH$_2$- and COOH-terminals, both PrPC and PrPSc consist of 209 residues. After limited proteolysis, the NH$_2$-terminus of PrPSc is truncated to form PrP 27-30, which is composed of approximately 142 amino acids.*

insoluble membranes that might be caveolae or early endosomes.[53-55] Within this cholesterol-rich, nonacidic compartment, GPI-anchored PrP^C can be either converted into PrP^{Sc} or partially degraded.[53] PrP^{Sc} is trimmed at the N-terminus in an acidic compartment in scrapie-infected cultured cells to form PrP 27-30.[56] In contrast, N-terminal trimming of PrP^{Sc} is minimal in brain, where little PrP 27-30 is found.[38]

Rodent Models of Prion Disease

Mice and hamsters are commonly used in experimental studies of prion disease. The shortest incubation times are achieved with intracerebral inoculation of homologous prions; under these conditions, all of the animals develop prion disease within a narrow interval for a particular dose [*see Figure 2*]. The term *homologous prion* indicates that the PrP gene of the donor in which the prion was previously passaged has the same sequence as that of the recipient. When the PrP sequence of the donor differs from that of the recipient, the incubation time is prolonged, the length of the incubation becomes quite variable, and often many of inoculated animals do not acquire disease.[57-60] This phenomenon is often called the prion species barrier.[10]

Kinetics of Prion Replication and Incubation Times

When the titer of prions reaches a critical threshold level, the animals show signs of neurologic dysfunction [*see Figure 2*]. The time from inoculation to reach this threshold level of prions (incubation period) can be modified by (1) the dose of prions, (2) the level of PrP^C expression, and (3) the strain of prion. The higher the dose of prions, the shorter the time required to reach the threshold level for PrP^{Sc} at which signs of illness appear. The higher the level of PrP^C, the more rapid the accumulation of PrP^{Sc} and, hence, the shorter the incubation time. At present, the mechanism by which strains modify the incubation time is not understood.

The most reliable clinical signs of prion disease in rodents consist of (1) limb and truncal ataxia, (2) rigidity of the tail, (3) forelimb flexion instead of extension when suspended by the tail, and (4) difficulty righting from a supine position. The diagnosis of probable prion disease requires progression of at least two of these signs over a period of 10 to 20 days; in addition, histopathologic examination is necessary in a portion of the animals. Spongiform degeneration and reactive astrocytic gliosis are obligatory features of prion disease. Nonobligatory neuropathologic features include PrP amyloid plaques and protease-resistant PrP^{Sc} on histoblotting.

A recent study purports to have separated prion infectivity in mice from PrP^{Sc}.[61] In this study, bovine prions were inoculated into mice, and between 300 and 700 days later, the mice acquired an illness characterized by hindlimb paralysis, tremors, hypersensitivity to stimulation, apathy, and a hunched posture. All of the mice were said to have prion disease based on this constellation of clinical signs, yet only about 50 percent showed the neuropathological changes of prion disease, that is, spongiform degeneration and reactive astrocytic gliosis. In mice with these neuropathological changes PrP^{Sc} was found by immunoblotting, whereas the mice lacking these changes did not have detectable PrP^{Sc}. Brain extracts prepared from mice

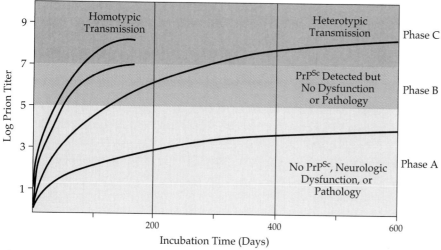

Figure 2 *Kinetics of experimental prion infection in rodents. Phases A and B are preclinical, whereas phase C is characterized by clinical signs of neurologic dysfunction. In phase A, only prion infectivity is detectable, and the titer is generally less than 10^5 ID_{50} units/ml of homogenate. In phase B, the titer is typically 10^5 to 10^7 ID_{50} units/ml, which is sufficiently high for PrP^Sc detection by immunoblotting; some neuropathological changes can also be seen in this phase. In phase C, the titer is greater than 10^7 ID_{50} units/ml, PrP^Sc is readily detected, frank neuropathology is seen, and clinical signs are obvious. The curves at the left depict transmission of prions between homologous animals, and the curves at the right are for heterologous transmission (i.e., crossing the species barrier), in which there is much more variation.*

with spongiform degeneration, reactive astrocytic gliosis, and PrP^Sc transmitted disease to recipient mice in approximately 150 days, whereas extracts from mice without these features required more than 250 days for transmission to recipient mice. Such results are consistent with the interpretation that the mice lacking spongiform degeneration, reactive astrocytic gliosis, and PrP^Sc carry low titers of prions. The clinical illness of the mice with low titers of prions was not attributable to prion disease, a contention supported by the lack of spongiform degeneration and reactive astrocytic gliosis. This argument is further supported by the prolonged incubation times that were found when the brains of the mice were subjected to bioassay. The apparent absence of PrP^Sc was expected because the prion titers were low; PrP^Sc could not be detected by the relatively insensitive immunoassay that was used.[43]

Why these investigators chose to equate seemingly apoptotic neurons in mice lacking spongiform degeneration and gliosis with prion disease is unclear. By doing so they were forced to conclude that the mice had a new form of prion disease, in which PrP^Sc could not be found.[61] As noted above, the absence of detectable PrP^Sc would be expected because the mice did not have bona fide prion disease and the titers of prions were low based on the prolonged incubation times for transmission to recipient animals. In fact, the neuropathological changes they describe can be attributed to artifacts from immersion fixation in formalin, including the hyperchromatic Purkinje cells (so-called spiky artifact) and the pseudoapoptotic appearance of the Purkinje cell nucleus by electron microscopy.

It is noteworthy that the lack of protease-resistant PrP has been reported in ill Tg mice expressing PrP with the P102L mutation of Gerstmann-Sträussler-Scheinker disease (GSS) at high levels.[62] When these GSS mice develop progressive neurologic dysfunction, they lack protease-resistant PrP but exhibit spongiform degeneration, astrocytic gliosis, and PrP amyloid plaques. This constellation of findings leaves no doubt about the diagnosis. Whereas prion disease was transmitted to Tg mice expressing the mutant transgene at low levels by inoculation of extracts from the foregoing Tg mice or from patients with GSS, these ill, recipient Tg mice likewise did not exhibit protease-resistant PrP.[33,34,59,62] The recipient Tg mice did not acquire prion disease spontaneously but showed spongiform degeneration, astrocytic gliosis, and PrP amyloid plaques after inoculation.

PrP Gene

The entire open reading frame of all known mammalian and avian PrP genes resides within a single exon,[29,63-65] which eliminates the possibility that PrP^{Sc} arises from alternative RNA splicing.[29,63,66] The two exons of the SHaPrP gene are separated by a 10 kb intron: exon 1 encodes a portion of the 5' untranslated leader sequence, whereas exon 2 encodes the ORF and 3' untranslated region.[29] Recently, a low-abundance SHa PrP mRNA containing an additional small exon in the 5' untranslated region was discovered that is encoded by the SHa PrP gene.[67] Mapping of PrP genes to the short arm of human (Hu) chromosome 20 and to the homologous region of mouse (Mo) chromosome 2 argues for the existence of PrP genes prior to the speciation of mammals.[68,69]

Expression of the PrP Gene

Although PrP mRNA is constitutively expressed in the brains of adult animals,[21,31] it is highly regulated during development. In the septum, levels of PrP mRNA and choline acetyltransferase were found to increase in parallel during development.[70] In other brain regions, PrP gene expression occurred at an earlier age. In situ hybridization studies show that the highest levels of PrP mRNA are found in neurons.[71]

Since no antibodies are currently available that distinguish PrP^{C} from PrP^{Sc} and vice versa, PrP^{C} is generally measured in tissues from uninfected control animals in which no PrP^{Sc} is found. PrP^{Sc} must be measured in tissues of infected animals after PrP^{C} has been hydrolyzed by digestion with a proteolytic enzyme. Immunostaining of PrP^{C} in the SHa brain is most intense in the stratum radiatum and stratum oriens of the CA1 region of the hippocampus and is virtually absent from the granule cell layer of the dentate gyrus and the pyramidal cell layer throughout Ammon's horn. PrP^{Sc} staining was minimal in these regions, which were intensely stained for PrP^{C}. A similar relationship between PrP^{C} and PrP^{Sc} was found in the amygdala. In contrast, PrP^{Sc} accumulated in the medial habenular nucleus, the medial septal nuclei, and the diagonal band of Broca; these areas were virtually devoid of PrP^{C}. In the white matter, bundles of myelinated axons contained PrP^{Sc} but were devoid of PrP^{C}. These findings suggest that prions are transported along axons, which is consistent with earlier findings in which scrapie infectivity migrated in a pattern consistent with retrograde transport.[48,72,73]

Overexpression of wtPrP Transgenes

Transgenic mice expressing different levels of the SHaPrP transgene Tg(SHaPrP) showed after inoculation with SHa prions an abrogation of the species barrier, resulting in abbreviated incubation times. Incubation time after inoculation with SHa prions was inversely proportional to the level of SHaPrP^C in the brains of Tg(SHaPrP) mice.[74] Bioassays of brain extracts from clinically ill Tg(SHaPrP) mice inoculated with Mo prions revealed that only Mo prions but no SHa prions were produced. Conversely, inoculation of Tg(SHaPrP) mice with SHa prions led to the synthesis of only SHa prions.

In these transgenetic studies, we discovered that uninoculated older mice harboring high copy numbers of wild type (wt)PrP transgenes derived from Syrian hamsters, sheep, and mice spontaneously exhibited truncal ataxia, hind-limb paralysis, and tremors.[75] These mice exhibited a profound necrotizing myopathy involving skeletal muscle, a demyelinating polyneuropathy, and focal vacuolation of the CNS.

PrP-Deficient Mice

Ablation of the PrP gene (*Prnp*$^{0/0}$) in mice (often called knockouts) did not affect development of these animals.[76,77] In fact, they generally remained healthy at almost two years, except for one report in which the Purkinje cell loss was accompanied by ataxia beginning at about 70 weeks of age.[78] Although brain slices from *Prnp*$^{0/0}$ mice were reported to show altered synaptic behavior,[79,80] these results could not be confirmed by other investigators.[81,82]

Although *Prnp*$^{0/0}$ mice are resistant to prions Sha,[82a,83] some investigators reported measurable titers of prions many weeks after inoculation of *Prnp*$^{0/0}$ mice with mouse-passaged prions.[84,85] Such findings have been cited by some investigators as indicating that prion infectivity replicates in the absence of PrP gene expression.[61,86,87]

Prnp$^{0/0}$ mice crossed with Tg(SHaPrP) mice are rendered susceptible to SHa prions but remain resistant to Mo prions.[83,84] Because the absence of PrP^C expression does not provoke disease, scrapie and other prion diseases likely are a consequence of PrP^{Sc} accumulation rather than an inhibition of PrP^C function.[76] Such an interpretation is consistent with the dominant inheritance of familial prion diseases by a so-called toxic gain of function.

Mice heterozygous (*Prnp*$^{0/+}$) for ablation of the PrP gene had prolonged incubation times when inoculated with Mo prions, showing signs of neurologic dysfunction 400 to 460 days after inoculation.[83,88] These findings are in accord with studies on Tg(SHaPrP) mice in which increased SHaPrP expression was accompanied by diminished incubation times.[74]

Species Variations in the PrP Sequence and Processing

PrP is posttranslationally processed to remove a 22 amino acid NH$_2$-terminal signal peptide. At the COOH-terminus, 23 residues are also removed during the addition of a GPI moiety that anchors the protein to the cell membrane.[89] Also contributing to the mass of the protein are two large oligosaccharides linked to Asn side chains with multiple structures that have been shown to be complex and diverse.[90] Although many species variants of PrP have now been sequenced,[91] only the chicken sequence is greatly different from the human sequence.[65,92] The alignment of the translated sequences from more than 40 PrP genes shows a striking degree of conservation among the mammalian sequences and is suggestive of the retention of some important function through evolution [*see Figure 3a*].

The NH_2-terminal domain of mammalian PrP contains five copies of a P(H/Q)GGG(G)WGQ octarepeat sequence, occasionally more, as in the case of one sequenced bovine allele that has six copies.[93,94] These repeats are remarkably conserved among species, implying a functionally important role. The chicken sequence contains a different repeat, PGYP(H/Q)N.[65,92] Although insertions of extra repeats have been found in patients with familial disease, naturally occurring deletions of single octarepeats do not appear to cause disease, and deletion of all these repeats does not prevent PrP^C from undergoing a conformational transition into PrP^{Sc}.[95,96]

The other region of notable conservation is in the sequence following the COOH-terminus of the last octarepeat. Here an unusual Gly- and Ala-rich region from A113 to Y128 is found [*see Figure 3a*]. Although no differences between species have been found in this part of the sequence, a single point mutation A117V is linked to GSS.[97] The conservation of structure suggests an important role in the function of PrP^C; in addition, this region is likely to be important in the conversion of PrP^C into PrP^{Sc}.

Structures of PrP Isoforms

A search for posttranslational chemical modifications that might explain the differences in the properties of PrP^C and PrP^{Sc} utilized mass spectrometry and Edman sequencing. No modifications differentiating PrP^C from PrP^{Sc} were found.[98] These observations forced consideration of the possibility that a conformational change distinguishes the two PrP isoforms.

Molecular Modeling of PrP Isoforms

In comparing the amino acid sequences of one avian and 11 mammalian PrPs, structure prediction algorithms identified four regions of putative secondary structure. Using a neural network algorithm, these four regions were predicted to be α-helices, which were designated H1, H2, H3, and H4.[99,100] When synthetic peptides corresponding to each of these domains were produced, the H1, H3, and H4 peptides readily adopted a β-sheet conformation in aqueous buffer.[99] Once we learned that PrP^C has a high α-helical content, three-dimensional models of PrP^C and PrP^{Sc} were developed [*see Figure 4*].[100,101]

Optical Spectroscopy of PrP Isoforms

PrP^C and PrP^{Sc} were examined by Fourier transform infrared (FTIR) and circular dichroism (CD) spectroscopy. These studies demonstrated that PrP^C has a high α-helix content (42 percent) and a low β-sheet content (three percent).[19] In contrast, the β-sheet content of PrP^{Sc} was 43 percent and the α-helix 30 percent.[102]

We and others have suggested that the conversion of PrP^C into PrP^{Sc} may proceed through an intermediate designated PrP*.[103,104] Possible intermediates have been identified by FTIR and CD spectroscopy.[105,106] Whether the conformation of PrP^C is altered or the molecule is simply protected from proteolysis by binding to PrP^{Sc} could not be established.

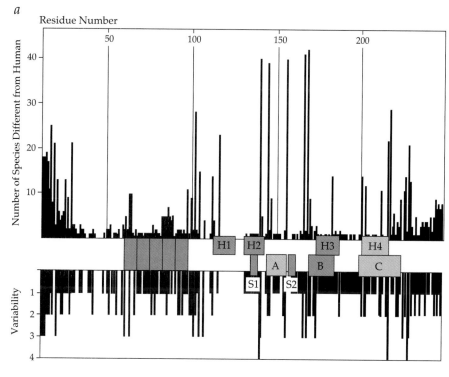

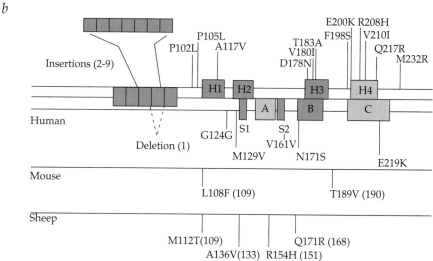

Figure 3 Species variations and mutations of the prion protein gene. (a) Species variations. The x-axis represents the human PrP sequence, with the five octarepeats and H1 through H4 shown as well as the three α-helices A, B, and C and the two β-strands S1 and S2. Vertical bars above the axis indicate the number of species whose sequence differs from the human sequence at each position. Below the axis, the length of the bars indicates the number of alternative amino acids at each position. (b) Mutations causing inherited human prion disease and polymorphisms in human, mouse, and sheep. Above the line of the human sequence are mutations that cause prion disease. Below the lines are polymorphisms, some but not all of which are known to influence the onset as well as the phenotype of disease.

Using a similar protocol, synthetic PrP peptides were also found to produce protease-resistant PrP[C].[107,108]

NMR Structural Studies of Recombinant PrP

Nuclear magnetic resonance (NMR) studies provide evidence supporting these molecular models. A recombinant protein (rPrP) of 142 residues corresponding to PrP 27-30 was expressed in *Escherichia coli* and purified.[109] The structure of rPrP, solved by multidimensional heteronuclear NMR [*see Figure 5*] and compared with two shorter PrP fragments,[110,111] shows that H1 and H2 regions, particularly H1, can form α-helices.[112] Like that of a smaller fragment, Mo PrP, composed of 111 residues,[111] the structure of rPrP is composed of three α-helices and a small, two-stranded, antiparallel β-sheet.[112]

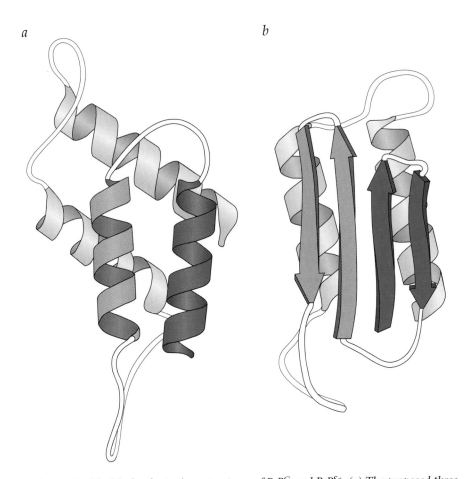

a   *b*

Figure 4   *Models for the tertiary structures of PrP^C and PrP^Sc. (a) The proposed three-dimensional structure of PrP^C. Helix 1 is shown in dark blue and helix 2 in light blue. Helices 1 and 2 are converted into β-sheet structures during the formation of PrP^Sc. The H₂ to H₃ loop corresponding to the S_{2b} to H₃ loop in PrP^Sc is shown at the bottom of the illustration. (b) The proposed three-dimensional structure of PrP^Sc. This structure correlates best with genetic data on residues involved in species barrier. It contains a four-strand, mixed β-sheet with two α-helices packed against one face of the sheet. Strands 1a and 1b (red) correspond to the helix 1 in PrP^C; strands 2a and 2b (green) correspond to the helix 2. Helices 3 and 4 in this model remain unchanged from the PrP^C model. The S_{2b} to H₃ loop (bottom right) connecting the β-sheet and helix 3 is implicated in the species barrier. This conformationally flexible loop could come into contact with PrP^C during the formation of the PrP^Sc-PrP^C complex. Therefore, the specific molecular recognition during prion replication might involve both the β-sheet as the primary binding site and the S_{2b} to H₃ loop as an additional site for interaction.*

NMR spectra reveal multiple signals for some nuclei, indicating that the NH₂-terminus and spatially proximate regions assume multiple discrete conformers. This conformational heterogeneity may be the basis for transformation of PrP^C into PrP^Sc.

Prion Species Barrier and Protein X

As noted previously, the prolonged incubation time and the apparent resistance of some animals to the inoculated prions from another species is often called the prion species barrier.[10] The species barrier concept is of practical importance in assessing the risk for humans of developing CJD after consumption of scrapie-infected lamb or BSE-infected beef.[113] The passage of prions between species is usually a stochastic process involving a randomly determined sequence of events that results in prolonged incubation times during the first passage into a new host.[10] On subsequent passage in a homologous host, the incubation time shortens to that recorded for all subsequent passages. Prions synthesized de novo reflect the sequence of the host PrP gene and not that of the PrPSc molecules in the inoculum.[114]

Attempts to abrogate the prion species barrier between humans and mice were initially unsuccessful. Mice expressing HuPrP transgenes did not develop signs of CNS dysfunction more frequently or rapidly than non-Tg controls.[57] Subsequently, success in breaking the species barrier between humans and mice came with mice expressing chimeric PrP transgenes derived from Hu and Mo PrP. HuPrP differs from MoPrP at 28 of 254 positions,[115] whereas chimeric MHu2MPrP differs at only nine residues. The mice expressing the MHu2M transgene acquire disease after injection with human prions and exhibit abbreviated incubation times of approximately 200 days.[57]

We believe that prion propagation involves the formation of a complex between PrPSc and the homotypic substrate PrPC.[74] However, such propagation may require the participation of other proteins, such as chaperones, which may be necessary to catalyze the conformational changes for the formation of PrPSc.[19] Notably, efficient transmission of HuCJD prions to Tg(HuPrP)/Prnp[0/0] mice was obtained when the endogenous MoPrP gene was inactivated, suggesting that MoPrPC competes with HuPrPC for binding to a cellular component.[57] In contrast, the sensitivity of Tg(MHu2M) mice to HuCJD prions was only minimally affected by the expression of MoPrPC. One explanation for the difference in susceptibility of Tg(MHu2M) and Tg(HuPrP) mice to Hu prions in mice is that MoPrPC binds more readily to

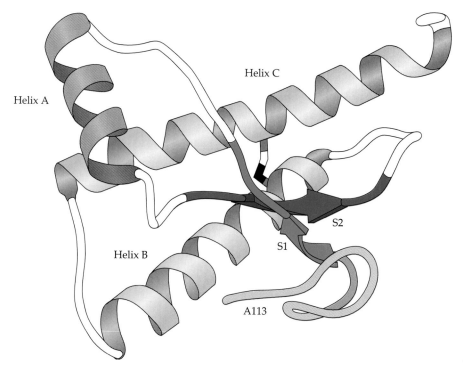

Figure 5 *Nuclear magnetic resonance (NMR) structure of a 142-residue recombinant PrP (rPrP) corresponding to the sequence of PrP 27-30 that is found in infectious prion preparations. The rPrP was produced in Escherichia coli, purified after denaturation and subsequently refolded into a predominantly α-helical form. Isotopically labeled amino acids were incorporated into rPrP during fermentation of the E. coli and used for determination of the three-dimensional structure by NMR.*

mouse chaperones that catalyze the refolding of PrP^C into PrP^{Sc} than to HuPrP^C; thus MoPrP^C inhibits conversion of HuPrP^{Sc} into PrP^{Sc} but has little effect on the conversion of chimeric MHu2M PrP^C into MHu2M PrP^{Sc}.

Because the conversion of PrP^C into PrP^{Sc} involves a profound conformational change, the participation of one or more molecular chaperones seems likely. We have referred to such chaperones as protein X. No direct identification of a molecular chaperone involved in prion formation in mammalian cells has been accomplished to date.

Spectrum of Human Prion Diseases

The human prion diseases present as infectious, genetic, and sporadic disorders.[116] Such an unprecedented spectrum of diseases demands a new mechanism: prions provide a conceptual framework within which this remarkably diverse spectrum can be explained.

Human prion disease should be considered in any patient who experiences a progressive subacute or chronic decline in cognitive or motor function. Typically, adults between 40 and 80 years of age are affected.[117] Some patients as young as 15 years of age have died of variant (v) CJD in the United Kingdom and France. The early onset age has raised the possibility that these patients were infected with bovine prions from contaminated beef products.[113,118,119] Ninety-one young adults have also been diagnosed with iatrogenic CJD four to 30 years after receiving human growth hormone (HGH) or gonadotropin derived from cadaveric pituitaries.[120,121] The longest incubation periods (20 to 30 years) are similar to those associated with more recent cases of kuru.[122,123]

Mutations of the PrP gene are not found in most cases of CJD and possibly not in some cases of GSS.[124] How prions arise in patients with these sporadic forms is unknown; hypotheses include horizontal transmission from humans or animals,[24] somatic mutation of the PrP gene ORF, and spontaneous conversion of PrP^C into PrP^{Sc}.[116,125] Numerous attempts to establish an infectious link between sporadic CJD and a preexisting prion disease in animals or humans have been unrewarding.[126,127]

Inherited Prion Diseases

The recognition that 10 percent of CJD cases are familial led to the suspicion that genetics plays a role in this disease.[128,129] Like sheep scrapie, the relative contributions of genetic and infectious etiologies in the human prion diseases remained puzzling. A mutation in the ORF or protein coding region of the PrP gene has been found in all reported kindreds with inherited human prion disease [see Figure 3b].

GSS and Genetic Linkage

The discovery that GSS, which was known to be a familial disease, could be transmitted to apes and monkeys was first reported when many still thought that scrapie, CJD, and related disorders were caused by viruses.[129] Only the discovery of a proline (P)→leucine (L) mutation at codon 102 of the human PrP gene that was genetically linked to GSS permitted the unprecedented conclusion that prion disease can have both genetic and infectious etiologies.[64,116] In that study, the codon 102 mutation was linked to development of GSS with a logarithm of the odds (LOD) score exceeding 3, demonstrating a tight association between the altered genotype and the disease phenotype [*see Figure 3b*].[64] This mutation may be caused by the deamination of a methylated CpG in a germ line PrP gene, resulting in the substitution of a thymine (T) for cytosine (C). The same mutation has been found in many families in numerous countries, including the original GSS family.[130-132]

fCJD Caused by Octarepeat Inserts

An insert of 144 base pairs (bp) containing six octarepeats at codon 53, in addition to the five that are normally present, was described in patients with CJD from four families residing in southern England.[133,134] Genealogic investigations have shown that all four families are related, arguing for a single founder born more than two centuries ago. The LOD score for this extended pedigree exceeds 11. Studies from several laboratories have demonstrated that inserts of two, four, five, six, seven, eight, or nine octarepeats in addition to the normal five are found in patients with inherited CJD [*see Figure 3b*].[133,135]

fCJD in Libyan Jews

The unusually high incidence of CJD among Israeli Jews of Libyan origin was thought to be caused by the consumption of lightly cooked sheep brain or eyeballs.[136] Molecular genetic investigations revealed that Libyan and Tunisian Jews with fCJD have a PrP gene point mutation at codon 200, resulting in a Glu→Lys substitution [*see Figure 3b*].[125,137] The E200K mutation has been genetically linked to the mutation with an LOD score exceeding 3,[138] and the same mutation has been found in patients from Orava in north-central Slovakia,[137] in a cluster of familial cases in Chile,[139] and in a large German family living in the United States.[140]

Most patients are heterozygous for the mutation and thus express both mutant and wtPrP^C. In the brains of patients who die of fCJD(E200K), the mutant PrP^{Sc} is both insoluble and protease resistant, whereas much of wt PrP differs from both PrP^C and PrP^{Sc} in that it is insoluble but readily digested by proteases. Whether this form of PrP is an intermediate in the conversion of PrP^C into PrP^{Sc} remains to be established.[141]

Penetrance of fCJD

Life-table analyses of carriers harboring the codon 200 mutation show complete penetrance for the mutation.[142,143] In other words, if the carriers live long enough, they will all eventually acquire prion disease. Some investigators have argued that the inherited prion diseases are not fully penetrant, and thus an environmental factor such as the ubiquitous scrapie virus is required for illness to be manifested. However, no viral pathogen has been found in prion disease.[144,145]

Fatal Familial Insomnia

Studies of inherited human prion diseases demonstrate that changing a single polymorphic residue at position 129 in addition to the D178N pathogenic mutation alters the clinical and neuropathological phenotype. The D178N mutation combined with a Met encoded at position 129 results in a prion disease called *fatal familial insomnia* (FFI) [*see Figure 3b*].[145,146] In this disease, adults generally older than 50 years acquire a progressive sleep disorder and usually die within about a year.[147] In their brains, deposition of PrP[Sc] is confined largely within the anteroventral and the dorsal medial nuclei of the thalamus. The D178N mutation has been linked to the development of FFI with an LOD score over 5.[148] More than 30 families worldwide with FFI have been recorded.[149] In contrast, the same D178N mutation with a Val encoded at position 129 produces fCJD, in which the patients have dementia, and widespread deposition of PrP[Sc] is found postmortem.[150] The first family to be recognized with CJD was recently found to carry the D178N mutation.[128,151]

Human PrP Gene Polymorphisms

The M→V polymorphism at position 129[152] appears able to influence prion disease expression not only in inherited forms but also in iatrogenic and sporadic forms of prion disease [*see Figure 3b*]. A second polymorphism, resulting in an amino acid substitution at codon 219 (E→K), has been reported to occur with a frequency of about 12 percent in the Japanese population but not in Caucasians.[153] A third polymorphism is the deletion of a single octarepeat (24 bp), which has been found in 2.5 percent of Caucasians.[154-156] In another study of over 700 individuals, this single octarepeat was found in one percent of the population.[157]

Studies of Caucasian patients with sCJD have shown that most are homozygous for Met or Val at codon 129.[158] In the general Caucasian population, frequencies for the codon 129 polymorphism are 12 percent V/V, 37 percent M/M, and 51 percent M/V.[159] In contrast, the frequency of the Val allele in the Japanese population is much lower,[160,161] and heterozygosity at codon 129 (M/V) is more frequent (18 percent) in CJD patients than in the general population, in which the polymorphism frequencies are zero percent V/V, 92 percent M/M, and eight percent M/V.[162]

De Novo Generation of Prions

The generation of prions in brain without introduction from an external source is referred to here as de novo generation. Introduction of the codon 102 point mutation found in GSS patients into the MoPrP gene resulted in Tg(MoPrP-P101L) mice that developed CNS degeneration indistinguishable from experimental murine scrapie; neuropathology consisted of widespread spongiform morphology, astrocytic gliosis, and PrP amyloid plaques.[33,34] Brain extracts prepared from spontaneously ill Tg(MoPrP-P101L) mice transmitted CNS degeneration to Tg196 mice.[33,34] The Tg196 mice expressed low levels of the mutant transgene MoPrP-P101L but did not develop spontaneous disease, whereas the Tg(MoPrP-P101L) mice expressing high levels of the mutant transgene product did develop CNS degeneration spontaneously. These studies, as well as transmission of prions from patients who died of GSS to apes and monkeys[129] and to Tg(MHu2M-P102L) mice,[59] argue persuasively that prions are generated de novo by mutations in PrP. In contrast to species-specific variations in PrP, all of the known point mutations in PrP occurred either within or adjacent to regions of putative secondary structure in PrP and, as such, appeared to destabilize the structure of PrP.[100,110,111] How the presence of a Leu at codon 102 might alter binding of mutant PrPSc to wtPrPC is unknown, but its place in the central region of PrP suggests that the position is critical. It is noteworthy that extracts from the brains of patients who died with GSS(P102L) have transmitted disease to Tg(MHu2M-P102L) mice but not to Tg(MHu2M) mice.[59] In contrast, human prions from patients with sCJD, fCJD(E200K), or FFI have all transmitted disease to Tg(MHu2M) mice.[22,59]

Why mutations of the PrP gene that produce seemingly unstable PrPC molecules require many decades in humans to be manifested as CNS dysfunction is unknown. In Tg(MoPrP-P101L) mice, the level of expression of the mutant transgene is inversely related to the age of disease onset. In addition, the presence of the wtMoPrP gene slows the onset of disease and diminishes the severity of the neuropathological changes.

Strains of Prions

Goats with scrapie may exhibit either of two different syndromes: one in which the animals become nervous and the other that causes scratching, raising the possibility that different strains of prions might exist.[163] Studies with mice also documented the existence of multiple strains.[164,165] Different prion strains were identified using two strains of mice: C57BL and VM. One group of prions exhibited short incubation times in C57BL mice and long ones in VM mice. Other prion strains showed unexpected behavior with respect to the length of the incubation time. These variations in apparent virulence of prions have not been explained.

Selective Neuronal Targeting of Prion Strains

New techniques of in situ detection of PrPSc made it possible to localize and quantify PrPSc as well as to determine whether different prion strains produce different but reproducible patterns of PrPSc accumulation. Patterns of PrPSc accumulation were different for each prion strain when the genotype of the host was held constant.[166,167] A single prion strain produced different patterns when inoculated into mice expressing various PrP transgenes. From these observations, we concluded that the pattern of PrPSc deposition is a manifestation of the particular prion strain but is not related to its propagation.[168,169]

Whereas studies with both mice and Syrian hamsters established that each strain has a specific signature as defined by a specific pattern of PrPSc accumulation in the brain,[166,167,170] such comparisons must be made on an isogenic background.[33,171] When a single strain was inoculated into mice expressing different PrP genes, variations in the patterns of PrPSc accumulation were found to be as great as those seen between two strains.[168] Based upon the initial studies performed in animals of a single genotype, we suggested that PrPSc synthesis occurs in specific populations of cells for a given distinct prion isolate.[116,166]

Isolation of New Strains

Additional evidence that implicates PrP in the phenomenon of prion strains is derived from studies on the transmission of strains from one species to another. Mice expressing chimeric SHa/MoPrP transgenes showed that new strains pathogenic for Syrian hamsters could be obtained from prion strains that had been previously cloned in mice.[172,173] Both the generation and the propagation of prion strains seem to be results of interactions between PrP^{Sc} and PrP^{C}. Additionally, strains once thought to be distinct that were isolated from different breeds of sheep with scrapie were shown to have indistinguishable properties. Such findings argue for the convergence of some strains and raise questions about the limits of prion diversity.

Comparisons of prion strains in mice have not revealed any biochemical or physical differences in PrP^{Sc}.[166] However, a difference in PrP^{Sc} related to strains was found when two prion strains were isolated from mink with transmissible encephalopathy.[174] One strain (HY) produced hyperactivity in Syrian hamsters, and the other (DY) produced lethargy, reminiscent of the scrapie strain differences first seen in goats.[174] PrP^{Sc} produced by the DY prions showed diminished resistance to proteinase K digestion and truncation of the N-terminus compared with HY,[175] providing evidence for the hypothesis that different strains might represent different conformers of PrP^{Sc}.[23] It was not possible to demonstrate the de novo synthesis of infectious prions using this system.

Enciphering Diversity through Protein Conformation

Although the notion that PrP^{Sc} tertiary structure might encrypt the information for each strain is consistent with all of the foregoing experimental data, such a hypothesis enjoyed little enthusiasm. Only with the transmission of two different inherited human prion diseases to mice expressing chimeric Hu/Mo PrP transgenes has firm evidence emerged supporting this concept. In FFI, the protease-resistant fragment of PrP^{Sc} after deglycosylation has a relative molecular mass (M_r) of 19 kd, whereas that from other inherited and sporadic prion diseases is 21 kd [see Figure 6].[176,177] Extracts from the brains of FFI patients transmitted disease to mice expressing a chimeric MHu2M PrP gene about 200 days after inoculation and induced formation of the 19 kd PrP^{Sc}, whereas fCJD(E200K) and sCJD produced the 21 kd PrP^{Sc} in these mice.[22] These findings argue that PrP^{Sc} acts as a template for the conversion of PrP^{C} into nascent PrP^{Sc}. Imparting the size of the protease-resistant fragment of PrP^{Sc} through conformational templating provides a mechanism for both the generation and the propagation of prion strains.

The Interplay Between the Species and Strains of Prions

The recent advances in our understanding of the role of the primary and tertiary structures of PrP in the transmission of disease have given new insights into the pathogenesis of the prion diseases. The amino acid sequence of PrP encodes the species of the prion [see Table 3],[59,178] and the prion derives its PrP^{Sc} sequence from the last mammal in which it was passaged. Whereas the primary structure of PrP is likely to be the most important or even the sole determinant of the tertiary structure of PrP^{C}, existing PrP^{Sc} seems to function as a template in determining the tertiary structure of nascent PrP^{Sc} molecules as they are formed from PrP^{C}.[23,104] In turn, prion diversity may be enciphered in the conformation of PrP^{Sc}, and prion strains may represent different conformers of PrP^{Sc}.[22,179]

As unlikely as the foregoing scenario seems, and as unprecedented as it is in biology, there is now considerable experimental data to support the concept. However, it is not yet known whether multiple conformers of PrP^C also exist that might serve as precursors for selective conversion into different PrP^{Sc}. In this light, it is useful to consider another phenomenon that is not yet understood: the selective targeting of neuronal populations in the CNS of the host mammal. Recent data suggest that variations in glycosylation of the Asn-linked sugar chains may influence the rate at which a particular PrP^C molecule is converted into PrP^{Sc}.[168] Because Asn-linked oligosaccharides are known to modify the conformation of some proteins,[180,181] it seemed reasonable to assume that variations in complex type sugars may alter the size of the energy barrier that must be traversed during formation of PrP^{Sc}. If this is the case, then regional variations in oligosaccharide structure in the CNS could account for selective targeting, that is, formation of PrP^{Sc} in particular areas of the brain. Such a mechanism could also explain the variations in the ratio of the various PrP^{Sc} glycoforms observed by some investigators.[182] However, such a mechanism, although accounting for specific patterns of PrP^{Sc} distribution, does not seem to influence to any measurable degree the properties of the resulting PrP^{Sc} molecule. Molecular

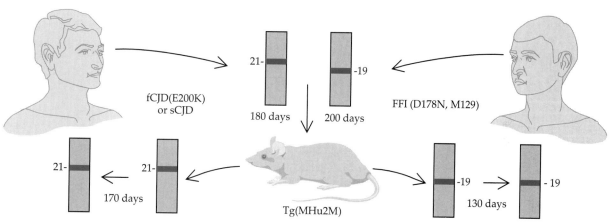

Figure 6 *Evidence for prion diversity being enciphered within the conformation of PrP^{Sc}. Mutant prions template-specific conformations onto PrP^{Sc} formed in Tg(MHu2M)Prnp$^{0/0}$ mice. Western blot analysis of PrP^{Sc} from the brains of patients who died of fCJD(E200K) or FFI(D178N, M129) showed PrP molecules of 21 or 19 kd, respectively. Homogenates were digested with proteinase K, denatured, and digested with PNGase F before sodium dodecylsulfate–polyacrylamide gel electrophoresis, followed by immunoblotting with α-PrP 3F4 mAbs. Aliquots of brain homogenates from patients were inoculated intracerebrally into Tg(MHu2M)Prnp$^{0/0}$ mice; the mice were sacrificed about 200 days later, when they developed signs of central nervous system dysfunction. Western blot analysis of PrP^{Sc} from the brains of Tg(MHu2M) mice showed MHu2M PrP^{Sc} molecules of 21 or 19 kd. If the mice were inoculated with homogenates prepared from the brains of the patients who died of FFI(D178N, M129), then after proteinase K and PNGase F digestions, MHu2M PrP^{Sc} migrated with a relative molecular mass (M$_r$) of 19 kd. In contrast, if the mice were inoculated with fCJD(E200K) prions, then MHu2M PrP^{Sc} migrated at 21 kd. On the second serial passage in Tg(MHu2M)Prnp$^{0/0}$ mice, aliquots of brain homogenates from first-passage Tg(MHu2M)Prnp$^{0/0}$ mice were inoculated intracerebrally into Tg(MHu2M)Prnp$^{0/0}$ mice, and the mice were sacrificed when they developed signs of CNS dysfunction. For mice receiving an inoculum that began with an fCJD(E200K) patient, the incubation time was about 170 days on second passage; for mice receiving an inoculum that began with an FFI patient, the incubation time was about 130 days. Western blot analysis of PrP^{Sc} from the brains of Tg(MHu2M) mice showed MHu2M PrP^{Sc} molecules of 21 or 19 kd. If the mice were initially inoculated with homogenates prepared from the brains of the patients who died of FFI(D178N, M129), then after proteinase K and PNGase F digestions, MHu2M PrP^{Sc} migrated with an M$_r$ of 19 kd. If the mice were inoculated with fCJD(E200K) prions, then MHu2M PrP^{Sc} migrated at 21 kd.*

Table 3 Influence of Prion Species and Strain on Transmission from Syrian Hamsters to Hamsters and Mice

Inoculum	Recipient	Strain Sc237		Strain 137H	
		Incubation Time*	n/n_0†	Incubation Time*	n/n_0†
SHa→SHa	SHa	77 ± 1	48/48	167 ± 1	94/94
SHa→SHa	non-Tg mice	>700	0/9	499 ± 15	11/11
SHa→SHa	Tg(SHaPrP)81 mice	75 ± 2	22/22	110 ± 2	19/19

Note. The species of prion is encoded by the primary structure of PrP^{Sc} and the strain of prion is enciphered by the tertiary structure of PrP^{Sc}. The primary structure as well as posttranslational chemical modifications determine the tertiary structure of PrP^C but the conformation of PrP^C is modified by PrP^{Sc} as it is refolded into a nascent molecule of PrP^{Sc}.[22]

*Days ± SEM.

†Number affected/number inoculated.

modeling and NMR structural studies may provide an explanation for such phenomena because the Asn-linked oligosaccharides appear to be on the opposite face of PrP from where PrP^C and PrP^{Sc} are expected to interact during the formation of nascent PrP^{Sc}.[100,101,110,111]

Prion Diseases of Animals

Sheep and Cattle PrP Gene Polymorphisms

Parry argued that host genes were responsible for the development of scrapie in sheep.[183,184] He was convinced that natural scrapie is a genetic disease that could be eradicated by proper breeding protocols. He considered its transmission by inoculation important primarily for laboratory studies and communicable infection of little consequence in nature. Other investigators viewed natural scrapie as an infectious disease and argued that host genetics only modulates susceptibility to an endemic infectious agent.[185]

In sheep, polymorphisms at codons 136, 154, and 171 of the PrP gene that produce amino acid substitutions have been studied with respect to the occurrence of scrapie [*see Figure 3b*].[186-189] Studies of natural scrapie in the United States have shown that approximately 85 percent of the afflicted sheep are of the Suffolk breed. Only those Suffolk sheep homozygous for Gln (Q) at codon 171 were found with scrapie, although healthy control subjects with this genotype were also found.[190-194] These results argue that susceptibility in Suffolk sheep is governed by the PrP codon 171 polymorphism. In Cheviot sheep, the PrP codon 171 polymorphism has a profound influence on susceptibility to scrapie as in Suffolks, and codon 136 seems to play a less pronounced role.[195,196]

In contrast to sheep, different breeds of cattle have no specific PrP polymorphisms. The only polymorphism recorded in cattle is a variation in the number of octarepeats: most cattle, like humans, have five octarepeats, but some have six.[93,94] Humans with seven octarepeats develop fCJD,[197] but six octarepeats do not seem to be overrepresented in BSE.[93,94,198]

Bovine Spongiform Encephalopathy

Beginning in 1986, an epidemic of a previously unknown prion disease, named bovine spongiform encephalopathy or so-called mad cow disease, appeared in cattle in Great Britain,[18] in which protease-resistant PrP was found in brains of ill cattle.[94,199] It has been proposed that BSE represents a massive common-source epidemic caused by meat and bone meal (MBM) fed primarily to dairy cows.[200] The MBM was prepared from the inedible parts or offal of sheep, cattle, pigs, and chickens as a high-protein nutritional supplement. In the late 1970s, the solvent extraction method used in rendering offal began to be abandoned, resulting in an MBM with a much higher fat content.[201] It is now thought that this change in the rendering process allowed scrapie prions from sheep to survive rendering and pass into cattle. Alternatively, bovine prions that had caused clinical CNS dysfunction at a level low enough to go undetected might have survived the rendering process and were passed back to cattle through the MBM. It now appears that the epidemic is disappearing as a result of the 1988 ban on feeding MBM.

Brain extracts from BSE cattle have transmitted disease to mice, cattle, sheep, pigs, and mink after intracerebral inoculation.[202,203] Transmissions to mice and sheep suggest that cattle preferentially propagate a single strain of prions.[204] Of particular significance is the transmission of BSE to the marmoset after intracerebral inoculation and, more recently, to the macaque.[205,206]

Have Bovine Prions Been Transmitted to Humans?

The recent cases of vCJD in three teenagers and 20 young adults in the United Kingdom and France raise the possibility of transmission of BSE to humans.[118,119] The average age of these individuals was 27 years, much younger than that of any other group of people who died of CJD, except for those who received pituitary HGH. Not only does age set these patients apart from others who died of prion disease but so does the neuropathology. The deposition of PrP in the brains of these patients is extreme, and numerous multinucleated PrP amyloid plaques surrounded by intense spongiform degeneration have been observed. These neuropathological changes seem to be unlike any observed in other forms of prion disease. Why such cases should be confined to young people is unclear. Whether the young CNS is more vulnerable to invasion by bovine prions or the dietary habits of these young persons exposed them to a greater dose of bovine prions is unknown.

Epidemiological studies over nearly three decades have failed to establish convincing evidence for transmission of sheep prions to humans.[126,127] Of interest is the high incidence of CJD among Libyan Jews, which was initially attributed to the consumption of lightly cooked sheep brain[136]; however, subsequent studies showed that this geographical cluster of CJD is attributable to the E200K mutation [*see Figure 3b*].[125,144]

Toward Effective Therapies for Prion Diseases

Although the study of prions has taken several unexpected directions over the past three decades, a rather unprecedented story of prion biology is emerging. Although learning the detailed structures of the prion proteins and deciphering the mechanism of PrP^C transformation into PrP^{Sc} remain important goals, the broader principles of prion biology have become reasonably clear. Although some investigators view the composition of the infectious prion particle as unresolved, a growing body of data supports the prion concept.

The discovery that prion diseases in humans are uniquely both genetic and infectious greatly strengthened and extended the prion concept. Twenty different mutations in the human PrP gene, each resulting in nonconservative substitutions, have been found either to be linked genetically to or to segregate with the inherited prion diseases [*see Figure 3b*]. Yet, the transmissible prion particle is composed largely, if not exclusively, of an abnormal PrP isoform designated PrP^{Sc}.[23]

Understanding how PrP^C unfolds and refolds into PrP^{Sc} will be of paramount importance in transferring advances in the prion diseases to studies of other degenerative illnesses. The mechanism by which PrP^{Sc} is formed must involve a templating process whereby existing PrP^{Sc} directs the refolding of PrP^C into a nascent PrP^{Sc} with the same conformation. Undoubtedly, molecular chaperones of some type participate in a process that seems confined to caveolalike domains of the cell.

Studies of prion-like proteins in yeast may prove particularly helpful in dissecting some of the events that are featured in PrP^{Sc} formation.[207] Conversion to the prionlike state in yeast requires the molecular chaperone Hsp104; however, no homologue of Hsp104 has been found in mammals.[208,209] It is notable that other examples of acquired inheritance in lower organisms may also occur through a prionlike mechanism.[210,211]

As our understanding of prion propagation increases, it should be possible to design effective therapeutics. Because people at risk for inherited prion diseases can now be identified decades before neurologic dysfunction is evident, the development of an effective therapy is imperative. Moreover, the possible transmission of bovine prions to humans in the United Kingdom and France raises concern regarding the number of people who might succumb to prion disease in the future.[113] Seeking an effective means of therapy now seems most prudent. On the basis of our current knowledge, the conversion of PrP^C into PrP^{Sc} seems to be the most attractive target.[104] Either stabilizing the structure of PrP^C by binding a drug, or interfering with the action of protein X, which presumably functions as a molecular chaperone, seems to be a reasonable strategy for the rational design of a pharmacotherapeutic.

Whether therapies designed to prevent the conversion of PrP^C into PrP^{Sc} will be effective in the more common neurodegenerative diseases, such as Alzheimer's disease, Parkinson's disease, and amyotrophic lateral sclerosis, is unknown. Alternatively, developing a therapy for the prion diseases might provide a blueprint for designing somewhat different drugs for these disorders. Like the inherited prion diseases, an important subset of Alzheimer's disease and amyotrophic lateral sclerosis is caused by mutations that result in nonconservative amino acid substitutions in proteins expressed in the CNS.

References

1. Sigurdsson B: Rida, a chronic encephalitis of sheep with general remarks on infections which develop slowly and some of their special characteristics. *Br Vet J* 110:341, 1954
2. Hadlow WJ: Scrapie and kuru. *Lancet* 2:289, 1959

3. Gajdusek DC, Gibbs CJ Jr, Alpers M: Experimental transmission of a kuru-like syndrome to chimpanzees. *Nature* 209:794, 1966

4. Klatzo I, Gajdusek DC, Zigas V: Pathology of kuru. *Lab Invest* 8:799, 1959

5. Gibbs CJ Jr, Gajdusek DC, Asher DM, et al: Creutzfeldt-Jakob disease (spongiform encephalopathy): transmission to the chimpanzee. *Science* 161:388, 1968

6. Alper T, Haig DA, Clarke MC: The exceptionally small size of the scrapie agent. *Biochem Biophys Res Commun* 22:278, 1966

7. Alper T, Cramp WA, Haig DA, et al: Does the agent of scrapie replicate without nucleic acid? *Nature* 214:764, 1967

8. Latarjet R, Muel B, Haig DA, et al: Inactivation of the scrapie agent by near monochromatic ultraviolet light. *Nature* 227:1341, 1970

9. Gibbs CJ Jr, Gajdusek DC, Latarjet R: Unusual resistance to ionizing radiation of the viruses of kuru, Creutzfeldt-Jakob disease. *Proc Natl Acad Sci USA* 75:6268, 1978

10. Pattison IH: Experiments with scrapie with special reference to the nature of the agent and the pathology of the disease. *Slow, Latent and Temperate Virus Infections, NINDB Monograph 2.* Gajdusek DC, Gibbs CJ Jr, Alpers MP, Eds. U.S. Government Printing Office, Washington, DC, 1965, p 249

11. Gibbons RA, Hunter GD: Nature of the scrapie agent. *Nature* 215:1041, 1967

12. Pattison IH, Jones KM: The possible nature of the transmissible agent of scrapie. *Vet Rec* 80:1, 1967

13. Hunter GD, Kimberlin RH, Gibbons RA: Scrapie: a modified membrane hypothesis. *J Theor Biol* 20:355, 1968

14. Field EJ, Farmer F, Caspary EA, et al: Susceptibility of scrapie agent to ionizing radiation. *Nature* 222:90, 1969

15. Hunter GD: Scrapie: a prototype slow infection. *J Infect Dis* 125:427, 1972

16. Prusiner SB: An approach to the isolation of biological particles using sedimentation analysis. *J Biol Chem* 253:916, 1978

17. Prusiner SB, Garfin DE, Baringer JR, et al: Evidence for multiple molecular forms of the scrapie agent. *Persistent Viruses.* Stevens J, Todaro G, Fox CF, Eds. Academic Press, New York, 1978, p 591

18. Wells GAH, Scott AC, Johnson CT, et al: A novel progressive spongiform encephalopathy in cattle. *Vet Rec* 121:419, 1987

19. Pan K-M, Baldwin M, Nguyen J, et al: Conversion of α-helices into β-sheets features in the formation of the scrapie prion proteins. *Proc Natl Acad Sci USA* 90:10962, 1993

20. Meyer RK, McKinley MP, Bowman KA, et al: Separation and properties of cellular and scrapie prion proteins. *Proc Natl Acad Sci USA* 83:2310, 1986

21. Oesch B, Westaway D, Wälchli M, et al: A cellular gene encodes scrapie PrP 27-30 protein. *Cell* 40:735, 1985

22. Telling GC, Parchi P, DeArmond SJ, et al: Evidence for the conformation of the pathologic isoform of the prion protein enciphering and propagating prion diversity. *Science* 274:2079, 1996

23. Prusiner SB: Molecular biology of prion diseases. *Science* 252:1515, 1991

24. Gajdusek DC: Unconventional viruses and the origin and disappearance of kuru. *Science* 197:943, 1977

25. Prusiner SB, Groth DF, Cochran SP, et al: Molecular properties, partial purification, and assay by incubation period measurements of the hamster scrapie agent. *Biochemistry* 19:4883, 1980

26. Prusiner SB: Novel proteinaceous infectious particles cause scrapie. *Science* 216:136, 1982

27. Bolton DC, McKinley MP, Prusiner SB: Identification of a protein that purifies with the scrapie prion. *Science* 218:1309, 1982

28. Prusiner SB, Groth DF, Bolton DC, et al: Purification and structural studies of a major scrapie prion protein. *Cell* 38:127, 1984

29. Basler K, Oesch B, Scott M, et al: Scrapie and cellular PrP isoforms are encoded by the same chromosomal gene. *Cell* 46:417, 1986

30. Gabizon R, McKinley MP, Groth DF, et al: Immunoaffinity purification and neutralization of scrapie prion infectivity. *Proc Natl Acad Sci USA* 85:6617, 1988

31. Chesebro B, Race R, Wehrly K, et al: Identification of scrapie prion protein-specific mRNA in scrapie-infected and uninfected brain. *Nature* 315:331, 1985

32. Locht C, Chesebro B, Race R, et al: Molecular cloning and complete sequence of prion protein cDNA from mouse brain infected with the scrapie agent. *Proc Natl Acad Sci USA* 83:6372, 1986

33. Hsiao KK, Groth D, Scott M, et al: Serial transmission in rodents of neurodegeneration from transgenic mice expressing mutant prion protein. *Proc Natl Acad Sci USA* 91:9126, 1994

34. Telling GC, Haga T, Torchia M, et al: Interactions between wild-type and mutant prion proteins modulate neurodegeneration in transgenic mice. *Genes Dev* 10:1736, 1996

35. Caughey B, Neary K, Butler R, et al: Normal and scrapie-associated forms of prion protein differ in their sensitivities to phospholipase and proteases in intact neuroblastoma cells. *J Virol* 64:1093, 1990

36. Prusiner SB, McKinley MP, Bowman KA, et al: Scrapie prions aggregate to form amyloid-like birefringent rods. *Cell* 35:349, 1983

37. Caughey BW, Dong A, Bhat KS, et al: Secondary structure analysis of the scrapie-associated protein PrP 27-30 in water by infrared spectroscopy. *Biochemistry* 30:7672, 1991

38. McKinley MP, Meyer R, Kenaga L, et al: Scrapie prion rod formation *in vitro* requires both detergent extraction and limited proteolysis. *J Virol* 65:1440, 1991

39. Gasset M, Baldwin MA, Fletterick RJ, et al: Perturbation of the secondary structure of the scrapie prion protein under conditions associated with changes in infectivity. *Proc Natl Acad Sci USA* 90:1, 1993

40. Safar J, Roller PP, Gajdusek DC, et al: Thermal-stability and conformational transitions of scrapie amyloid (prion) protein correlate with infectivity. *Protein Sci* 2:2206, 1993

41. Muramoto T, Scott M, Cohen F, et al: Recombinant scrapie-like prion protein of 106 amino acids is soluble. *Proc Natl Acad Sci USA* 93:15457, 1996

42. Riesner D, Kellings K, Post K, et al: Disruption of prion rods generates 10-nm spherical particles having high α-helical content and lacking scrapie infectivity. *J Virol* 70:1714, 1996

43. Lasmézas CI, Deslys J-P, Demaimay R, et al: Strain specific and common pathogenic events in murine models of scrapie and bovine spongiform encephalopathy. *J Gen Virol* 77:1601, 1996

44. DeArmond SJ, Mobley WC, DeMott DL, et al: Changes in the localization of brain prion proteins during scrapie infection. *Neurology* 37:1271, 1987

45. Taraboulos A, Jendroska K, Serban D, et al: Regional mapping of prion proteins in brains. *Proc Natl Acad Sci USA* 89:7620, 1992

46. Czub M, Braig HR, Diringer H: Pathogenesis of scrapie: study of the temporal development of clinical symptoms of infectivity titres and scrapie-associated fibrils in brains of hamsters infected intraperitoneally. *J Gen Virol* 67:2005, 1986

47. Xi YG, Ingrosso L, Ladogana A, et al: Amphotericin B treatment dissociates *in vivo* replication of the scrapie agent from PrP accumulation. *Nature* 356:598, 1992

48. Jendroska K, Heinzel FP, Torchia M, et al: Proteinase-resistant prion protein accumulation in Syrian hamster brain correlates with regional pathology and scrapie infectivity. *Neurology* 41:1482, 1991

49. McKenzie D, Kaczkowski J, Marsh R, et al: Amphotericin B delays both scrapie agent replication and PrP-res accumulation early in infection. *J Virol* 68:7534, 1994

50. Stahl N, Borchelt DR, Hsiao K, et al: Scrapie prion protein contains a phosphatidylinositol glycolipid. *Cell* 51:229, 1987

51. Borchelt DR, Scott M, Taraboulos A, et al: Scrapie and cellular prion proteins differ in their kinetics of synthesis and topology in cultured cells. *J Cell Biol* 110:743, 1990

52. Caughey B, Raymond GJ: The scrapie-associated form of PrP is made from a cell surface precursor that is both protease- and phospholipase-sensitive. *J Biol Chem* 266:18217, 1991

53. Taraboulos A, Scott M, Semenov A, et al: Cholesterol depletion and modification of COOH-terminal targeting sequence of the prion protein inhibit formation of the scrapie isoform. *J Cell Biol* 129:121, 1995

54. Vey M, Pilkuhn S, Wille H, et al: Subcellular colocalization of the cellular and scrapie prion proteins in caveolae-like membranous domains. *Proc Natl Acad Sci USA* 93:14945, 1996

55. Kaneko K, Vey M, Scott M, et al: COOH-terminal sequence of the cellular prion protein directs subcellular trafficking and controls conversion into the scrapie isoform. *Proc Natl Acad Sci USA* 94:2333, 1997

56. Caughey B, Raymond GJ, Ernst D, et al: N-terminal truncation of the scrapie-associated form of PrP by lysosomal protease(s): implications regarding the site of conversion of PrP to the protease-resistant state. *J Virol* 65:6597, 1991

57. Telling GC, Scott M, Hsiao KK, et al: Transmission of Creutzfeldt-Jakob disease from humans to transgenic mice expressing chimeric human-mouse prion protein. *Proc Natl Acad Sci USA* 91:9936, 1994

58. Carlson GA, Westaway D, DeArmond SJ, et al: Primary structure of prion protein may modify scrapie isolate properties. *Proc Natl Acad Sci USA* 86:7475, 1989

59. Telling GC, Scott M, Mastrianni J, et al: Prion propagation in mice expressing human and chimeric PrP transgenes implicates the interaction of cellular PrP with another protein. *Cell* 83:79, 1995

60. Tateishi J, Kitamoto T, Hoque MZ, et al: Experimental transmission of Creutzfeldt-Jakob disease and related diseases to rodents. *Neurology* 46:532, 1996

61. Lasmézas CI, Deslys J-P, Robain O, et al: Transmission of the BSE agent to mice in the absence of detectable abnormal prion protein. *Science* 275:402, 1997

62. Hsiao KK, Scott M, Foster D, et al: Spontaneous neurodegeneration in transgenic mice with mutant prion protein. *Science* 250:1587, 1990

63. Westaway D, Goodman PA, Mirenda CA, et al: Distinct prion proteins in short and long scrapie incubation period mice. *Cell* 51:651, 1987

64. Hsiao K, Baker HF, Crow TJ, et al: Linkage of a prion protein missense variant to Gerstmann-Sträussler syndrome. *Nature* 338:342, 1989

65. Gabriel J-M, Oesch B, Kretzschmar H, et al: Molecular cloning of a candidate chicken prion protein. *Proc Natl Acad Sci USA* 89:9097, 1992

66. Westaway D, Mirenda CA, Foster D, et al: Paradoxical shortening of scrapie incubation times by expression of prion protein transgenes derived from long incubation period mice. *Neuron* 7:59, 1991

67. Li G, Bolton DC: A novel hamster prion protein mRNA contains an extra exon: increased expression in scrapie. *Brain Res* 751:265, 1997

68. Robakis NK, Devine-Gage EA, Kascsak RJ, et al: Localization of a human gene homologous to the PrP gene on the p arm of chromosome 20 and detection of PrP-related antigens in normal human brain. *Biochem Biophys Res Commun* 140:758, 1986

69. Sparkes RS, Simon M, Cohn VH, et al: Assignment of the human and mouse prion protein genes to homologous chromosomes. *Proc Natl Acad Sci USA* 83:7358, 1986

70. Mobley WC, Neve RL, Prusiner SB, et al: Nerve growth factor increases mRNA levels for the prion protein and the beta-amyloid protein precursor in developing hamster brain. *Proc Natl Acad Sci USA* 85:9811, 1988

71. Kretzschmar HA, Prusiner SB, Stowring LE, et al: Scrapie prion proteins are synthesized in neurons. *Am J Pathol* 122:1, 1986

72. Kimberlin RH, Field HJ, Walker CA: Pathogenesis of mouse scrapie: evidence for spread of infection from central to peripheral nervous system. *J Gen Virol* 64:713, 1983

73. Fraser H, Dickinson AG: Targeting of scrapie lesions and spread of agent via the retinotectal projection. *Brain Res* 346:32, 1985

74. Prusiner SB, Scott M, Foster D, et al: Transgenetic studies implicate interactions between homologous PrP isoforms in scrapie prion replication. *Cell* 63:673, 1990

75. Westaway D, Cooper C, Turner S, et al: Structure and polymorphism of the mouse prion protein gene. *Proc Natl Acad Sci USA* 91:6418, 1994

76. Büeler H, Fischer M, Lang Y, et al: Normal development and behaviour of mice lacking the neuronal cell-surface PrP protein. *Nature* 356:577, 1992

77. Manson JC, Clarke AR, Hooper ML, et al: 129/Ola mice carrying a null mutation in PrP that abolishes mRNA production are developmentally normal. *Mol Neurobiol* 8:121, 1994

78. Sakaguchi S, Katamine S, Nishida N, et al: Loss of cerebellar Purkinje cells in aged mice homozygous for a disrupted PrP gene. *Nature* 380:528, 1996

79. Collinge J, Whittington MA, Sidle KC, et al: Prion protein is necessary for normal synaptic function. *Nature* 370:295, 1994

80. Whittington MA, Sidle KCL, Gowland I, et al: Rescue of neurophysiological phenotype seen in PrP null mice by transgene encoding human prion protein. *Nat Genet* 9:197, 1995

81. Herms JW, Kretzschmar HA, Titz S, et al: Patch-clamp analysis of synaptic transmission to cerebellar Purkinje cells of prion protein knockout mice. *Eur J Neurosci* 7:2508, 1995

82. Lledo P-M, Tremblay P, DeArmond SJ, et al: Mice deficient for prion protein exhibit normal neuronal excitability and synaptic transmission in the hippocampus. *Proc Natl Acad Sci USA* 93:2403, 1996

82a. Sailer A, Bueler H, Fischer M, et al: No propagation of prions in mice devoid of PrP. *Cell* 77: 967, 1994

83. Prusiner SB, Groth D, Serban A, et al: Ablation of the prion protein (PrP) gene in mice prevents scrapie and facilitates production of anti-PrP antibodies. *Proc Natl Acad Sci USA* 90:10608, 1993

84. Büeler H, Aguzzi A, Sailer A, et al: Mice devoid of PrP are resistant to scrapie. *Cell* 73:1339, 1993

85. Sakaguchi S, Katamine S, Shigematsu K, et al: Accumulation of proteinase K-resistant prion protein (PrP) is restricted by the expression level of normal PrP in mice inoculated with a mouse-adapted strain of the Creutzfeldt-Jakob disease agent. *J Virol* 69:7586, 1995

86. Caughey B, Chesebro B: Prion protein and the transmissible spongiform encephalopathies. *Trends Cell Biol* 7:56, 1997

87. Chesebro B, Caughey B: Scrapie agent replication without the prion protein? *Curr Biol* 3:696, 1993

88. Büeler H, Raeber A, Sailer A, et al: High prion and PrP^{Sc} levels but delayed onset of disease in scrapie-inoculated mice heterozygous for a disrupted PrP gene. *Mol Med* 1:19, 1994

89. Stahl N, Baldwin MA, Burlingame AL, et al: Identification of glycoinositol phospholipid linked and truncated forms of the scrapie prion protein. *Biochemistry* 29:8879, 1990

90. Endo T, Groth D, Prusiner SB, et al: Diversity of oligosaccharide structures linked to asparagines of the scrapie prion protein. *Biochemistry* 28:8380, 1989

91. Schätzl HM, Da Costa M, Taylor L, et al: Prion protein gene variation among primates. *J Mol Biol* 245:362, 1995

92. Harris DA, Falls DL, Walsh W, et al: Molecular cloning of an acetylcholine receptor-inducing protein. *Soc Neurosci* 15:70, 1989

93. Goldmann W, Hunter N, Martin T, et al: Different forms of the bovine PrP gene have five or six copies of a short, G-C-rich element within the protein-coding exon. *J Gen Virol* 72:201, 1991

94. Prusiner SB, Fuzi M, Scott M, et al: Immunologic and molecular biological studies of prion proteins in bovine spongiform encephalopathy. *J Infect Dis* 167:602, 1993

95. Rogers M, Yehiely F, Scott M, et al: Conversion of truncated and elongated prion proteins into the scrapie isoform in cultured cells. *Proc Natl Acad Sci USA* 90:3182, 1993

96. Fischer M, Rülicke T, Raeber A, et al: Prion protein (PrP) with amino-proximal deletions restoring susceptibility of PrP knockout mice to scrapie. *EMBO J* 15:1255, 1996

97. Hsiao KK, Cass C, Schellenberg GD, et al: A prion protein variant in a family with the telencephalic form of Gerstmann-Sträussler-Scheinker syndrome. *Neurology* 41:681, 1991

98. Stahl N, Baldwin MA, Teplow DB, et al: Structural analysis of the scrapie prion protein using mass spectrometry and amino acid sequencing. *Biochemistry* 32:1991, 1993

99. Gasset M, Baldwin MA, Lloyd D, et al: Predicted α-helical regions of the prion protein when synthesized as peptides form amyloid. *Proc Natl Acad Sci USA* 89:10940, 1992

100. Huang Z, Gabriel J-M, Baldwin MA, et al: Proposed three-dimensional structure for the cellular prion protein. *Proc Natl Acad Sci USA* 91:7139, 1994

101. Huang Z, Prusiner SB, Cohen FE: Scrapie prions: a three-dimensional model of an infectious fragment. *Folding Design* 1:13, 1996

102. Safar J, Roller PP, Gajdusek DC, et al: Conformational transitions, dissociation, and unfolding of scrapie amyloid (prion) protein. *J Biol Chem* 268:20276, 1993

103. Jarrett JT, Lansbury PT Jr: Seeding "one-dimensional crystallization" of amyloid: a pathogenic mechanism in Alzheimer's disease and scrapie? *Cell* 73:1055, 1993

104. Cohen FE, Pan K-M, Huang Z, et al: Structural clues to prion replication. *Science* 264:530, 1994

105. Safar J, Roller PP, Gajdusek DC, et al: Scrapie amyloid (prion) protein has the conformational characteristics of an aggregated molten globule folding intermediate. *Biochemistry* 33:8375, 1994

106. Kocisko DA, Come JH, Priola SA, et al: Cell-free formation of protease-resistant prion protein. *Nature* 370:471, 1994

107. Kaneko K, Peretz D, Pan K-M, et al: Prion protein (PrP) synthetic peptides induce cellular PrP to acquire properties of the scrapie isoform. *Proc Natl Acad Sci USA* 32:11160, 1995

108. Kaneko K, Wille H, Mehlhorn I, et al: Complexes of the prion protein and synthetic peptides implicate multiple intermediates in the formation of the scrapie isoform. *J Mol Biol* 270:574, 1997

109. Mehlhorn I, Groth D, Stöckel J, et al: High-level expression and characterization of a purified 142-residue polypeptide of the prion protein. *Biochemistry* 35:5528, 1996

110. Zhang H, Kaneko K, Nguyen JT, et al: Conformational transitions in peptides containing two putative α-helices of the prion protein. *J Mol Biol* 250:514, 1995

111. Riek R, Hornemann S, Wider G, et al: NMR structure of the mouse prion protein domain PrP(121-231). *Nature* 382:180, 1996

112. James TL, Liu H, Ulyanov NB, et al: Solution structure of a 142-residue recombinant prion protein corresponding to the infectious fragment of the scrapie isoform. *Proc Natl Acad Sci USA* 94:10086, 1997

113. Cousens SN, Vynnycky E, Zeidler M, et al: Predicting the CJD epidemic in humans. *Nature* 385:197, 1997

114. Bockman JM, Prusiner SB, Tateishi J, et al: Immunoblotting of Creutzfeldt-Jakob disease prion proteins: host species-specific epitopes. *Ann Neurol* 21:589, 1987

115. Kretzschmar HA, Stowring LE, Westaway D, et al: Molecular cloning of a human prion protein cDNA. *DNA* 5:315, 1986

116. Prusiner SB: Scrapie prions. *Annu Rev Microbiol* 43:345, 1989

117. Roos R, Gajdusek DC, Gibbs CJ Jr: The clinical characteristics of transmissible Creutzfeldt-Jakob disease. *Brain* 96:1, 1973

118. Chazot G, Broussolle E, Lapras CI, et al: New variant of Creutzfeldt-Jakob disease in a 26-year-old French man. *Lancet* 347:1181, 1996

119. Will RG, Ironside JW, Zeidler M, et al: A new variant of Creutzfeldt-Jakob disease in the UK. *Lancet* 347:921, 1996

120. Koch TK, Berg BO, DeArmond SJ, et al: Creutzfeldt-Jakob disease in a young adult with idiopathic hypopituitarism: possible relation to the administration of cadaveric human growth hormone. *N Engl J Med* 313:731, 1985

121. PHS Interagency Coordinating Committee: *Report on Human Growth Hormone and Creutzfeldt-Jakob Disease.* 14:1, 1997

122. Gajdusek DC, Gibbs CJ Jr, Asher DM, et al: Precautions in medical care of and in handling materials from patients with transmissible virus dementia (CJD). *N Engl J Med* 297:1253, 1977

123. Klitzman RL, Alpers MP, Gajdusek DC: The natural incubation period of kuru and the episodes of transmission in three clusters of patients. *Neuroepidemiology* 3:3, 1984

124. Masters CL, Harris JO, Gajdusek DC, et al: Creutzfeldt-Jakob disease: patterns of worldwide occurrence and the significance of familal and sporadic clustering. *Ann Neurol* 5:177, 1978

125. Hsiao K, Meiner Z, Kahana E, et al: Mutation of the prion protein in Libyan Jews with Creutzfeldt-Jakob disease. *N Engl J Med* 324:1091, 1991

126. Malmgren R, Kurland L, Mokri B, et al: The epidemiology of Creutzfeldt-Jakob disease. *Slow Transmissible Diseases of the Nervous System,* Vol. 1. Prusiner SB, Hadlow WJ, Eds. Academic Press, New York, 1979, p 93

127. Cousens SN, Harries-Jones R, Knight R, et al: Geographical distribution of cases of Creutzfeldt-Jakob disease in England and Wales 1970-84. *J Neurol Neurosurg Psychiatry* 53:459, 1990

128. Meggendorfer F: Klinische und genealogische Beobachtungen bei einem Fall von spastischer Pseudosklerose Jakobs. *Z Ges Neurol Psychiatr* 128:337, 1930

129. Masters CL, Gajdusek DC, Gibbs CJ Jr: Creutzfeldt-Jakob disease virus isolations from the Gerstmann-Sträussler syndrome. *Brain* 104:559, 1981

130. Doh-ura K, Tateishi J, Sasaki H, et al: Pro→Leu change at position 102 of prion protein is the most common but not the sole mutation related to Gerstmann-Sträussler syndrome. *Biochem Biophys Res Commun* 163:974, 1989

131. Goldgaber D, Goldfarb LG, Brown P, et al: Mutations in familial Creutzfeldt-Jakob disease and Gerstmann-Sträussler-Scheinker's syndrome. *Exp Neurol* 106:204, 1989

132. Kretzschmar HA, Honold G, Seitelberger F, et al: Prion protein mutation in family first reported by Gerstmann, Sträussler, and Scheinker. *Lancet* 337:1160, 1991

133. Owen F, Poulter M, Lofthouse R, et al: Insertion in prion protein gene in familial Creutzfeldt-Jakob disease. *Lancet* 1:51, 1989

134. Poulter M, Baker HF, Frith CD, et al: Inherited prion disease with 144 base pair gene insertion. 1. Genealogical and molecular studies. *Brain* 115:675, 1992

135. Goldfarb LG, Brown P, McCombie WR, et al: Transmissible familial Creutzfeldt-Jakob disease associated with five, seven, and eight extra octapeptide coding repeats in the *PRNP* gene. *Proc Natl Acad Sci USA* 88:10926, 1991

136. Kahana E, Milton A, Braham J, et al: Creutzfeldt-Jakob disease: focus among Libyan Jews in Israel. *Science* 183:90, 1974

137. Goldfarb LG, Mitrova E, Brown P, et al: Mutation in codon 200 of scrapie amyloid protein gene in two clusters of Creutzfeldt-Jakob disease in Slovakia. *Lancet* 336:514, 1990

138. Gabizon R, Rosenmann H, Meiner Z, et al: Mutation and polymorphism of the prion protein gene in Libyan Jews with Creutzfeldt-Jakob disease. *Am J Hum Genet* 33:828, 1993

139. Goldfarb LG, Brown P, Mitrova E, et al: Creutzfeldt-Jacob disease associated with the PRNP codon 200^{Lys} mutation: an analysis of 45 families. *Eur J Epidemiol* 7:477, 1991

140. Bertoni JM, Brown P, Goldfarb L, et al: Familial Creutzfeldt-Jakob disease with the PRNP codon 200^{lys} mutation and supranuclear palsy but without myoclonus or periodic EEG complexes (abstr). *Neurology* 42(suppl 3):350, 1992

141. Gabizon R, Telling G, Meiner Z, et al: Insoluble wild-type and protease-resistant mutant prion protein in brains of patients with inherited prion disease. *Nat Med* 2:59, 1996

142. Chapman J, Ben-Israel J, Goldhammer Y, et al: The risk of developing Creutzfeldt-Jakob disease in subjects with the *PRNP* gene codon 200 point mutation. *Neurology* 44:1683, 1994

143. Spudich S, Mastrianni JA, Wrensch M, et al: Complete penetrance of Creutzfeldt-Jakob disease in Libyan Jews carrying the E200K mutation in the prion protein gene. *Mol Med* 1:607, 1995

144. Goldfarb L, Korczyn A, Brown P, et al: Mutation in codon 200 of scrapie amyloid precursor gene linked to Creutzfeldt-Jakob disease in Sephardic Jews of Libyan and non-Libyan origin. *Lancet* 336:637, 1990

145. Goldfarb LG, Petersen RB, Tabaton M, et al: Fatal familial insomnia and familial Creutzfeldt-Jakob disease: disease phenotype determined by a DNA polymorphism. *Science* 258:806, 1992

146. Medori R, Montagna P, Tritschler HJ, et al: Fatal familial insomnia: a second kindred with mutation of prion protein gene at codon 178. *Neurology* 42:669, 1992

147. Lugaresi E, Medori R, Montagna P, et al: Fatal familial insomnia and dysautonomia with selective degeneration of thalamic nuclei. *N Engl J Med* 315:997, 1986

148. Petersen RB, Tabaton M, Berg L, et al: Analysis of the prion protein gene in thalamic dementia. *Neurology* 42:1859, 1992

149. Gambetti P, Parchi P, Petersen RB, et al: Fatal familial insomnia and familial Creutzfeldt-Jakob disease: clinical, pathological and molecular features. *Brain Pathol* 5:43, 1995

150. Goldfarb LG, Haltia M, Brown P, et al: New mutation in scrapie amyloid precursor gene (at codon 178) in Finnish Creutzfeldt-Jakob kindred. *Lancet* 337:425, 1991

151. Kretzschmar HA, Neumann M, Stavrou D: Codon 178 mutation of the human prion protein gene in a German family (Backer family): sequencing data from 72-year-old celloidin-embedded brain tissue. *Acta Neuropathol* 89:96, 1995

152. Owen F, Poulter M, Collinge J, et al: Codon 129 changes in the prion protein gene in Caucasians. *Am J Hum Genet* 46:1215, 1990

153. Kitamoto T, Tateishi J: Human prion diseases with variant prion protein. *Philos Trans R Soc Lond B* 343:391, 1994

154. Laplanche J-L, Chatelain J, Launay J-M, et al: Deletion in prion protein gene in a Moroccan family. *Nucleic Acids Res* 18:6745, 1990

155. Vnencak-Jones CL, Phillips JA: Identification of heterogeneous PrP gene deletions in controls by detection of allele-specific heteroduplexes (DASH). *Am J Hum Genet* 50:871, 1992

156. Cervenáková L, Brown P, Piccardo P, et al: 24-nucleotide deletion in the *PRNP* gene: analysis of associated phenotypes. *Transmissible Subacute Spongiform Encephalopathies: Prion Diseases.* Court L, Dodet B, Eds. Elsevier, Paris, 1996, p 433

157. Palmer MS, Mahal SP, Campbell TA, et al: Deletions in the prion protein gene are not associated with CJD. *Hum Mol Genet* 2:541, 1993

158. Palmer MS, Dryden AJ, Hughes JT, et al: Homozygous prion protein genotype predisposes to sporadic Creutzfeldt-Jakob disease. *Nature* 352:340, 1991

159. Collinge J, Palmer MS, Dryden AJ: Genetic predisposition to iatrogenic Creutzfeldt-Jakob disease. *Lancet* 337:1441, 1991

160. Doh-ura K, Kitamoto T, Sakaki Y, et al: CJD discrepancy. *Nature* 353:801, 1991

161. Miyazono M, Kitamoto T, Doh-ura K, et al: Creutzfeldt-Jakob disease with codon 129 polymorphism (valine): a comparative study of patients with codon 102 point mutation or without mutations. *Acta Neuropathol* 84:349, 1992

162. Tateishi J, Kitamoto T: Developments in diagnosis for prion diseases. *Br Med Bull* 49:971, 1993

163. Pattison IH, Millson GC: Scrapie produced experimentally in goats with special reference to the clinical syndrome. *J Comp Pathol* 71:101, 1961

164. Dickinson AG, Meikle VMH, Fraser H: Identification of a gene which controls the incubation period of some strains of scrapie agent in mice. *J Comp Pathol* 78:293, 1968

165. Fraser H, Dickinson AG: Scrapie in mice. Agent-strain differences in the distribution and intensity of grey matter vacuolation. *J Comp Pathol* 83:29, 1973

166. Hecker R, Taraboulos A, Scott M, et al: Replication of distinct prion isolates is region specific in brains of transgenic mice and hamsters. *Genes Dev* 6:1213, 1992

167. DeArmond SJ, Yang S-L, Lee A, et al: Three scrapie prion isolates exhibit different accumulation patterns of the prion protein scrapie isoform. *Proc Natl Acad Sci USA* 90:6449, 1993

168. DeArmond SJ, Sánchez H, Yehiely F, et al: Selective neuronal targeting in prion disease. *Neuron* 19:1337, 1997

169. Carp RI, Meeker H, Sersen E: Scrapie strains retain their distinctive characteristics following passages of homogenates from different brain regions and spleen. *J Gen Virol* 78:283, 1997

170. Carlson GA, Ebeling C, Yang S-L, et al: Prion isolate specified allotypic interactions between the cellular and scrapie prion proteins in congenic and transgenic mice. *Proc Natl Acad Sci USA* 91:5690, 1994

171. Scott M, Groth D, Foster D, et al: Propagation of prions with artificial properties in transgenic mice expressing chimeric PrP genes. *Cell* 73:979, 1993

172. Dickinson AG, Meikle VM, Fraser H: Genetical control of the concentration of ME7 scrapie agent in the brain of mice. *J Comp Pathol* 79:15, 1969

173. Scott MR, Safar J, Telling G, et al: Identification of a prion protein epitope modulating transmission of bovine spongiform encephalopathy prions to transgenic mice. *Proc Natl Acad Sci USA* 94:14279, 1997

174. Bessen RA, Marsh RF: Identification of two biologically distinct strains of transmissible mink encephalopathy in hamsters. *J Gen Virol* 73:329, 1992

175. Bessen RA, Marsh RF: Distinct PrP properties suggest the molecular basis of strain variation in transmissible mink encephalopathy. *J Virol* 68:7859, 1994

176. Monari L, Chen SG, Brown P, et al: Fatal familial insomnia and familial Creutzfeldt-Jakob disease: different prion proteins determined by a DNA polymorphism. *Proc Natl Acad Sci USA* 91:2839, 1994

177. Parchi P, Castellani R, Capellari S, et al: Molecular basis of phenotypic variability in sporadic Creutzfeldt-Jakob disease. *Ann Neurol* 39:767, 1996

178. Scott M, Foster D, Mirenda C, et al: Transgenic mice expressing hamster prion protein produce species-specific scrapie infectivity and amyloid plaques. *Cell* 59:847, 1989

179. Bessen RA, Marsh RF: Biochemical and physical properties of the prion protein from two strains of the transmissible mink encephalopathy agent. *J Virol* 66:2096, 1992

180. Otvos L Jr, Thurin J, Kollat E, et al: Glycosylation of synthetic peptides breaks helices. *Int J Peptide Protein Res* 38:476, 1991

181. O'Connor SE, Imperiali B: Modulation of protein structure and function by asparagine-linked glycosylation. *Chem Biol* 3:803, 1996

182. Collinge J, Sidle KCL, Meads J, et al: Molecular analysis of prion strain variation and the aetiology of "new variant" CJD. *Nature* 383:685, 1996

183. Parry HB: Scrapie: a transmissible and hereditary disease of sheep. *Heredity* 17:75, 1962

184. Parry HB: *Scrapie Disease in Sheep.* Academic Press, New York, 1983

185. Dickinson AG, Young GB, Stamp JT, et al: An analysis of natural scrapie in Suffolk sheep. *Heredity* 20:485, 1965

186. Goldmann W, Hunter N, Foster JD, et al: Two alleles of a neural protein gene linked to scrapie in sheep. *Proc Natl Acad Sci USA* 87:2476, 1990

187. Goldmann W, Hunter N, Manson J, et al: The PrP gene of the sheep, a natural host of scrapie (abstr). VIIIth International Congress of Virology, Berlin, Aug. 26–31, 1990, p 284

188. Laplanche J-L, Chatelain J, Beaudry P, et al: French autochthonous scrapied sheep without the 136Val PrP polymorphism. *Mamm Genome* 4:463, 1993

189. Clousard C, Beaudry P, Elsen JM, et al: Different allelic effects of the codons 136 and 171 of the prion protein gene in sheep with natural scrapie. *J Gen Virol* 76:2097, 1995

190. Westaway D, Zuliani V, Cooper CM, et al: Homozygosity for prion protein alleles encoding glutamine-171 renders sheep susceptible to natural scrapie. *Genes Dev* 8:959, 1994

191. Ikeda T, Horiuchi M, Ishiguro N, et al: Amino acid polymorphisms of PrP with reference to onset of scrapie in Suffolk and Corriedale sheep in Japan. *J Gen Virol* 76:2577, 1995

192. O'Rourke KI, Melco RP, Mickelson JR: Allelic frequencies of an ovine scrapie susceptibility gene. *Anim Biotechnol* 7:155, 1996

193. Hunter N, Moore L, Hosie BD, et al: Association between natural scrapie and PrP genotype in a flock of Suffolk sheep in Scotland. *Vet Rec* 140:59, 1997

194. Hunter N, Cairns D, Foster JD, et al: Is scrapie solely a genetic disease? *Nature* 386:137, 1997

195. Goldmann W, Hunter N, Benson G, et al: Different scrapie-associated fibril proteins (PrP) are encoded by lines of sheep selected for different alleles of the *Sip* gene. *J Gen Virol* 72:2411, 1991

196. Hunter N, Foster JD, Benson G, et al: Restriction fragment length polymorphisms of the scrapie-associated fibril protein (PrP) gene and their association with susceptiblity to natural scrapie in British sheep. *J Gen Virol* 72:1287, 1991

197. Goldfarb LG, Brown P, Little BW, et al: A new (two-repeat) octapeptide coding insert mutation in Creutzfeldt-Jakob disease. *Neurology* 43:2392, 1993

198. Hunter N, Goldmann W, Smith G, et al: Frequencies of PrP gene variants in healthy cattle and cattle with BSE in Scotland. *Vet Rec* 135:400, 1994

199. Hope J, Reekie LJD, Hunter N, et al: Fibrils from brains of cows with new cattle disease contain scrapie-associated protein. *Nature* 336:390, 1988

200. Wilesmith JW, Ryan JBM, Atkinson MJ: Bovine spongiform encephalopathy: epidemiologic studies on the origin. *Vet Rec* 128:199, 1991

201. Wilesmith JW, Ryan JBM, Hueston WD: Bovine spongiform encephalopathy: case-control studies of calf feeding practices and meat and bonemeal inclusion in proprietary concentrates. *Res Vet Sci* 52:323, 1992

202. Fraser H, McConnell I, Wells GAH, et al: Transmission of bovine spongiform encephalopathy to mice. *Vet Rec* 123:472, 1988

203. Robinson MM, Hadlow WJ, Huff TP, et al: Experimental infection of mink with bovine spongiform encephalopathy. *J Gen Virol* 75:2151, 1994

204. Bruce M, Chree A, McConnell I, et al: Transmission of bovine spongiform encephalopathy and scrapie to mice: strain variation and the species barrier. *Philos Trans R Soc Lond B* 343:405, 1994

205. Baker HF, Ridley RM, Wells GAH: Experimental transmission of BSE and scrapie to the common marmoset. *Vet Rec* 132:403, 1993

206. Lasmézas CI, Deslys J-P, Demaimay R, et al: BSE transmission to macaques. *Nature* 381:743, 1996

207. Wickner RB: [URE3] as an altered URE2 protein: evidence for a prion analog in *Saccharomyces cerevisiae*. *Science* 264:566, 1994

208. Chernoff YO, Lindquist SL, Ono B, et al: Role of the chaperone protein Hsp104 in propagation of the yeast prion-like factor [*psi*$^+$]. *Science* 268:880, 1995

209. Patino MM, Liu J-J, Glover JR, et al: Support for the prion hypothesis for inheritance of a phenotypic trait in yeast. *Science* 273:622, 1996

210. Landman OE: The inheritance of acquired characteristics. *Annu Rev Genet* 25:1, 1991

211. Deleu C, Clavé C, Bégueret J: A single amino acid difference is sufficient to elicit vegetative incompatibility in the fungus *Podospora anserina*. *Genetics* 135:45, 1993

Acknowledgments

I thank Fred E. Cohen, Stephen J. DeArmond, and Michael Scott for many stimulating discussions and for their help in preparing this manuscript. This work was supported by grants from the National Institutes of Health (NS14069, AG08967, AG02132, NS22786, and AG10770) and the American Health Assistance Foundation, as well as by gifts from the Sherman Fairchild Foundation, the G. Harold and Leila Y. Mathers Foundation, the Bernard Osher Foundation, and Centeon.

Figures 1 and 2   Marcia Kammerer.

Figure 3   Marcia Kammerer. Data were compiled by Paul Bamborough and Fred E. Cohen.

Figures 4 through 6   Dimitry Schidlovsky.

CHAPTER 11

The Molecular Pathogenesis of Multiple Sclerosis

Jorge R. Oksenberg, Ph.D., Stephen L. Hauser, M.D.

Multiple sclerosis (MS), the prototypic demyelinating disease in humans, affects more than one million in people in North America and Western Europe. Symptoms of MS are caused by immune-mediated injury to the insulating myelin sheath within the central nervous system; peripheral nerve myelin is spared. Transmission of action potentials along exposed axons is impaired, resulting in varied neurologic deficits, including sensory loss, weakness, visual loss, vertigo, incoordination, sphincter disturbances, and cognitive deficits.

MS generally begins in early or middle adulthood, although onset as early as two years or as late as 80 years of age may rarely occur. Women are affected twice as often as men. Onset may be insidious or explosive, and the disease course, although unpredictable, ultimately disables 90 percent or more of those affected. The course may consist of recurrent attacks, each followed by variable recovery (*relapsing MS*), by attacks followed by or superimposed upon steady progressive worsening (*relapsing-progressive MS*), or by a progressive course from onset (*primary progressive MS*). Total societal costs in the United States are an estimated $150 billion annually. No curative therapy exists for MS, although the Food and Drug Administration has recently approved three different drugs (interferon [IFN] beta-1a, IFN beta-1b, and Copolymer 1) that appear to be modestly effective in reducing the frequency and severity of relapses.

A large body of immunologic, epidemiological, and genetic data indicate that tissue injury in MS results from an abnormal immune response to one or more myelin antigens that occurs in genetically susceptible persons after exposure to an as yet undefined causal agent. The past few years have seen real progress in defining the molecular basis of MS and have set the stage for new therapeutic approaches based on correction of specific underlying disease mechanisms. In this chapter, we review current understanding of the immunology of MS, the nature of genetic susceptibility, and the exciting prospects for effective therapy based on targeting of specific molecules.

The Nature of the Lesion

The pathological hallmark of MS is the plaque—a well-demarcated gray or pink lesion characterized histologically by inflammation, demyelination, and gliosis (scarring). MS plaques are multiple and generally asymmetric and tend to concentrate in deep white matter near the lateral ventricles, corpus callosum, floor of the fourth ventricle, deep periaqueductal region, optic nerves and tracts, corticomedullary junction, and cervical spinal cord.

Inflammation may be the initial event in plaque formation following a breach in the blood-brain barrier (BBB).[1] The acute MS lesion consists of perivascular and parenchymal infiltration by mononuclear cells, both thymus-derived (T) cells and macrophages, and myelin breakdown that appears to be mediated by the infiltrating cells. Only rare B cells and plasma cells are present. Preservation of axon cylinders in the presence of demyelination is characteristic, although this finding is relative rather than absolute; in approximately 10 percent of lesions, there is significant axonal destruction, and in rare cases, complete destruction of the neuropil and cavitation. Vesicular disruption of the myelin membrane is an early pathological feature, a finding that can be simulated in vitro by the application of cytokines, autoantibodies, or calcium ionophores. As lesions evolve, axons traversing the lesions show marked irregular beading, astrocytes proliferate, and lipid-laden macrophages containing myelin debris become prominent. Progressive fibrillary gliosis ensues, and mononuclear cells gradually disappear. In some MS lesions but not others, oligodendrocytes appear to proliferate initially, but these cells are apparently destroyed as the gliosis progresses.

In chronic MS lesions, complete or nearly complete demyelination, dense gliosis (more severe than in most other neuropathological conditions), and loss of oligodendroglia are found. In some chronic active MS lesions, gradations in the histologic findings from the center to the lesion edge suggest that lesions expand by concentric outward growth.

Multiple Sclerosis as an Autoimmune Disease

Normally, the immune system recognizes and reacts against foreign substances (antigens) but tolerates the body's own components. If tolerance to self-antigens fails to develop, the result may be an autoimmune disease; that is, a person's immune system is triggered to mount a specific pathological response against self. This response can be generalized, as in systemic lupus erythematosus, or targeted to a particular tissue or organ, as in MS. Ten percent of the population is thought to suffer from some form of autoimmune disease.[2]

The Molecular Basis of Antigen Recognition

To maintain immune homeostasis, the immune system must be capable of responding specifically and appropriately to a wide array of antigenic challenges. These capacities are mediated by the generation and expression of diverse antigen receptors on T and B cells. Diversity is generated through genetic rearrangement within the antigen receptor gene complexes, a process in which different gene segments are apposed to create rearranged receptor gene transcripts with an extraordinarily diverse array of sequence possibilities. A single functional antigen receptor molecule is expressed by each T and B cell and its progeny, and the rearrangement generates as many as 10^9 or more different populations (clones) of immune cells, each with unique antigen-binding characteristics.

Unlike immunoglobulins (Ig), which are capable of recognizing free antigen, the T cell antigen receptor (TCR) only recognizes antigen fragments that are noncovalently bound to major histocompatibility complex (MHC) molecules expressed on the surface of specialized antigen-presenting cells (APCs) [*see Figure 1*]. APCs are a heterogeneous group of cells able to directly bind free antigen (i.e., B cells) or to process and present antigen after internalization (i.e., macrophages, Langerhans' cells, endothelial cells, and fibroblasts). MHC molecules are encoded by a cluster of genes located in humans on the short arm of chromosome 6 [*see Figure 2*]. The class I and class II MHC genes are members of the immunoglobulin supergene family and encode highly polymorphic cell surface molecules responsible for antigen presentation. MHC class I molecules engage CD8+ T cells, whereas MHC class II molecules interact with CD4+ T cells. Multiple cell surface molecules interact during antigen presentation to T cells, but the trimolecular complex (MHC/peptide/TCR) is the focal point for initiation and propagation of most immune responses to protein antigens.

Why has the immune system evolved this complex process of antigen recognition by T cells? T cell recognition of antigen is the key initiating event for most immune responses and is prerequisite for triggering of effector T cells, for provision of help for B cell activation and differentiation into antibody-forming cells, and for activation of macrophages. Many foreign antigens share structural homologies with self-molecules, yet removal of all TCRs that recognize cross-reactive determinants would cripple the ability of the system to respond appropriately. By narrowly restricting T cell recognition to specific microenvironments in which APCs are present and antigen processing occurs, the underlying activity of the immune system can be regulated without sacrificing its diversity and specificity. Thus, naturally occurring autoreactive T cells are retained in the adult immune system without deleterious consequences because under normal circumstances, self-antigens are not efficiently presented to such cells. Clearly, however, the potential for autoimmunity exists.

During intrauterine life, stem cells migrate into the thymus, where they undergo differentiation. In the thymus, two competitive selection processes influence the final composition of the mature T cell repertoire: (1) positive selection of T cells bearing receptors that have modest affinity for self-MHC and that will be useful to the host and (2) negative selection by clonal deletion or inactivation of T cells bearing receptors with high affinity for self-MHC or

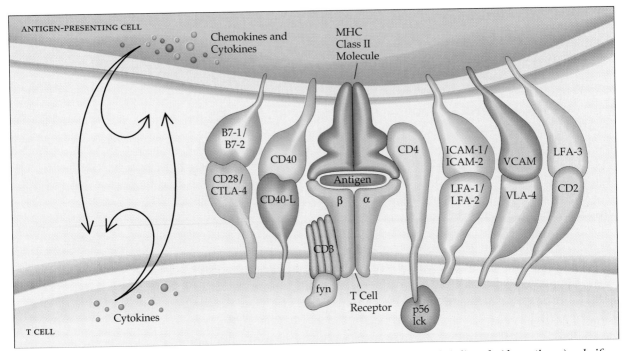

Figure 1   *The molecular basis of antigen recognition. T cell antigen receptors bind their ligands (the antigens) only if presented in the framework of MHC molecules. The T cell receptors (TCRs) of circulating lymphocytes are glycoprotein heterodimers (either αβ, present in over 95 percent of peripheral blood lymphocytes, or γδ) noncovalently associated with five other invariant molecules collectively known as the CD3 complex. Other cell surface molecules, including members of the integrin and immunoglobulin supergene families, participate in the process; some of these ligand–receptor interactions have a pure anchoring effect. Studies estimate relatively low-affinity binding values for the MHC peptide:TCR complex, in the order of the lowest known antibody binding reactions. T cells initially bind the antigen-presenting cell through low-affinity leukocyte function–associated antigen–intercellular adhesion molecule (LFA-ICAM) interactions. Conformational changes in LFA-1 increase affinity and prolong cell-cell contact. Other molecules participate in signal transduction to one or both cells. For example, the TCR-CD3 complex delivers transduction signals after the interaction with the MHC-antigen complex. A second signal delivered by the B7-CD28 interaction is necessary to induce T cell activation. The absence of this second signal may result in anergy. Similarly, the CD2 molecule on T cells is involved in activation once engaged with LFA-3, which is widely distributed on APCs. CD40 is the most potent activating signal during T cell–B cell interactions. T cell activation is expressed as proliferation, differentiation to effector functions, and cytokine production.*

MHC bound to antigens present in the thymus. Negative selection will remove or inactivate most T cells with autoreactive potential. Thymic selection involves interactions between TCRs, self-antigens, MHC, and accessory molecules. Notably, some myelin antigens are expressed in the thymus. Regulation of autoreactive T cells is extended to the periphery during adult life in the form of (1) paralysis or anergy of autoreactive cells that renders them unable to respond, (2) apoptosis (programmed cell death) in the presence of high concentrations of antigen, and/or (3) specific (idiotypic) T cell–T cell interactions that actively suppress autoimmune cells.[3] Clonal ignorance has also been demonstrated and refers to the observation that self-reactive lymphocytes simply ignore self-antigens, such as when the antigens are in tissues sequestered from the circulation. Perhaps the fundamental problem in autoimmunity is the loss of self-tolerance and the escape of harmful T cells from surveillance.

The complex cellular communications between T cells and other cells of the immune system are mediated by a group of signaling molecules called *cytokines*. Cytokines regulate the activation, differentiation, and proliferation of T cells. Under their influence, these cells differentiate into two major pathways [*see Figure 3*]. T helper type 1 (Th1) cells produce primarily interleukin-2

(IL-2), tumor necrosis factor (TNF), and IFN gamma and participate in inflammatory responses such as macrophage activation, cell-mediated immunity, and delayed type hypersensitivity (DTH) responses. In addition, Th1-type cytokines influence B cell differentiation into IgG2a-producing plasma cells. T helper type 2 (Th2) cells produce IL-3, IL-4, IL-5, IL-6, and IL-10. They facilitate humoral immunity and promote synthesis of IgG1, IgG4, and IgE antibody isotypes. Th2 cells are also involved in the induction of T cell tolerance and differentiation of mast cells and eosinophils. In many pathological states, cytokines have been shown to induce and regulate numerous critical cellular processes, including activation and differentiation, recruitment and migration, and proliferation and death. Elucidation of specific roles for individual cytokines in the MS disease process may provide opportunities to use them as potential starting points for therapeutic intervention.[4]

Identification of Autoantigens in Multiple Sclerosis

An essential prerequisite to defining the molecular basis of an autoimmune disease is knowledge of the responsible autoantigen(s). The role of individual autoantigens as inducers of demyelination has proved difficult to investigate in humans; thus, investigators have turned to animal models. Beginning in the 1930s, Rivers and others defined experimental allergic (or autoimmune) encephalomyelitis (EAE).[5] EAE can be induced in a variety of animal species, including nonhuman primates, by immunization with myelin proteins or their peptide derivatives. In genetically susceptible animals, immunization induces brain inflammation accompanied by various signs of neurologic disease. EAE and MS share common clinical, histologic, immunologic, and genetic features; hence, EAE is widely considered a relevant model for the human disease. Demonstration in the early 1960s that EAE could be adoptively transferred by myelin-sensitized T cells[6] inaugurated the era of T cell immunology in MS research, an approach that in many respects dominates the field to this day.

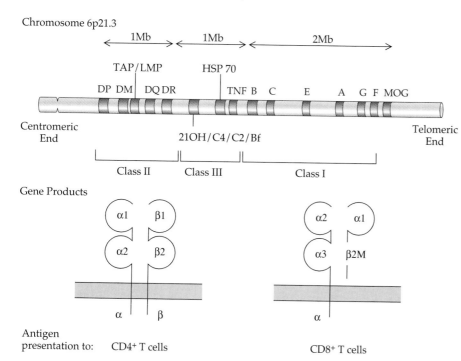

Figure 2 *Genetic organization of the major histocompatibility complex. The diagram show the relative positions of the major class I and class II loci involved in antigen presentation, as well as other examples of the more than 200 genes encoded within this 3-million base pair (Mb) complex.*

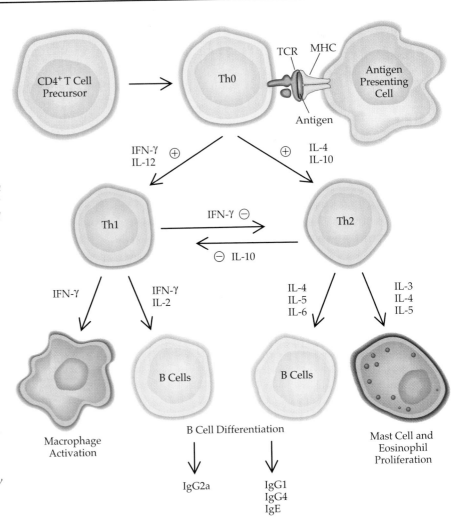

Figure 3 *Immune responses mediated by T helper type 1 (Th1) and T helper type 2 (Th2) cells. The types of cytokines released by different populations of CD4+ T cells are thought to influence different facets of the immune response. Th1 cells secrete cytokines associated with cellular immunity, whereas Th2 cells secrete cytokines associated with B cell activation. The two populations of Th cells are in some respects mutual antagonists. Interferon gamma (IFN-γ) produced by Th1 cells and interleukin-10 (IL-10) produced by Th2 cells antagonize each other's effects. A third subset of T helper cells, Th0, with a pattern of cytokine production overlapping both Th1 and Th2, has been identified and may represent a precursor population. Preferential activation of one T cell type or the other may explain why the immune response may be predominantly cellular in some circumstances (Th1), and humoral (Th2) in others.*

The two quantitatively major myelin proteins, myelin basic protein (MBP) and proteolipid protein (PLP), which make up about 30 and 50 percent of myelin proteins by weight, respectively, have received the most attention as potential T cell autoantigens in MS [*see Figure 4*]. Both can effectively induce EAE. Both are thought to participate in myelin compaction—MBP between intracytoplasmic surfaces (the major dense line by electron microscopy) and PLP between adjacent myelin lamellae (the intraperiod line). Several lines of evidence point to the importance of an anti-MBP T cell immune response in MS. MBP-reactive T cells have been found to undergo chronic stimulation in the peripheral blood in MS[7] and appear to migrate selectively to the CNS, where they are detected in MS lesions[8] and CSF.[9] Because MBP is located not only in the CNS but also in the peripheral nervous system (PNS), it is difficult to explain the specificity of MS for the CNS on the basis of autoimmunity to MBP. Until recently, PLP was thought to be expressed exclusively in the CNS but is now known to be expressed in the PNS as well.[10] Direct evidence in support of a primary role for PLP in MS is scant; circulating PLP-reactive T cells home to the CSF in MS, but this occurs to a lesser degree than for MBP-reactive cells.

The most promising new MS autoantigen to be identified is myelin oligodendrocyte glycoprotein (MOG). MOG has been shown to be a critical antigen in a primate model of EAE.[11] The common marmoset *(Callithrix jacchus)* is a small New World monkey (about the size of a guinea pig) that, when immunized against MOG, develops a chronic relapsing form of EAE that is unique because the histologic lesions are identical to those of MS. These primates develop in utero as genetically distinct (nonidentical) twins or triplets that share bone-marrow–derived cells through a common placental circulation. As a consequence, a permanent, stable bone-marrow chimerism develops that permits adoptive transfer (passage) of functional T cells between genetically distinct siblings without tissue rejection. This model has clarified the immune components of an MS-like lesion and has produced several surprising findings that have modified thinking about potential pathogenic events in MS. One such finding is that the MS-like lesion is not exclusively T cell mediated but that a synergistic T cell and antibody response is required to produce demyelination; a second finding is that MOG is a dominant autoantigen of myelin. In the marmoset model, T cells that recognize MBP, MOG, or both enter the CNS and disrupt the BBB; they cause inflammation but not demyelination. Demyelination requires the presence of antibodies against MOG that gain access to the CNS across the disrupted barrier.[12-15] The finding that MS-like demyelination occurs only in the presence of autoantibodies against MOG clarifies to a great extent why typical EAE in rodents, mediated exclusively by T cells, is an inflammatory but not a demyelinating disease.

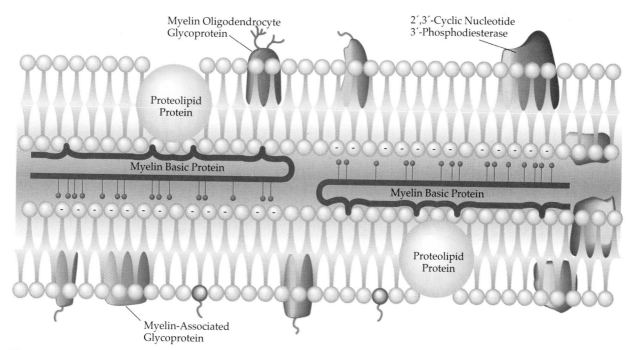

Figure 4 *The molecular architecture of the central myelin sheath at the myelin surface-associated zone. Myelin is formed by membrane extensions of oligodendrocytes that wrap around the axon. Central myelin consists of about 75 to 80 percent lipid and 20 to 25 percent protein. Proteolipid protein (PLP) accounts for about 50 percent of total myelin protein and myelin basic protein (MBP) for another 30 percent. Myelin-associated glycoprotein (MAG) and myelin oligodendrocyte glycoprotein (MOG) are minor constituents of whole myelin protein. PLP is integrated in the myelin membrane. MBP is a cytosolic protein. MAG is located in the periaxonal space of the myelin sheath. MOG is located at the surface, so it is the myelin protein most exposed to humoral and cellular immune responses. PLP, MBP, and MOG are encephalitogenic in sensitive animals. CNP is 2′,3′-cyclic nucleotide 3′-phosphodiesterase.*

MOG is a quantitatively minor constituent of CNS myelin proteins; it is located exclusively on oligodendrocyte surfaces and on the outermost lamellae of the myelin sheath, making it readily accessible to an immune attack, particularly to attack by an autoantibody. One hypothesis is that MOG may function to terminate myelin synthesis.[16] In keeping with a possible role for MOG as an autoantigen in MS, MOG is exclusively a CNS protein not known to be present in the PNS. Furthermore, studies suggest that peripheral blood T cells from MS patients are specifically sensitized to MOG. Other minor components of myelin as well as other antigens, including heat shock proteins, β-arrestin and arrestin, glial fibrillary acidic protein (GFAP), and astrocyte-derived calcium-binding protein (S1000β), have also been proposed as candidate antigens in MS.[17]

In most MS patients, an elevated level of intrathecally synthesized immunoglobulins can be detected in the CNS. Although the specificity of these antibodies is mostly unknown, anti-MBP specificities have been reported.[18,19] The antibodies may coat the myelin sheath and make it more available for phagocytosis by macrophages. CNS immunoglobulins may also induce direct myelinolysis via activation of a Ca^{2+}-dependent protease acting on MBP.[20]

In summary, it appears likely that a multifaceted cellular and humoral immune response against multiple different proteins underlies autoimmunity in MS and that therapies directed against any single antigen are unlikely to succeed.

Triggering an MS Attack

Magnetic resonance imaging studies have provided invaluable contributions to our understanding of the dynamics of the MS disease process. MRI measurements of BBB breakdown, a marker of inflammation in MS, indicate that very frequent bursts of inflammation occur in patients with active disease. These bursts are often multifocal and appear approximately monthly, or seven to 10 times more frequently than do clinical attacks of MS; thus, most bursts are clinically silent.

What triggers these bursts of multifocal inflammation in MS? The most plausible theories propose that recurrent exposures to exogenous pathogens activate encephalitogenic T cells in peripheral blood. This could occur via nonspecific polyclonal activation of T and B cells by bacterial or viral determinants, by molecular mimicry, by an innocent-bystander mechanism, or by superantigens. Molecular mimicry refers to induction of an autoimmune response because of a structural homology between a self-protein and a protein in a viral or bacterial pathogen. MBP shares extensive amino acid homologies with proteins of measles, influenza, and adenovirus. For example, residues 91 to 101 of MBP (a region known to be commonly recognized by human MBP-reactive T cells) share identical stretches of four to six amino acids with adenovirus. Homology may be necessary at only a few amino acids for efficient T cell recognition to occur.[2] Amino acid homology may not even be required for cross-reactivity as long as the three-dimensional structure of the determinant that is recognized by T cells (the epitope) is similar.[21]

The innocent-bystander hypothesis proposes that activation of autoimmune T cells occurs as a consequence of viral infections of the CNS; such infections may be asymptomatic. As examples, immune responses against MBP are generated during measles encephalitis and human T cell lymphotropic virus type I (HTLV-I) infection in humans, and coronavirus infection in rodents. Thus, a neurotropic virus may infect the nervous system and, in doing so, stimulate an immune response not only to the virus but also to normal nervous system proteins. This may occur via release of cytokines that amplify autoreactive T cell responses or an increase in the efficiency of presentation of autoantigens on APCs.

Another potential mechanism implicates superantigens in the etiology of autoimmunity. The term *superantigen* defines antigens that, at very low concentrations (in the picomolar range), can stimulate subsets of T cells.[22] Superantigens bind with high affinity to class II MHC molecules outside the antigen-binding groove. They interact with the variable region of the β chain (Vβ) of the TCR in the region of the β-pleated sheet, away from the antigen-binding site (the CDR3 region). In a non-MHC restricted manner, without need for antigen processing, the class II superantigen complexes trigger proliferation of T cells expressing particular TCR-Vβ chains. A notable feature of superantigenic stimulation is that the responding T cells initially mount a vigorous response but then are deleted or lose the ability to respond to antigen (i.e., they become anergic). Exogenous superantigens include the toxins of many common bacteria and possibly components of certain viruses. Superantigens are associated with numerous human diseases, including food poisoning, toxic-shock syndrome, and scalded skin syndrome.[23] In one study,[24] the ability of staphylococcal toxins to stimulate human T cells specific to MBP or PLP was examined; all myelin-specific T cells responded in proliferation studies to at least one superantigen. In some experiments, the superantigenic toxins were more than 10^5 times more potent in stimulating the T cells than were the myelin antigens to which the T cells were initially sensitized! Hence, superantigen stimulation accompanying infection clearly has the potential to activate preexisting myelin-specific T cells and perhaps to induce exacerbations.

Multiple Sclerosis as a Genetic Disease

A strong genetic influence on susceptibility to MS has been identified in studies of different ethnic groups and in family, adoption, and twin studies.[25,26] MS is a common disease among whites, and it has long been known that MS tends to aggregate in certain families. One measure of familial aggregation in MS is the estimate of the disease risk in siblings of affected patients, determined by evaluating the ratio of lifetime risk to siblings (Ks) versus the population prevalence (K) of the disease (λ_S= Ks/K). For MS, the total λ_S is between 20 (0.02/0.001) and 40 (0.04/0.001).[27] These estimates are roughly similar to λ_S values for type 1 diabetes mellitus. Recent studies of adoption and half siblings and of offspring of two parents with MS support the concept that familial aggregation in MS is determined primarily by genetics rather than environment.[28] Twin studies provide the most powerful evidence for a genetic etiology for MS; concordance estimates of 30 percent in monozygotic twins compared with five percent in dizygotic twins (the same as nontwin siblings) have been consistently reported.[29] Recurrence risk estimates (that is, the risk to a relative of a patient with known MS) in first-, second-, and third-degree relatives, combined with twin data, have been used to predict the nature of genetic transmission; this type of modeling indicates that multiple genes must interact to result in an MS-prone genotype. Thus, MS is a *polygenic disorder*. It is also possible (but unproved) that *genetic heterogeneity* exists, meaning that different genes influence susceptibility in some individuals but not in others.

Genetic studies in EAE have been extraordinarily useful in defining the complex interplay of multiple genes that can result in brain inflammation; these studies have also been useful in validating EAE as an MS disease model.[26] The best-characterized EAE susceptibility gene resides within the *MHC*.[30] Induction and full clinical manifestations of EAE are strongly influenced by inherited differences (polymorphisms) of MHC class II region genes; certain *MHC* haplotypes are permissive for EAE, whereas others are resistant. Classic genetics and whole-genome screening have been used to identify several additional genetic regions that participate in conferring EAE susceptibility. These studies provide compelling evidence for the hypothesis that susceptibility to demyelination is largely genetically determined. In both EAE and MS, a multiple-locus model is applicable. Each locus may contribute to a specific stage of EAE pathology, although some loci are probably involved in several steps of the autoimmune process [*see Figure 5*]. However, no locus seems to be an absolute requirement for the susceptible

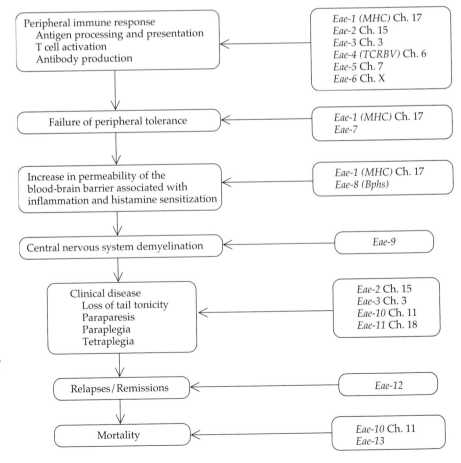

Figure 5 *Model of genetic contributions to experimental allergic (or autoimmune) encephalomyelitis (EAE) pathology. Autoimmune demyelination depends on multiple genes with complex interactions.*

phenotype; that is, a susceptible EAE phenotype can be achieved by different combinations of genotypes. As the genes that contribute to EAE are identified, such genes will represent strong candidates for testing in MS.

Following Strong Leads: the MHC and Multiple Sclerosis

The association of MS with *MHC* genes has been known for more than 20 years. However, despite great progress in understanding the immunobiology of the MHC, much remains to be learned about the underlying mechanism responsible for its genetic association with many autoimmune disease states. MHC class I and class II molecules are highly polymorphic cell surface glycoproteins whose primary role in an immune response is to display and present short antigenic peptide fragments to antigen-specific T cells. The importance of *MHC* compatibility in clinical transplantation stimulated the development of reagents and methods to fully comprehend the degree of genetic variability in the system. As a result, an impressive body of knowledge has been accumulated on the function and population genetics of the human *MHC* (the human leukocyte antigen [*HLA*] system) [*see Figure 2*]. The ability to respond to an antigen, whether foreign or self, and the nature of that response, is partly determined by the unique amino acid sequences of *MHC* molecules, an observation that provided the rationale for focusing on associations between *HLA* genotypes and susceptibility to autoimmune disease.

Most *MHC* association studies in MS have focused on Caucasians of northern European descent, in whom predisposition to disease has been consistently associated with the class II *HLA-DRB1*1501-DQA1*0102-DQB1*0602* haplotype (the molecular designation for the serologically defined DR2 haplotype or combination of alleles). Attempts to further localize a susceptibility gene within the *DR* or *DQ* region of the *HLA* have not resulted in consensus. The close proximity of the *DR* and *DQ* gene clusters (300,000 base pairs apart) and the fact that *DRB1*1501* and *DQB1*0602* are found exclusively in combination in Caucasians of northern European descent have prevented clear resolution of the relative contribution of each gene.

The mechanism(s) underlying the association of *HLA-DRB1*1501-DQA1*0102-DQB1*0602* with MS are not yet fully understood. Possible explanations include the following:

1. Determinant model. *HLA-DRB1*1501* and/or *DQA1*0102-DQB1*0602* genes encode a class II recognition molecule with a propensity to bind peptide antigens of myelin and stimulate encephalitogenic T cells.
2. Thymic selection model. *HLA-DRB1*1501* and/or *DQA1*0102-DQB1*0602* determinants fail to negatively select (delete) autoreactive T cells within the embryonic thymic microenvironment.
3. Cytokine regulation model. *HLA-DRB1*1501* and/or *DQA1*0102-DQB1*0602* synergize with other genes to preferentially stimulate Th1 cells (producing IFN gamma, TNF-α, IL-12) rather than Th2 cells (IL-4, IL-6, IL-10, transforming growth factor–β [TGF-β]) in response to encephalitogenic stresses, trauma, or immunologic insults. It has been postulated that tolerance to self can be restored in MS by inducing a shift to Th2 response.
4. Molecular mimicry or superantigen model. *HLA-DRB1*1501* and/or *DQA1*0102-DQB1*0602* bind with high affinity to a bacterial- or viral-derived antigenic peptide with sequence or structural homologies to a

CNS protein. Alternatively, sensitivity to massive superantigen stimulation with infection may trigger activation of myelin-specific T cells with no need for molecular homologies with the autoantigen.[31]

5. Aberrant expression model. Polymorphisms in promoter regions of *MHC* genes may induce local overexpression of MHC molecules within the CNS; such overexpression could facilitate recognition of myelin antigens by T cells that normally passage through the CNS.

6. Linkage disequilibrium model. Most individuals who carry the *MHC* genes *DRB1*1501* and *DQA1*0102-DQB1*0602* never develop MS. It is possible that these genes do not themselves predispose to MS but are located close to another unidentified gene that confers susceptibility. For some closely linked genes, certain combinations of alleles (alternative variants of a single gene, each with a unique DNA sequence) preferentially occur together on the same chromosome. For example, the *1501* form of the *DRB1** gene preferentially occurs with the *0602* form of the *DQB1** gene. Such genes are said to be in linkage disequilibrium. The MHC region is characterized by extensive linkage disequilibrium across the 3 million base pairs and more than 200 genes that constitute this region. Linkage disequilibrium makes it difficult, and at times impossible, to determine exactly which gene is responsible for an observed genetic effect.[32] Other genes mapped in this region include the gene for hemochromatosis and genes that encode proteins that control antigen processing and transport, the complement proteins C2, factor B and C4, heat shock protein 70, steroid 21-hydroxylase, the gene for hemochromatosis, genes for the cytokine tumor necrosis factor, and the myelin protein MOG.

A Full Genomic Screen Underscores a Role for the MHC

The genetic analysis of MS has been based mainly on association studies of candidate genes, by which the frequencies of marker alleles in groups of patients and healthy controls are compared, and the difference is subjected to statistical analysis. The so-called association is often expressed as the relative risk of an individual to acquire the disorder if he or she carries the particular allele or marker, compared with an individual who does not carry the allele or marker. Candidate genes are defined as genes that could logically play a role in the disease, such as cytokines and immune receptors. Association studies have met with only modest success in identifying disease-causing genes in MS, in part because of difficulty in selecting from among the many candidate gene possibilities, and in identifying ideal control populations. An alternative approach is the whole-genome screen; in this approach, no assumptions are made as to which genes are involved or how they are inherited [*see Chapter 1*]. The strategy involves scanning the entire genome in multiply affected families through the systematic analysis of the inheritance of discrete chromosomal segments that cosegregate with the disease, in an attempt to establish the chromosomal locations of MS disease genes.

In contrast to monogenic traits that account for diseases such as Huntington's disease, Alzheimer's disease, and familial amyotrophic lateral sclerosis, linkage analysis in complex disorders such as MS must be performed on an extremely large group of individuals if small genetic effects are to be detected or if genetic heterogeneity is present. The smaller the genetic contribution, the larger is the required screen, and more stringent are the required inclusion criteria to minimize effects of genetic heterogeneity. Furthermore, even when linkage to a discrete genomic region is found, the screening is pursued throughout the entire genome. A multiple-stage, whole-genome screen in multiplex MS families collected in the United States and selected according to rigorous clinical criteria was completed in 1996.[33] The initial study used 443 markers on all chromosomes with an average spacing of 9.6 centimorgans to genotype 471 individuals belonging to families with two to eight affected members. The data, which represented close to 210,000 experiments (genotypes), were analyzed with a combination of traditional and novel statistical methods. This multianalytical strategy identified 19 susceptibility regions, including 5q13-q23, 7q21-q22, 19q13, and the *MHC* on 6p21. Follow-up screening in a larger, confirmatory dataset of multiplex families is in progress. In addition to the MHC, the putative chromosome 19 locus is of particular interest as it is located in the region of apolipropotein E, a susceptibility gene for Alzheimer's disease.[34] Parallel efforts in the United Kingdom[35] and Canada[36] identified additional suggestive MS loci.

These data support a model of MS inheritance that involves multiple interacting susceptibility loci, each with a relatively small contribution to the overall risk. Additional studies in progress, with extended data sets and dense collection of markers, should enable the fine mapping and cloning of MS susceptibility genes. Their identification and characterization should help us understand the pathogenesis of demyelination, improve diagnosis and, hopefully, influence therapy. By analogy to emerging data on the genetic basis of EAE, it will be of particular interest to determine whether some loci are involved in the initial pathogenic events and others influence the later development and progression of the disease.

A Model for Disease Pathogenesis

Because of the likely role of an undefined environmental exposure, our partial knowledge of the full set of genes involved in conferring susceptibility, and the clinical heterogeneity of the disease, it has been difficult to formulate a unifying mechanism that explains the pathogenesis of MS. Nevertheless, it is likely that lymphocytes activated in the periphery home to the CNS and become attached to receptors on endothelial cells; they then pass across the BBB through the endothelium and the subendothelial basal lamina directly into the interstitial matrix [*see Figure 6*]. Interestingly, in EAE, this process has been shown to be dependent on the activation state of the T cells but is independent of their antigen specificity. Thus, MBP-reactive T cells cross the BBB with no greater efficiency than do, for example, insulin-reactive T cells.

After traversing the BBB, pathogenic T cells are reactivated by fragments of myelin antigens presented in the framework of MHC class II molecules on the surface of APCs (macrophages, microglia, and perhaps astrocytes). Possibly, enhanced or dysregulated MHC class II expression by APCs within the CNS predisposes to autoimmunity in MS, as it does in EAE. Reactivation induces release of proinflammatory cytokines that open the BBB and stimulate chemotaxis, resulting in a second, larger wave of inflammatory cell recruitment and leakage of antibody and other plasma proteins into the nervous system. Pathogenic T cells may not be capable of inducing tissue injury in the absence of this secondary leukocyte recruitment. For example, in EAE mediated by adoptive transfer of MBP-reactive encephalitogenic T cells, these cells are among the first to infiltrate the CNS, but they constitute only a minor component of the total infiltrate in the full-blown lesion.

The resident microglia, lying within the parenchyma, also become activated as a result of locally released cytokines.[37] Microglia act as scavengers that remove debris and as APCs that present processed antigens to T cells, contributing to their local clonal expansion. Mutual interactions between T cells and macrophages induce proliferation of both cell types through mediation of molecules such as IL-2 and colony-stimulating factors. Furthermore, endothelial and T cells provide colony-stimulating factors that maintain macrophage activation and prevent apoptosis and cell death. Microglia are also likely to directly induce myelin damage through the release of mediators such as free radicals (nitric oxide [NO] and superoxide [O_2^-]), vasoactive amines, complement, proteases, cytokines (IL-1, TNF-α), and eicosanoids.[38] Finally, autoantibodies may also play key roles in macrophage activation and demyelination.

This model provides a useful conceptual framework for understanding the mechanisms of action of existing therapies for MS and the rationale behind drugs currently under development.[39] Interference with one or several steps in the disease process—activation of T cells in the periphery, adhesion to brain vasculature, migration across endothelia, antigen recognition and reactivation within the CNS, opening of the BBB, and tissue damage mediated by Th1 cytokines and/or antibody—is the goal of all MS therapies. IFN beta appears to act by downregulating MHC expression on APCs, altering the pattern of cytokine response to a Th2 pattern and blocking migration across endothelia. Copolymer 1 (a synthetic protein designed as an analogue of MBP) may induce active T cell suppression against MBP and saturate MHC molecules on APCs, preventing presentation of autoantigens. Glucocorticoids are potent inhibitors of APC function. The chemotherapeutic drug cyclophosphamide is lympholytic and stimulates production of Th2 cytokines. Most experimental therapies focus on interference with antigen presentation to encephalitogenic T cells (altered peptide ligand, intravenous antigen), induction of a Th2 response (oral tolerance), blockade of adhesion molecules (anti-VLA4 antibody), administration of anti-inflammatory cytokines (IL-10, TGFβ), or neutralization of proinflammatory cytokines (anti-TNF antibody, soluble TNF receptor).

One exciting approach to experimental therapy targets proteases expressed by activated T cells and responsible for lysis of the dense subendothelial basal lamina by T cells as they migrate across the BBB to reach the CNS parenchyma. Although macrophages are a rich source of enzymes that will disrupt the endothelium and allow traffic into the subendothelial basal lamina, T cells may have their own arsenal of proteases. We demonstrated in 1995 that highly purified normal peripheral blood T cells express two matrix metalloproteinases: gelatinases A (72 kd) and B (92 kd).[40] Functionally, the secretion of gelatinases by T cells correlates with the migratory and cytotoxic capacity of the cells. Both gelatinases are structurally related and share the proteolytic selectivity for basal lamina collagens. Remarkably, the

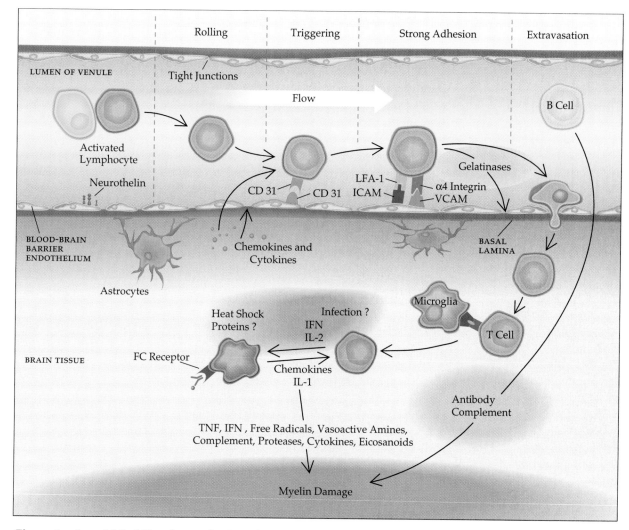

Figure 6 *A model for MS pathogenesis. Crucial steps for disease initiation and progression include peripheral activation of preexisting autoreactive T cells; homing to the CNS and extravasation across the blood-brain barrier; reactivation of T cells by exposed autoantigens; secretion of cytokines; activation of microglia and astrocytes and recruitment of a secondary inflammatory wave; and immune-mediated myelin destruction. ICAM, intercellular adhesion molecule; LFA-1, leukocyte function–associated antigen–1; VCAM, vascular cell adhesion molecule; IFN, interferon; IL, interleukin; TNF, tumor necrosis factor.*

sequence in the putative cleavage of the TNF-α precursor reveals homologies with peptide sequences known to be cleaved by metalloproteinase-like enzymes.[41,42] Thus, metalloproteinases not only may act as a mediator of cell traffic across the BBB but also may increase the inflammatory and homing reactions through TNF processing.

The clinical relevance of metalloproteinases is underlined by the observation that gelatinase B is present in the CSF of patients with MS and acute encephalitis but not in the CSF of normal controls. Inhibition of gelatinases by enzyme inhibitors results in suppression of cell migration across a model BBB in vitro and amelioration of clinical symptoms in a rodent EAE model of MS. Furthermore, IFN beta-1b, the first approved drug for treatment of MS, down-regulates gelatinase production in T cells.[43] This effect is associated with down-regulation of the IL-2 receptor α chain.

An additional appeal of gelatinase inhibition as a strategy in MS is that an essential step in brain inflammation, T cell trafficking, can be retarded by targeting a small family of enzymes with unique function. By contrast, alternative approaches—such as targeting an individual antigen, adhesion molecule, or cytokine—may be less effective because of the complex and overlapping roles of many molecules at other stages of the disease process.

In conclusion, the development of new molecular tools to define the immune response in MS and to characterize its genetic basis is opening a window into the intricacies of genetic and environmental interactions involved in this disorder. This new knowledge has clarified the likely mechanism of action of therapies currently in use against MS. More significant, elucidation of the sequence of events that leads to MS is likely to result in improved diagnosis and opportunities for primary prevention or selective immunotherapy targeted against underlying disease mechanisms.

References

1. Raine CS: The Dale E. McFarlin memorial lecture: the immunology of the multiple sclerosis lesion. *Ann Neurol* 36:S61, 1994

2. Steinman L: Escape from "horror autotoxicus": pathogenesis and treatment of autoimmune disease. *Cell* 80:7, 1995

3. Schwartz RH: Models of T cell anergy: is there a common molecular mechanism? *J Exp Med* 184:1, 1996

4. Steinman L: Some misconceptions about understanding autoimmunity through experiments with knockouts. *J Exp Med* 185:2039, 1997

5. Rivers TM, Schwenther FF: Encephalomyelitis accompanied by myelin destruction experimentally produced in monkeys. *J Exp Med* 61:689, 1935

6. Paterson PY: Transfer of allergic encephalomyelitis in rats by means of lymph node cells. *J Exp Med* 111:119, 1960

7. Allegretta M, Nicklas JA, Sriram S, et al: T cells responsive to myelin basic protein in patients with multiple sclerosis. *Science* 247:718, 1990

8. Oksenberg JR, Panzara MA, Begovich AB, et al: Selection for T-cell receptor Vβ-Dβ-Jβ gene rearrangements with specificity for a myelin basic protein peptide in brain lesions of multiple sclerosis. *Nature* 362:68, 1993

9. Hafler DA, Duby AD, Lee SJ, et al: Oligoclonal T lymphocytes in the cerebrospinal fluid of patients with multiple sclerosis. *J Exp Med* 167:1313, 1988

10. Garbern JY, Cambi F, Tang X-M, et al: Proteolipid protein is necessary in peripheral as well as central myelin. *Neuron* 19:205, 1997

11. Massacesi L, Genain CP, Lee-Parritz D, et al: Active and passively induced experimental autoimmune encephalomyelitis in common marmosets: a new model for multiple sclerosis. *Ann Neurol* 37:519, 1995

12. Genain CP, Hauser SL: Creation of a model for multiple sclerosis in *Callithrix jacchus* marmosets. *J Mol Med* 75:187, 1997

13. Genain CP, Nguyen M-H, Letvin NL, et al: Antibody facilitation of multiple sclerosis-like lesions in a nonhuman primate. *J Clin Invest* 96:2966, 1995

14. Genain CP, Abel K, Belmar N, et al: Late complications of immune deviation therapy in a non-human primate. *Science* 274:2054, 1996

15. Schluesener HJ, Sobel RA, Linington C, et al: Monoclonal antibodies against a myelin oligodendrocyte glycoprotein induces relapses and demyelination in central nervous system autoimmune disease. *J Immunol* 139:4016, 1987

16. Brunner C, Lassmann H, Waehneldet TV, et al: Differential ultrastructural localization of mylein basic protein, myelin oligodendroglia glycoprotein, and 2', 3'-cyclic nucleotide 3'-phosphodiesterase in the CNS of adult rats. *J Neurochem* 52:296, 1989

17. Oksenberg JR, Hauser SL: Pathogenesis of multiple sclerosis: relationship to therapeutic strategies. *Treatment of Multiple Sclerosis*, 2nd ed. Goodkin D, Rudick R, Eds. Springer-Verlag, London, 1996, p 17

18. Bernard CCA, Randell VB, Horvath L, et al: Antibody to myelin basic protein in extracts of multiple sclerosis brain. *Immunology* 43:447, 1981

19. Warren KG, Catz I, Johnson E, et al: Anti-myelin basic protein and anti-proteolipid protein specific forms of multiple sclerosis. *Ann Neurol* 35:280, 1994

20. Kerlero de Rosbo N, Bernard CCA: Multiple sclerosis brain immunoglobulins stimulate myelin basic protein degradation in human myelin: a new cause of demyelination. *J Neurochem* 53:513, 1989

21. Wucherpfennig WW, Strominger JL: Molecular mimicry in T cell-mediated autoimmunity: viral peptides activate human T cell clones specific for myelin basic protein. *Cell* 80:695, 1995

22. Chatila T, Geha RS: Superantigens. *Curr Opin Immunol* 4:74, 1992

23. Zumla A: Superantigens, T cells, and microbes. *Clin Infect Dis* 15:313, 1992

24. Burns J, Littlefield K, Gill J, et al: Bacterial toxin superantigens activate human T lymphocytes reactive with myelin autoantigens. *Eur J Immunol* 32:352, 1992

25. Oksenberg JR, Seboun E, Hauser SL: Genetics of demyelinating diseases. *Brain Pathol* 6:289, 1996

26. Seboun E, Oksenberg JR, Hauser SL: Molecular and genetic aspects of multiple sclerosis. *The Molecular and Genetic Basis of Neurological Disease*, 2nd ed. Rosenberg RG, Prusiner SB, DiMauro C et al, Eds. Butterworth-Heinemann, Boston, 1997, p 631

27. Ebers GC, Sadovnick AD: The role of genetic factors in multiple sclerosis susceptibility. *J Neuroimmunol* 54:1, 1994

28. Sadovnick AD, Ebers GC, Dyment DA, et al: Evidence for genetic basis of multiple sclerosis. *Lancet* 347:1278, 1996

29. Mumford C, Wood NW, Kellar-Wood H: The British Isles survey of multiple sclerosis in twins. *Neurology* 44:11, 1994

30. Blankenhorn EP, Stranford SA: Genetic factors in demyelinating diseases: genes that control demyelination due to experimental allergic encephalomyelitis and Theiler's murine encephalitis virus. *Reg Immunol* 4:331, 1992

31. Brocke S, Veromaa T, Weissman IL, et al: Infection and multiple sclerosis: a possible role for superantigens? *Trends Microbiol* 2:250, 1994

32. Feder JN, Gnirke A, Thomas W, et al: A novel MHC class I-like gene is mutated in patients with hereditary haemochromatosis. *Nature Genet* 13:399, 1996

33. The Multiple Sclerosis Genetic Group: A complete genomic screen for multiple sclerosis underscores a role for the major histocompatibility complex. *Nature Genet* 13:469, 1996

34. Saunders AM, Strittmatter WJ, Schemchel D, et al: Association of apolipoprotein allele e4 with late onset familial and sporadic Alzheimer's disease. *Neurology* 43:1467, 1993

35. Sawcer S, Jones HB, Feakes R, et al: A genome screen in multiple sclerosis reveals susceptibility loci on chromosome 6p21 and 17q22. *Nature Genet* 13:464, 1996

36. Ebers GC, Kukay K, Bulman DE, et al: A full genome search in multiple sclerosis. *Nature Genet* 13:472, 1996

37. Sriram S, Rodriguez M: Indictment of the microglia as the villain in multiple sclerosis. *Neurology* 48:464, 1997

38. Merrill JE, Benveniste EN: Cytokines in inflammatory brain lesions: helpful and harmful. *Trends Neurosci* 19:331, 1996

39. Hohlfeld R: Biotechnological agents for immunotherapy of multiple sclerosis: principles, problems and perspectives. *Brain* 120:865, 1997

40. Leppert D, Waubant E, Galardy R, et al: T cell gelatinases mediate basement membrane transmigration in vitro. *J Immunol* 154:4379, 1995

41. Gearing AJH, Beckett P, Christodoulou M: Processing of tumor necrosis factor-alpha precursor by metalloproteinases. *Nature* 370:555, 1994

42. McGeehan GM, Becherer JD, Bast RC, et al: Regulation of tumor necrosis factor-alpha processing by a metalloproteinase inhibitor. *Nature* 370:558, 1994

43. Leppert D, Waubant E, Burk MR, et al: Interferon beta 1-b inhibits gelatinase secretion and in vitro migration of human T cells: a possible mechanism for treatment efficacy in multiple sclerosis. *Ann Neurol* 40:846, 1996

Acknowledgment

Figures 1, 3, and 6   Dimitry Schidlovsky.

Figures 2 and 5   Marcia Kammerer.

Figure 4   Stephen L. Hauser. Adapted by Dimitry Schidlovsky.

CHAPTER 12

Amyotrophic Lateral Sclerosis and the Inherited Motor Neuron Diseases

Robert H. Brown Jr., D. Phil., M.D.

Few diseases are as devastating as amyotrophic lateral sclerosis (ALS), also known as Lou Gehrig's disease or motor neuron disease. The hallmark of this disease is progressive paralysis that is uniformly lethal, largely because of respiratory weakness.

Clinical Features

The disease usually begins in the midadult years, heralded in most cases by focal muscle weakness. It may appear initially in a distal limb (causing footdrop or hand weakness) or in the muscles of chewing and swallowing. It then spreads over two to three years to involve all muscle groups except those subserving eye movements and bowel and bladder function. The process arises because of progressive loss of motor neurons in the brain, brain stem, and spinal cord. When cortical motor (upper) neurons are involved, the weakness is associated with spasticity and stiffness of movement. When spinal (lower) motor neurons are first involved, the weakness arises from denervation of the affected muscles and is associated with muscle twitching (fasciculations) and then muscle atrophy. Regardless of the site of onset, the striking and devastating aspect of the illness is that it migrates ultimately to involve virtually all motor neurons. Although ALS patients rarely show symptoms or signs outside of the motor system, the disease may infrequently be accompanied by dementia or sensory loss.

With rare exceptions, the epidemiological features of ALS are invariant worldwide. The mean age of onset is 55 years, and the mean duration is three to five years, with a more rapid course in bulbar forms of the disease. About 10 percent of cases are inherited as an autosomal dominant trait (familial ALS or FALS).[1] The incidence of new cases is approximately 1:100,000, and the total number of cases is about 5:100,000. Thus, in the United States, there are an estimated 10,000 to 20,000 cases.

The pathological hallmark of ALS is motor neuron degeneration. In most cases, motor neurons appear to undergo a process of involution or shrinkage, losing volume in both cytoplasm and nucleus. Among the earliest pathological changes in the dying motor neurons is the appearance of fine wisps of ubiquitin-positive threads or skeins.[2] There may then be other forms of cytoskeletal pathology, including formation of cytoskeletal inclusions in both the cytoplasm and the proximal axons. In general, the pathology in ALS is largely confined to motor neurons. Some subsets of motor neurons are spared, including the motor nuclei that innervate extraocular muscles and the sphincteric muscles. Some patients show pathology outside the motor system, and some familial cases show deterioration of the posterior columns and neurons within Clarke's column. In exceptional instances, an ALS-like process arises with diffuse cerebral gliosis and microvacuolization, often in the setting of frontotemporal dementia.

In the past decade, two forms of juvenile-onset ALS have been described. One is recessively inherited and hence is encountered in regions where consanguinity is commonplace. This disorder begins within the first decade of life and progresses extremely slowly; affected individuals survive for decades. The clinical hallmark is simultaneous involvement of spinal and corticospinal motor neurons, with complete sparing of other neuron systems. No autopsy data are available for this entity. In a large Tunisian kindred, the disease was linked to chromosome 2q.[3] A dominantly inherited form of juvenile-onset ALS is also described; it, too, has a striking pattern of slowly progressive upper and lower motor neuron degeneration and shows genetic linkage to human chromosome 9. Other inherited disorders also selectively afflict the spinal motor neurons (the recessively inherited spinal muscular atrophies) and the corticospinal neurons (the dominantly inherited hereditary spastic paraplegias) [see Table 1].

Biologic Properties of Motor Neurons

In considering possible causes of ALS, it is useful to briefly review some of the salient anatomic and biologic properties of motor neurons. Some studies have defined a complex set of events that trigger motor neuron development and diversification. Critical to early motor neuron formation is the protein Sonic hedgehog that activates transcription factors to induce formation of motor neurons and ventral interneurons.[4] Additional proteins, such as Isl-1, are critical for initial spinal motor neuron formation and identity.[5] Cells expressing Isl-1 may subsequently be differentiated in columnar subsets distinguished by expression of the proteins Isl-1, Isl-2, Lim-1, Lim-3, and Gsh4.[5]

As development progresses, axons elongate from newly specified motor neurons to seek and innervate target muscles. Axonal elongation and guidance is driven by a combination of extrinsic, local chemical cues and endogenously expressed molecules. The latter include the integrins, which interact with extracellular matrix, and the calcium-dependent cadherins that interact with proteins on the surfaces of adjacent cells. Synapse formation with muscle is a complex process. Initially, muscle cells may be innervated

Table 1 Inherited Motor Neuron Disorders

Disease	Defect	Locus	Reference
Amyotrophic lateral sclerosis (upper and lower neurons)			
Dominant*	Superoxide dismutase 1	21q	11
	Neurofilament subunit H	11	57
Dominant†		9q	
Recessive†		2q	3
Hereditary spastic paraplegia (upper motor neurons)			
Dominant			61
		2p	62
		14q	63
		15q	64
Recessive		8q	65
X-linked	Proteolipid protein	Xq22	66
	LICAM	Xq28	67
Lower motor neurons			
Recessive			
Spinal muscular atrophy	Survival motor neuron	5q	68
GM$_2$ gangliosidoses			
Tay-Sachs disease	Hexosaminidase A deficiency (α-subunit)	15q	69
Sandhoff disease	Hexosaminidase A and B deficiency (β-subunit)	5q	69
AB variant	GM$_2$ activator protein	5q	70
Arthrogryposis		9	71
X-linked			
Spinobulbar muscular atrophy	Androgen receptor	Xq11	53
Arthrogryposis		Xp11-q11	72
			73

*Adult onset.
†Juvenile onset.

by multiple motor neurons. However, synapses are eliminated over time, such that all neuromuscular junctions are unisynaptic by the time of birth. Although the factors that govern synapse elimination have not been identified in full detail, it is evident that maintenance of a synapse depends on activity of the junction, integrity of the basal lamina within the junction, and factors secreted by motor neurons to induce synthesis of postsynaptic proteins.[6] At about the time that synaptic structures are formed, nearly half of all spinal motor neurons undergo a dramatic process of natural, programmed cell death. This step in spinal cord development may precisely match numbers of motor neurons to the available target muscle cells.[6]

At more than 100 μm in diameter, fully differentiated, mature motor neurons are among the largest cells in the body. Moreover, they often have extraordinarily long axons. In the case of lumbosacral neurons innervating a toe muscle in an adult, for example, the length of the axon may be several thousand times greater than the diameter of the cell body. This has striking practical implications: by far the most cell cytoplasmic volume is in the motor neuron axon, despite the large cell soma. More than a year may be required to transport some proteins along the length of the axon from the soma to the motor neuron junction.

Several biochemical features distinguish the lower motor neuron from other neural cells. The spinal and bulbar motor neurons release acetylcholine as the transmitter at the neuromuscular junction. This is synthesized by choline acetyltransferase, an intracellular enzyme that is a biochemical marker for these cells in the cord. The lower motor neurons are also distinguished by cell surface molecules, such as the low-affinity NGF receptor (p75).[7] Some subsets of spinal motor neurons are further identified by particular surface molecules. Thus, the largest, ALS-susceptible neurons in the anterior horns of the spinal cord bear androgen receptors. By contrast, smaller motor neurons that innervate extraocular muscles and are largely spared in ALS do not express surface androgen receptors. Motor neurons are also distinguished by relatively low levels of calcium-buffering proteins such as calbindin.[8]

Hypotheses of the Pathogenesis of ALS

The cause of the motor neuron diseases has been elusive. Several categories of defects have been proposed, either as primary triggers or as processes that may sustain the disease once it is initiated. Because lead toxicity can selectively injure motor neurons, heavy-metal poisoning has long been invoked as a possible cause of ALS. However, studies have failed to show any consistent evidence of metal toxicity. The most devastating motor neuron disease is poliomyelitis, caused by an infection by an RNA virus that primarily targets motor neurons. This has given rise to the theory that ALS is a form of latent or slow viral infection from an agent like poliovirus. However, no infectious agents have consistently been identified in ALS tissues using any of several methods of detection.

An autoimmune disorder of motor neurons has been proposed as a pathogenic mechanism. Studies from the laboratory of Stanley Appel have demonstrated, in sporadic but not familial ALS, the presence of elevated titers of antibodies to voltage-sensitive calcium channels (VSCC) in the distal terminals of motor neurons.[9] An autoimmune disorder of motor neurons can be produced in rodents by inoculation with preparations of motor neurons. Generalization of these findings remains to be established. Unfortunately, several trials of immunosuppressive therapy in ALS have not been beneficial.

There has also been extensive investigation of the possibility that excitotoxicity may be one component of the cell death process in motor neuron disease and other neurodegenerative disorders. Brain tissue levels of glutamate are elevated in ALS, although levels in cerebrospinal fluid (CSF) are controversial; some studies show elevated levels, others show normal levels. Glutamate transport is defective in synaptosomes from motor and sensory cortex of ALS patients.[10] Consistent with this hypothesis is the report that abnormally spliced messenger RNA (mRNA) transcripts of the astrocytic glutamate transporter EAAT2 are detected selectively in the motor cortex of ALS patients. Therapeutic studies lend credence to the glutamate hypothesis. In both human and mouse ALS, an inhibitor of presynaptic glutamate release, riluzole, produces subtle but statistically significant slowing of the disease course.

By contrast with these somewhat conjectural hypotheses about ALS, in a subset of cases the existence of an underlying gene defect is incontrovertible. In some pedigrees, the disease is clearly monogenic, dominantly inherited, and highly penetrant. The penetrance in one subgroup of otherwise typical adult-onset FALS is more than 90 percent by 70 years of age.

Genetic Linkage Analysis in FALS

To better understand the pathogenesis of ALS, a multicenter group began an analysis of the genetic forms of ALS in 1985. This study was predicated on the fact that familial and sporadic forms of ALS are clinically indistinguishable. It therefore seemed likely that insights into the pathogenesis of familial forms would pertain to sporadic forms. In 1991, this group identified an FALS locus on the long arm of chromosome 21, and in 1993, the same collaborators found that a subset of FALS families linked to this locus had mutations in the gene encoding the cytosolic protein superoxide dismutase type 1 (SOD1).[11] Expressed in every cell in every eukaryotic organism, SOD1 is highly conserved during evolution and is evidently important to some aspect of cellular function.[12] Eukaryotes have three genes for different SOD molecules. SOD1 is a protein of 153 amino acids encoded by five exons. A second form, SOD2, is abundant in mitochondria, whereas SOD3 is principally an extracellular enzyme[12]; mutations in the genes encoding SOD2 and SOD3 have not been detected in sporadic or familial ALS. SOD1 functions as a homodimer. Each monomer contains one atom of zinc, which maintains folding of the enzyme, and one of copper, which catalyzes the enzyme's primary reaction: the conversion or dismutation of the superoxide anion (O_2) to hydrogen peroxide (H_2O_2) [*see Figure 1*]. Superoxide anion can react with nitric oxide to form peroxynitrite, which, like H_2O_2, can form the highly toxic hydroxyl radical [*see Chapter 7*]. Thus, SOD1 is implicated in the regulation of levels of several critical intracellular free radicals. For this reason, the discovery of mutations in SOD1 in ALS patients raised

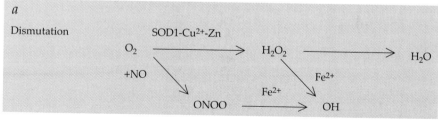

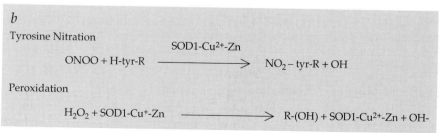

Figure 1 *Selected reactions of superoxide dismutase. (a) The normal dismutation activity of SOD1. (b) Two proposed aberrant reactions showing tyrosine nitration and peroxidation (see text). ONOO, peroxynitrite; OH, hydroxyl radical; tyr-R, tyrosine-containing protein.*

the hypothesis that one element in the pathogenesis of this disease is some perturbation or alteration of free radical homeostasis.

Mechanism of Toxicity of Mutant SOD1 Protein

More than 50 different mutations are reported in the SOD1 gene in FALS families, although it remains unclear how these trigger motor neuron death. Three important points have emerged. First, the mutations render the SOD1 protein less stable, as assessed by metal binding and the half-life of the mutant protein in cells in culture. Second, although many mutations partially reduce the dismutation activity of SOD1, these reductions do not, by themselves, cause motor neuron death.[13] Indeed, mice completely devoid of SOD1 activity (knockout mice) survive at least 18 months, although their motor neurons have subnormal resistance to some forms of injury.[14] Although SOD1 knockout mice may show evidence of subtle motor dysfunction in the second year, they do not develop a fulminant, rapid motor neuron disease. Third, the toxicity of the mutant SOD1 protein is mediated by an acquired cytotoxic property. By contrast with the knockout animals, some strains of mice (FALS mice) that express high levels of the mutant protein with no reduction in SOD1 activity do develop a rapid, fatal motor neuron disease.[15] To a remarkable degree, this mouse model shares many clinical and pathological features of human ALS. The disorder begins focally (usually in a hind limb) and spreads. It is strikingly specific to spinal motor neurons. It begins at an age comparable to human adulthood and follows a rapidly progressive time course. The subtle but reproducible benefit seen in human ALS from the drug riluzole, the inhibitor of glutamate release, is also seen in mouse ALS.

The mutant SOD1 protein likely is injurious to motor neurons through some property acquired (or accentuated) through the mutation (a gain of function mutation). However, the toxic function of the mutation in the SOD1 molecule has not been clearly established. Several hypotheses are under active investigation. All share the premise that protein folding is less stable in mutant than in wild-type SOD1. Consequently, the mutant enzyme binds metals subnormally and can accept in the active channel substrates whose dimensions would normally be excluded. One hypothesis is that peroxynitrite has enhanced access to copper in the mutant active channel and thereby acts as a nitrate donor, leading to nitration of tyrosines.[16] It is not difficult to envision this as a toxic influence, particularly if the nitrated tyrosines are functionally important (e.g., in neurofilaments or tyrosine kinase receptors). A second hypothesis is that the mutant proteins have a heightened capacity to interact with hydrogen peroxide, normally the product of the dismutation. It has been recognized for more than 20 years that normal SOD1 can act as a peroxidase, reacting with hydrogen peroxide to form hydroxyl radicals. Two groups have published data arguing that this peroxidation reaction is accelerated in the mutant SOD1 molecule,[17,18] perhaps because of an altered conformation of the active channel[18] and greater accessibility of copper to reducing agents.[19]

A third hypothesis is that the mutant molecule activates one or more programmed cell death pathways. Why the mutation should be proapoptotic is not clear. Emerging evidence indicates that reductions in levels of wild-type SOD1 promote apoptosis in neurons in cell culture[20-22]; this appears to be mediated by reactive oxygen intermediates that may act as signals upstream of the apoptotic program. Moreover, forced expression of the FALS-associated SOD1 mutations induces apoptosis in neurons, whether injected into the neuronal nucleus or administered to the cytoplasm (e.g., via transfection or infection using viral vectors).[23,24] This effect is not mediated by loss of SOD1 activity. Intriguingly, it seems to be particularly pronounced in neurons. It is blocked by inhibition of enzymes of the caspase family, such as interleukin-1β–converting enzyme (ICE), and by copper chelators.[24]

The hypothesis that the mutant SOD1 molecule induces apoptosis has received support from two recent studies of apoptotic cell death pathways in FALS mice expressing human SOD1 with a substitution of alanine for glycine at codon 93. In one pilot study, inactivation of ICE by a dominant negative caspase inhibitor prolonged survival in the FALS mice by approximately two weeks, although it did not alter the timing of disease onset.[25] In another study, FALS mice were bred to express high levels of *bcl-2*, the human proto-oncogene that inhibits programmed cell death. The resulting double transgenic animals showed a delay in disease onset of 19 percent and a prolongation in time to death of 15 percent.[26] The rate at which the disease progressed, once started, was not affected by *bcl-2*. The *bcl-2* transgene did not increase the numbers of motor neurons surviving at birth. Rather, it appeared to reduce spinal motor neuron death as indicated by cell counts and patterns of expression of two markers of cell injury, c-jun and ubiquitin.[26] Whether these findings will be pertinent to human ALS is not known. However, apoptotic cells have been described in ALS spinal cord at autopsy.[27] Further, it has been reported that levels of bax protein, which promotes apoptosis, are elevated in ALS spinal motor neurons.[28]

A fourth hypothesis is that the mutant SOD1 molecule adheres to and disrupts function of one or more cellular constituents. In one study, mutant but not wild-type SOD1 was found to bind avidly to a membrane constituent of endoplasmic reticulum and to lysl transfer RNA (tRNA) synthetase.[29] Also tenable is the possibility that the toxicity of the altered SOD1 protein is mediated through its diminished buffering of copper or zinc or its precipitation to form protein aggregates.[23] SOD1-positive aggregates have been observed within the neuronal cytoplasm in both sporadic and familial ALS. Of course, it must be emphasized that the mutant SOD1 peptide may promote more than one acquired toxic chemical reaction. Moreover, the importance of different chemical reactions may depend on the stage of the illness or the stage of the death process in any given cell.

Free Radical Pathology in ALS

Several of these hypotheses invoke mechanisms that involve free radical species. An analysis of brains from patients with SOD1 mutations is therefore predicted to show chemical signatures of oxidative reactions. Moreover, to the extent the pathogenesis of sporadic ALS overlaps that of familial ALS, sporadic cases may also show evidence of oxidative pathology. It has recently been proposed, for example, that age-dependent biochemical modifications of the SOD1 molecule in sporadic ALS patients may produce the same types of aberrant chemical reactivity seen with SOD1 mutations in FALS.[30] Consistent with this general expectation, it has been reported that carbonyl protein and free nitrotyrosine levels are elevated in sporadic ALS spinal cords and brains at autopsy.[31,32] Free nitrotyrosine levels are also elevated in the ALS mice, although hydroxyl radical production is not.[33] A 1996 report describes oxidative defects in DNA in two sporadic ALS cases.[34] One measure of oxidative stress is the tissue level of transition metals that may predispose to oxidative cytotoxicity. In this light, it is of interest that levels of iron have been reported to be elevated in ALS brains.[35,36] This is consistent with magnetic resonance imaging analyses showing increased deposition of iron in ALS brains.[37,38]

There have been few studies of levels of antioxidant defenses in ALS. At least one report suggests that levels of α-tocopherol are reduced in ALS patients.[39] On the other hand, data concerning glutathione peroxidase activity are conflicting. In a study of nine ALS brains, glutathione peroxidase activity was reduced in the precentral gyrus but not in the cerebellar cortex.[40] By contrast, the activity levels of Cu/Zn superoxide dismutase (SOD1), manganese SOD (SOD2), and catalase were normal. Two analogous studies documented unchanged[41] or increased[35] activity of glutathione peroxidase in ALS spinal cord. A single report describes increased binding sites for glutathione in five ALS brains; this observation was interpreted to indicate a deficiency of glutathione in these tissues.[42]

At least two antioxidant compounds (vitamin E and buckminsterfullerene) have been found to produce a mildly palliative effect in the mouse ALS model.[43] A trend toward improved survival was seen in a trial of the antioxidant N-acetylcysteine in 35 ALS patients.[44] Unfortunately, vitamin E administration did not prolong Lou Gehrig's life.[45]

Molecular Pathogenesis in Other Motor Neuron Disorders

In the past several years, remarkable progress has occurred in delineating the molecular defects underlying three other inherited motor neuron diseases. In each, the clinically significant pathology is confined to the lower or spinal motor neurons. Two are inherited as autosomal recessive traits; the third is X-linked. The most severe of these is spinal muscular atrophy (SMA). With a carrier frequency of one in 40 and an affected frequency of one in 10,000 births, this disorder is among the most common of the genetic neurologic diseases. In its acute form, SMA type 1 (SMA1 or Werdnig-Hoffmann disease), the problem begins prenatally, in the third trimester. The presenting feature may be a subtle diminution of intrauterine movements. Newborn babies with SMA1 are extremely feeble, with slack muscle tone, limb weakness, and poor sucking and crying. Milder forms of SMA (types 2 and 3) occur with later onset and slower disease progression. The defect causing SMA illness is loss of a protein encoded on the proximal long arm of chromosome 5, labeled survival motor neuron, or SMN. Present in a duplicated region of chromosome 5, the SMN gene is represented in centromeric (SMNC) and telomeric (SMNT) copies that differ only by single base differences in exons 7 and 8.[46] The SMN protein is expressed in most types of cells. It is present in nuclei in dense spots known as GEMS, in which SMN is twinned (gemini) with nuclear coiled bodies.[47] SMN is also present in the cytoplasm, particularly in large neurons such as motor neurons. The common feature in nuclear and cytoplasmic SMN may be association with ribosomal RNA. This raises the hypothesis that this molecule participates in RNA trafficking between the nucleus and cell body.

The spectrum of genetic defects in SMN causing SMA is only partially understood. About 95 percent of SMA patients, have deletions of the SMNT gene.[46] In most of the remaining patients, several other types of mutations have been identified, including minor (e.g., five base pair) deletions and missense mutations.[48-50] Some cases with apparent deletions of SMNT may represent single base pair mutations that convert SMNT to SMNC.[50]

Factors determining the severity of SMA have been difficult to define. Initially problematic were cases of types 1 and 2 SMA with the same apparent molecular defect—deletions of SMNT. It now appears that an important modifying factor is the number of copies of SMNC present in an SMA patient. In general, as the copy number of SMNC increases, the phenotype is milder. Thus, obligate carriers of SMA types 2 and 3 have more copies of SMNC than type 1 carriers.[51] It may also be that SMA patients with intact copies of a second gene in the locus, NAIP, have a milder phenotype. The NAIP gene encodes a protein (neuronal apoptosis inhibitory protein) with homology to viral antiapoptotic proteins.[52]

In X-linked spinobulbar muscular atrophy (SBMA), slowly progressive motor neuron loss begins in affected males in adulthood. The hallmark of SBMA is bulbar weakness with abundant facial fasciculations, arising concurrently with progressive limb weakness. Gynecomastia and subtle sensory loss may also occur. The molecular defect in SBMA is an expanded CAG repeat within the coding region of the androgen receptor gene[53] [*see Chapter 3*]. SBMA therefore belongs to the growing family of neurodegenerative diseases attributed to CAG expansions. It is likely that these diseases share some common pathogenic mechanism related to either the expanded CAG tract or the resulting expanded polyglutamine tract in the expressed protein. The nature of this mechanism is poorly understood. It may entail an aberrant affinity of the polyglutamine tracts for some molecular targets that are important in neuronal viability. It is also conceivable that either the expanded CAG or the expanded polyglutamine tract interferes with transcription or translation. This possibility seems particularly likely in view of recent reports that CAG repeat expansions in mice bearing a transgene for huntingtin cause this protein to be translocated to the nucleus, where it forms distinctive, ubiquitin-positive intraneuronal inclusions.[54] These structures, which have some features of amyloid,[55] appear before any behavioral disturbances in the mice occur. That the same defect (expanded CAG repeats) in different genes causes different neurodegenerative diseases indicates that (1) the pathogenic process is neuron specific; (2) the specific type of neurons targeted is determined by the expression patterns of the protein bearing the expanded polyglutamine tracts; and (3) in the long-term, a therapy that blocks the effect of the expansion on neuron survival in one of these neurodegenerative diseases may be effective for all of them.

The third inherited disease targeting predominantly lower motor neurons is adult-onset GM_2 gangliosidosis, or adult Tay-Sachs disease. In this disease, slowly progressive proximal muscle weakness often begins in early adulthood. The progression occurs over decades and may involve some spasticity and both dysarthria and cerebellar atrophy. The underlying defect is an accumulation of the GM_2 ganglioside caused by mutations in the gene for the enzyme N-acetylhexosaminidase A or B, or its associated activator protein.

Thus, each of these three selective, lower motor neuron disorders involves a defective gene that is expressed in multiple cell types throughout the neuraxis. Nonetheless, the clinical problem in each initially targets the motor neuron. This presumably reflects the fact that motor neurons are large, anatomically complex cells. These cells must be metabolically very active and are potentially very susceptible to subtle protein defects, which are not toxic to smaller neurons or nonneuronal cells.

Overview of Cell Death in the Inherited Motor Neuron Diseases

Insights into these motor neuron diseases (sporadic and familial ALS, SMA, SBMA, and Tay-Sachs disease) allow one to begin to distinguish major pathways leading to motor neuron cell death [*see Figure 2*]. In the ALS pathway, a central upstream event is presumed to be exposure to glutamate. This is potentially injurious either if glutamate levels are elevated or if the cell is sensitized to glutamate by some other defect. The major glutamate uptake transporter is EAAT2, whose activity is defective in some cases of ALS. Also shown at the top of this pathway are antibodies to VSCC. Either proximal stimulus (glutamate or the autoantibodies) mediates increases in intracellular calcium levels which, in turn, trigger parallel events. These include activation of nitric oxide synthase and xanthine oxidase, activation of proteases and nucleases, and increased mitochondrial calcium loading. Each of these has downstream consequences. Nitric oxide synthase and xanthine oxidase activation generate increased levels of nitric oxide and superoxide anion. These can interact to form toxic reactive oxygen species; formation of hydroxyl radical, peroxynitrite, and nitrated tyrosines may be enhanced in the presence of mutant SOD1. Increased mitochondrial calcium loading augments release of hydroxyl radicals from mitochondria and impairs adenosine triphosphate generation. Reciprocally, the latter enhances the sensitivity of the neuron to glutamate, amplifying this sequence of events.[56] Analogously, activation of proteases and the increased abundance of reactive species may injure the cellular membrane, particularly if lipid peroxidation occurs. This provides a leak pathway for additional ingress of calcium, which will also amplify the toxicity of glutamate.

The exact mechanisms whereby these activities are lethal to the motor neuron are not precisely defined. It has been proposed that nitration of tyrosines in neurofilaments prevents normal assembly of the cytoskeleton, leading to neurofilament aggregation and ultimately to disruption of axonal functions, including axonal transport. In this context, it is relevant that mutations in the gene encoding the neurofilament heavy subunit (NF-H) have been found in some patients with ALS.[57] It is not yet estab-

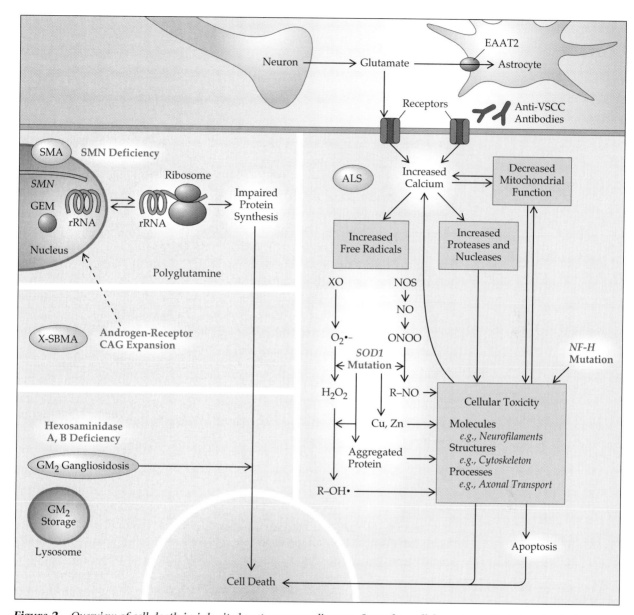

Figure 2 *Overview of cell death in inherited motor neuron diseases. Several parallel events are likely to arise during motor neuron death in ALS and the spinal muscular atrophies. This hypothetical cascade of events is triggered by glutamate, mediated by elevated levels of cytosolic calcium, and influenced downstream by superoxide dismutase (see text). Four major disease categories are amyotrophic lateral sclerosis (ALS), spinal muscular atrophy (SMA), X-linked spinobulbar muscular atrophy (X-SBMA), and GM$_2$ gangliosidosis (adult Tay-Sachs disease). Five genetic defects implicated in these diseases are SOD1 (Cu, Zn cytosolic superoxide dismutase) mutation, NF-H (neurofilament heavy subunit) mutation, SMN (survival motor neuron) deficiency, androgen receptor CAG expansion, and hexosaminidase A and B deficiency. EAAT2, astrocytic glutamate transporter; VSCC, voltage-sensitive calcium channel; Gem, guanosine triphosphate–binding protein.*

lished whether these mutations are causal, however. Indeed, some persons with these mutations do not have ALS. This may reflect the fact that they are still relatively young adults; an alternative possibility is that the NF-H defects are rare polymorphisms that do not cause disease. Extensive screening studies of populations of ALS patients have not identified mutations in NF subunits.[58]

The potential importance of neurofilaments in long-term motor neuron viability is indicated by the observation that mice expressing mutant neurofilament genes may develop axonal pathology in motor and sensory neurons and weakness.[59] It is important to note that even abnormal levels of wild-type neurofilament subunits can trigger motor neuron pathology, perhaps because of an abnormal stoichiometry of different NH subunits. Thus, mice expressing high levels of wild-type neurofilament light subunit (NF-L) acquire a type of motor neuron disease.[60] Like the mice with mutant NF subunits, the animals overexpressing human NF-L reveal neurofilament accumulation in both motor and sensory neurons. These pathological findings distinguish these mice from human ALS patients, in whom sensory neurons are not involved.

Limited experimental data illuminate events that signal cell death in the recessively inherited motor neuron diseases. At this time, it is at least plausible that the absence of SMN will compromise the trafficking of ribosomal RNA (rRNA) and possibly its ability to complex normally with ribosomes. This could impair protein synthesis and ultimately cause significant cellular dysfunction. The mechanism of neurotoxicity of the expanded CAG repeats in SBMA is not well understood. It is speculated that either the CAG itself or polyglutamine can, like the absence of SMN, impair protein synthesis [see Figure 2]. In adult Tay-Sachs disease, the cellular pathology likely reflects, in part, the failure to recycle neuronal gangliosides in the absence of hexosaminidase function. However, a direct toxic effect of accumulated GM_2 gangliosides is also possible, which might, for example, significantly impair normal lysosomal function.

Concepts in Therapy for ALS

Although the biochemical pathways in ALS [see Figure 2] are speculative, they are broadly useful. They suggest new hypotheses for further investigation, including the likely possibility that the death process will have multiple components. Presumably, different components will converge on final common pathways that lead to motor neuron death. As a corollary, some components will probably be more significant than others at different stages of the illness. Factors triggering the death process, for example, may not be the same as those sustaining it once it is under way.

If parallel death pathways are differentially active at different points in the illness, it will be necessary to devise treatment strategies accordingly. In the most general terms, it is useful to discern four distinct phases of the death process [see Table 2]. An initiation phase likely is acting relatively upstream to signal activity in the various death pathways. This appears to be followed by a propagation phase [see Figure 2], which may become self-sustaining because the progressive cellular impairment may enhance the initially triggering events. Strong evidence favoring this view is seen in studies of ALS mice. In this model, vitamin E, a potent lipid-penetrating antioxidant, retards the onset of ALS but has no impact on overall survival.[43] By contrast, the antiglutamate compound riluzole slows the disease course but has no effect on age of onset. This suggests that oxidative toxicity is important in the early triggering process, whereas glutamate-mediated toxicity

Table 2 Potential Approaches to Drug Therapy in Amyotrophic Lateral Sclerosis

Phase	Mechanism/Agent	Possible Therapy
Initiation	Lipid peroxidation	Membrane-soluble antioxidants
Propagation	1. Excitability cascade	↓ glutamate release Block glutamate receptors
	2. ↑ Cytosolic Ca^{2+}	↑ buffering of cytosolic Ca^{2+}
	3. NO, H_2O_2 toxicity	NOS inhibitors
	4. Mitochondrial impairment	? CoQ_{10}, ? Vitamin K
	5. Impaired axonal transport	?
Dissemination	1. NOS Activation/NO	NOS inhibitors
	2. ? H_2O_2	Zinc ↑ activity of glutathione peroxidase and catalase
Cell death	? Apoptosis	? ICE inhibitors ? Bcl-2
	Other	Neurotrophic factors

sustains the death process thereafter. For a given neuron, the next and final stage is cell death. In a population of neurons, it is probably also useful to distinguish a fourth stage, dissemination. This term encompasses events whereby the pathological process advances within the spinal cord and brain motor neurons. A body of data suggests that in most ALS patients, the disease begins focally and migrates, as if advancing in both the longitudinal and transverse axes within the spinal cord. Therapeutically, dissemination may be as important as initiation. Because most patients are first encountered with focal but progressive disease, knowledge of factors mediating dissemination and methods of blocking their activity would be invaluable. Therapies that block dissemination could convert ALS into a nonlethal disease. There are some broad strategies for treating these different phases of ALS [*see Table 2*]. Given the emerging complexities of these death pathways in motor neurons, it is a reasonable prediction that treatment of the motor neuron diseases will require simultaneous use of multiple therapies.

Conclusion

Since 1990, dramatic progress has been made in defining genetic loci linked to different motor neuron diseases. For many, the resulting molecular and protein defects are known. Some of the proteins, such as SMN, are novel; study of their functions will undoubtedly provide insight into both motor neuron biology and the motor neuron diseases. For others, such as SOD1, the normal function is well defined. Nonetheless, study of the role of the mutant protein in motor neuron death is extending our understanding of the protein itself and the associated disease. Because many neurodegenerative diseases share common features, insights into motor neuron death in ALS may enhance our understanding of diseases such as Alzheimer's and Parkinson's. Perhaps most important, it is already apparent that these studies will improve efforts to treat these problems, through both the generation of useful animal models and new perspectives on the complex pathogenesis of motor neuron cell death.

References

1. Mulder D, Kurland L, Offord K, et al: Familial adult motor neuron disease: amyotrophic lateral sclerosis. *Neurology* 36:511, 1986

2. Leigh PN, Anderton BH, Dodson A, et al: Ubiquitin deposits in anterior horn cells in motor neurone disease. *Neurosci Lett* 93:197, 1988

3. Hentati A, Bejaoui K, Pericak-Vance MA, et al: Linkage of recessive familial amyotrophic lateral sclerosis to chromosome 2q33-35. *Nature Genet* 7:425, 1994

4. Tanabe Y, Jessell T: Diversity and pattern in the developing spinal cord. *Science* 274:1115, 1996

5. Pfaff S, Mendelsohn M, Stewart C, et al: Requirement for LIM homeobox gene Isl1 in motor neuron generation reveals a dependent step in interneuron differentiation. *Cell* 84:309, 1996

6. Hall Z: *Introduction to Molecular Biology*. Sinauer, Sunderland, MA, 1992

7. McKay S, Garner A, Caldero J, et al: The expression of trkB and p75 and the role of BDNF in the developing neuromuscular system of the chick embryo. *Development* 122:715, 1996

8. Alexianu M, Ho B, Mohamed A, et al: The role of calcium-binding proteins in selective motoneuron vulnerability in amyotrophic lateral sclerosis. *Ann Neurol* 36:846, 1994

9. Appel S, Smith R, Engelhardt J, et al: Evidence of autoimmunity in amyotrophic lateral sclerosis (review). *J Neurol Sci* 124(suppl):14, 1994

10. Rothstein J, Kammen M, Levey A, et al: Selective loss of glial glutamate transporter GLT-1 in amyotrophic lateral sclerosis. *Ann Neurol* 38:73, 1995

11. Rosen DR, Siddique T, Patterson D, et al: Mutations in Cu/Zn superoxide dismutase gene are associated with familial amyotrophic lateral sclerosis. *Nature* 362:59, 1993

12. Fridovich I: Superoxide dismutases. *Adv Enzymol* 58:61, 1986

13. Borchelt DR, Guarnieri M, Wong P-C, et al: Superoxide dismutase 1 subunits with mutations linked to familial amyotrophic lateral sclerosis do not affect wild-type subunit function. *J Biol Chem* 270:1, 1995

14. Reaume A, Elliott J, Hoffman E, et al: Motor neurons in Cu/Zn superoxide dismutase-deficient mice develop normally but exhibit enhanced cell death after axonal injury. *Nature Genet* 13:43, 1996

15. Gurney ME, Pu H, Chiu AY, et al: Motor neuron degeneration in mice that express a human Cu, Zn superoxide dismutase mutation. *Science* 264:1772, 1994

16. Beckman JS, Carson M, Smith CD, et al: ALS, SOD, and peroxynitrite. *Nature* 364:584, 1993

17. Wiedau-Pazos M, Goto J, Rabizadeh S, et al: Altered reactivity of superoxide dismutase in familial amyotrophic lateral sclerosis. *Science* 271:515, 1996

18. Yim H-S, Kang J-H, Chock PB, et al: A familial amyotrophic lateral sclerosis-associated A4V Cu, Zn dismutase mutant has a lower Km for hydrogen peroxide. *J Biol Chem* 272:8861, 1997

19. Lyons T, Liu H, Goto J, et al: Mutations in copper zinc superoxide dismutase that cause amyotrophic lateral sclerosis alter the zinc binding site and the redox behavior of the protein. *Proc Natl Acad Sci USA* 93:12240, 1996

20. Rothstein JD, Bristol LA, Hosler BA, et al: Chronic inhibition of superoxide dismutase produces apoptotic death of spinal neurons. *Proc Natl Acad Sci USA* 91:4155, 1994

21. Troy CM, Shelanski M: Down-regulation of copper/zinc superoxide dismutase causes apoptotic death in PC12 neuronal cells. *Proc Natl Acad Sci USA* 91:6384, 1994

22. Greenlund L, Deckwerth T, Johnson E: Superoxide dismutase delays neuronal apoptosis: a role for reactive oxygen species in programmed neuronal death. *Neuron* 14:303, 1995

23. Durham H, Roy J, Dong L, et al: Aggregation of mutant Cu/Zn superoxide dismutase proteins in a culture model of ALS. *J Neuropathol Exp Neurol* 56:523, 1997

24. Roos R, Lee J, Bindokas V, et al: Gene delivery by replication-deficient recombinant adenoviruses (AdVs) in the study of Cu,Zn superoxide dismutase type 1 (SOD-1)-linked familial amyotrophic lateral sclerosis (FALS). *Neurology* 48:A150, 1997

25. Friedlander R, Brown J, Gagliardini V, et al: Inhibition of ICE slows ALS in mice. *Nature* 388:31, 1997

26. Kostic V, Jackson-Lewis V, Bilbao FD, et al: Bcl-2: prolonging life in a transgenic mouse model of familial amyotrophic lateral sclerosis. *Science* 277:559, 1997

27. Troost D, Aten J, Morsink F, et al: Apoptosis is not restricted to motoneurons: Bcl-2 expression is increased in post-central cortex, adjacent to affected motor cortex. *J Neurol Sci* 129(suppl):79, 1995

28. Mu X, He J, Anderson D, et al: Altered expression of bcl-2 and bax mRNA in amyotrophic lateral sclerosis spinal cord motor neurons. *Ann Neurol* 40:379, 1996

29. Kunst C, Mezey E, Brownstein M, et al: Mutations in SOD1 associated with familial ALS cause novel protein interactions. *Nature Genet* 15:91, 1997

30. Bredesen D, Ellerby L, Hart P, et al: Do post-translational modifications of CuZnSOD lead to sporadic amyotrophic lateral sclerosis? *Ann Neurol* 42:135, 1997

31. Bowling AC, Schulz JB, Brown RH Jr, et al: Superoxide dismutase activity, oxidative damage and mitochondrial energy metabolism in familial and sporadic amyotrophic lateral sclerosis. *J Neurochem* 61:2322, 1993

32. Shaw P, Ince P, Falkous G, et al: Oxidative damage to protein in sporadic motor neuorone disease spinal cord. *Ann Neurol* 38:691, 1995

33. Bruijn L, Beal M, Becher M, et al: Elevated free nitrotyrosine levels but not protein-bound nitrotyrosine or hydroxyl radicals, throughout amyotrophic lateral sclerosis (ALS)-like disease implicate tyrosine nitration as an aberrant in vivo property of one familial ALS-like superoxide dismutase 1 mutant. *Proc Natl Acad Sci USA* 94:7606, 1997

34. Fitzmaurice PS, Shaw IC, Kleiner HE, et al: Evidence for DNA damage in amyotrophic lateral sclerosis. *Muscle Nerve* 19:797, 1996

35. Ince P, Shaw P, Candy J, et al: Iron, selenium and glutathione peroxidase activity are elevated in sporadic motor neuron disease. *Neurosci Lett* 182:87, 1994

36. Karsarskis E, Tandon L, Lovell M, et al: Aluminum, calcium, and iron in the spinal cord of patients with sporadic amyotrophic lateral sclerosis using laser microprobe mass spectroscopy: a preliminary study. *J Neurol Sci* 130:203, 1995

37. Imon Y, Yamaguchi S, Yamamura Y, et al: Low intensity areas observed on T2-weighted magnetic resonance imaging of the cerebral cortex in various neurological diseases. *J Neurol Sci* 134:27, 1995

38. Oba H, Araki T, Ohtomo K, et al: Amyotrophic lateral sclerosis: T2 shortening in motor cortex at MR imaging. *Radiology* 189:843, 1993

39. Tohgi H, Abe T, Saheki M, et al: α-Tocopherol quinone level is remarkably low in the cerebrospinal fluid of patients with sporadic amyotrophic lateral sclerosis. *Neurosci Lett* 207:5, 1996

40. Przedborski S, Donaldson D, Jakowec M, et al: Brain superoxide dismutase, catalase and glutathione peroxidase activities in amyotrophic lateral sclerosis. *Ann Neurol* 39:158, 1996

41. Shaw I, Fitzmaurice PS, Mitchell J, et al: Studies on cellular free radical protection mechanisms in anterior horn from patients with amyotrophic lateral sclerosis. *Neurodegeneration* 4:391, 1995

42. Lanius R, Krieger C, Wagey R, et al: Increased 35-S-glutathione binding sites in spinal cords from patients with sporadic amyotrophic lateral sclerosis. *Neurosci Lett* 163:89, 1993

43. Gurney M, Cutting F, Zhai P, et al: Benefit of vitamin E, riluzole and gabapentin in a transgenic model of familial ALS. *Ann Neurol* 39:147, 1996

44. Louwerse E, Weverling G, Bossuyt P, et al: Randomized, double-blind, controlled trial of N-acetylcysteine in amyotrophic lateral sclerosis. *Arch Neurol* 52:559, 1995

45. Reider C, Paulson G: Lou Gehrig and amyotrophic lateral sclerosis: is vitamin E to be revisited? *Arch Neurol* 54:527, 1997

46. Lefebvre S, Bürglen L, Reboullet S, et al: Identification and characterization of a spinal muscular atrophy-determining gene. *Cell* 80:155, 1995

47. Liu Q, Dreyfuss G: A novel structure containing the survival motoneuron protein. *EMBO J* 15:3555, 1996

48. Brahe C, Clermont O, Zappata S, et al: Frameshift mutation in the survival motor neuron gene in a severe case of SMA type I. *Hum Mol Genet* 5:1971, 1996

49. Hahnen E, Schonling J, Rudnik-Schoneborn S, et al: Missense mutations in exon 6 of the survival motor neuron gene in patients with spinal muscular atrophy (SMA). *Hum Mol Genet* 6:821, 1997

50. Parsons D, McAndrew P, Monani U, et al: An 11 base pair duplication in exon 6 of the SMN gene produces a type I spinal muscular atrophy (SMA) phenotype: further evidence for SMN as the primary SMA-determining gene. *Hum Mol Genet* 5:1727, 1996

51. Velasco E, Valero C, Valero A, et al: Molecular analysis of the SMN and NAIP genes in Spanish spinal muscular atrophy (SMA) families and correlation between number of copies of cBCD541 and SMA phenotype. *Hum Mol Genet* 5:257, 1996

52. Roy N, Mahadevan MS, McLean M, et al: The gene for neuronal apoptosis inhibitory protein is partially deleted in individuals with spinal muscular atrophy. *Cell* 80:167, 1995

53. LaSpada AR, Wilson EM, Lubahn DB, et al: Androgen receptor gene mutations in X-linked spinal and bulbar muscular atrophy. *Nature* 352:77, 1991

54. Davies S, Turmaine M, Cozens B, et al: Formation of neuronal intranuclear inclusions underlies the neurological dysfunction in mice transgenic for the HD mutation. *Cell* 90:537, 1997

55. Scherzinger E, Lurz R, Turmaine M, et al: Huntingtin-encoded polyglutamine expansions form amyloid-like protein aggregates in vitro and in vivo. *Cell* 90:549, 1997

56. Beal M: Aging, energy, and oxidative stress in neurodegenerative diseases (review). *Ann Neurol* 38:357, 1995

57. Figlewicz DA, Krizus A, Martinoli MG, et al: Variants of the heavy neurofilament subunit are associated with the development of amyotrophic lateral sclerosis. *Hum Mol Genet* 3:1757, 1994

58. Vechio JD, Bruijn L, Xu Z-S, et al: Sequence variants in human neurofilament genes: absence of linkage to familial amyotrophic lateral sclerosis. *Ann Neurol* 40:603, 1996

59. Lee MK, Cleveland DW: A mutant neurofilament subunit causes massive, selective motor neuron death: implications for the pathogenesis of human motor neuron disease. *Neuron* 13:975, 1994

60. Xu Z, Cork LC, Griffin JW, et al: Increased expression of neurofilament subunit NF-L produces morphological alterations that resemble the pathology of human motor neuron disease. *Cell* 73:23, 1993

61. Fink J, Heiman-Patterson T, Bird T, et al: Hereditary spastic paraplegia: advances in genetic research. *Neurology* 46:1507, 1996

62. Hazan J, Fontaine B, Bruyn R, et al: Linkage to a new locus for autosomal dominant familial spastic paraplegia to chromosome 2p. *Hum Mol Genet* 3:1569, 1994

63. Hazan J, Lamy C, Melki J, et al: Autosomal dominant familial spastic paraplegia is genetically heterogeneous and one locus maps to chromosome 14q. *Nature Genet* 5:163, 1993

64. Fink JK, Wu CT, Jones SM, et al: Autosomal dominant familial spastic paraplegia: tight linkage to chromosome 15q. *Am J Hum Genet* 56:188, 1995

65. Hentati A, Pericak-Vance MA, Hung WY, et al: Linkage of "pure" autosomal recessive familial spastic paraplegia to chromosome 8 markers and evidence of genetic locus heterogeneity. *Hum Mol Genet* 3:1263, 1994

66. Saugier-Veber P, Munnich A, Bonneau D, et al: X-linked spastic paraplegia and Pelizaeus-Merzbacher disease are allelic disorders at the proteolipid protein locus. *Nature Genet* 6:257, 1994

67. Jouet M, Rosenthal A, Armstrong G, et al: X-linked spastic paraplegia (SPG1), MASA syndrome, and X-linked hydrocephalus result from mutations in the L1 gene. *Nature Genet* 7:402, 1994

68. Brzustowicz LM, Lehner T, Castilla LH, et al: Genetic mapping of chronic childhood-onset spinal muscular atrophy to chromosome 5q11.2-13.3. *Nature* 344:540, 1990

69. Takeda K, Nakai H, Hagiwara H, et al: Fine assignment of beta-hexosaminidase A alpha subunit on 15q23-24 by high resolution in situ hybrdization. *Tohoku J Exp Med* 60:203, 1990

70. Burg J, Conzelmann E, Sandhoff K, et al: Mapping of the gene coding for the human GM2-activator protein to chromosome 5. *Ann Hum Genet* 49:41, 1985

71. Bamshad M, Watkins W, Zenger R, et al: A gene for distal arthrogryposis type I maps to the pericentromeric region of chromosome 9. *Am J Hum Genet* 55:1153, 1994

72. Kobayashi H, Baumbach L, Matise T, et al: A gene for a severe, lethal form of X-linked arthrogryposis (X-linked infantile spinal muscular atrophy) maps to human chromosome Xp11.3-q11.2. *Hum Mol Genet* 4:1213, 1995

73. Hall J, Reed S, Scott C, et al: Three distinct types of X-linked arthrogryposis seen in 6 families. *Clin Genet* 21:81, 1982

Acknowledgments

Figure 1   Marcia Kammerer.
Figure 2   Seward Hung.

CHAPTER 13

Molecular Genetics of Peripheral Neuropathies

James R. Lupski, M.D., Ph.D.

Inherited disorders of human peripheral nerve are a heterogeneous collection of entities with distinct yet sometimes overlapping clinical features. They usually present with an insidious, slowly progressive course that begins in early adulthood, although more severe cases may manifest themselves in childhood or even infancy. These inherited peripheral neuropathies sometimes share clinical features with acquired conditions, compounding the diagnostic dilemma presented to the clinician, particularly in a sporadic case with no family history.

The elucidation of the molecular genetic bases of inherited peripheral neuropathies[1-10] has uncovered a novel genetic mechanism: a large DNA duplication affecting a dosage-sensitive gene, resulting in a commonly inherited trait. Furthermore, the investigations have resulted in the identification of three genes essential to myelin function: *PMP22* encoding peripheral myelin protein-22, *MPZ* encoding myelin protein zero (P_0), and *Cx32* encoding connexin-32. The study of these genes and their protein products provides new insights into the biology and structure of peripheral nerve. The identification of these genes has confirmed the long-held suspicion that inherited peripheral neuropathies represent a manifestation of peripheral nerve dysfunction resulting from abnormalities in Schwann cells and their myelin sheath. An understanding of the molecular etiology of inherited neuropathies has prompted some authors to advocate reclassification on the basis of the molecular defect involved.[11,12] Perhaps the most substantive effect of these molecular findings is the development of diagnostic tools that permit precise and secure diagnosis, accurate recurrence risk estimates, prognostic information, and the potential to design therapeutic approaches.

Clinical Features of Inherited Neuropathy

Charcot-Marie-Tooth (CMT) disease is the most common inherited peripheral neuropathy; it is characterized by progressive muscular atrophy of the distal muscles, foot deformities including *pes cavus* and *pes planus*, gait disturbances, and absent deep tendon reflexes. These findings are usually evident within the first two decades of life.[13,14] A tendency to walk on the toes during childhood sometimes provides early warning of the development of CMT. Two major forms of the disorder are delineated based on electrophysiologic and neuropathic features. Type 1 (CMT1), the hypertrophic form, displays uniformly slow motor nerve conduction velocity (NCV), usually less than 40 m/sec. Sural nerve biopsies show onion bulb formation that occurs when demyelination is followed by Schwann cell division and renewed myelination.[14] This process is nonspecific and may result from either genetic or acquired causes. Type 2 (CMT2), the neuronal form, exhibits normal or mildly reduced NCV with decreased amplitudes. CMT2 results from axonal dysfunction [*see Figure 1a*]. CMT1 and CMT2 have alternatively been referred to as hereditary motor and sensory neuropathy I and II (HMSNI and HMSNII).

Other inherited peripheral neuropathies[1-14] that primarily affect the myelin include Dejerine-Sottas syndrome (DSS), congenital hypomyelinating neuropathy (CHN), and hereditary neuropathy with liability to pressure palsies (HNPP) [*see Figure 1a*]. Dejerine-Sottas syndrome, also referred to as HMSNIII, was originally described as a severe neuropathy that manifests itself in early childhood, usually with delayed motor milestones, hypertrophic nerves, onion bulbs noted on nerve biopsy, and an apparent autosomal recessive inheritance pattern. Subsequently, markedly decreased NCVs, usually less than 10 m/sec, was also associated with Dejerine-Sottas syndrome. Congenital hypomyelinating neuropathy is characterized clinically by onset in infancy of hypotonia, areflexia, distal muscle weakness, and very slow NCV. In its extreme form, it may present with severe joint contractions or arthrogryposis multiplex congenita. Less severe cases of congenital hypomyelinating neuropathy can overlap clinically with Dejerine-Sottas syndrome.[15] Pathological features on sural nerve biopsy can often differentiate between these disorders; patients with congenital hypomyelinating neuropathy lack both active myelin breakdown and well-organized onion bulbs and show hypomyelination of all or most fibers.

HNPP was initially described in a European family with three generations of affected members who had recurrent peroneal neuropathy after digging potatoes in a kneeling position. Clinical features include periodic episodes of numbness and muscular weakness after relatively minor compression or trauma to the peripheral nerves. The axillary, median, radial, ulnar, and peroneal nerves are the most commonly affected. Entrapment neuropathies, including carpal tunnel syndrome, are frequent manifestations of HNPP. Electrophysiologic studies in affected patients and asymptomatic carriers sometimes show mildly decreased NCV with conduction blocks. A specific neuropathologic feature observed is tomacula, or sausagelike focal thickening of the myelin sheaths. Electron microscopy of tomacula reveals loss of myelin compaction, suggesting a potential structural and perhaps an adhesive role for the proteins affected in HNPP.[16]

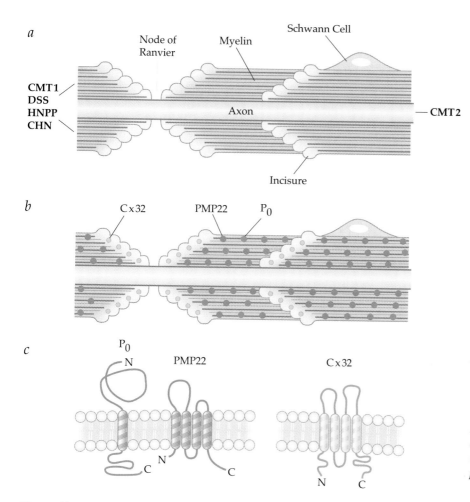

Figure 1 *Inherited peripheral nerve disorders and involved proteins. (a) Structure of a typical peripheral nerve with axon and surrounding myelin made by the supporting glia or Schwann cell. The node of Ranvier and the Schmidt-Lanterman incisure (an area of uncompacted myelin) are depicted. On the left are listed the disorders affecting myelin: Charcot-Marie-Tooth disease type 1 (CMT1), Dejerine-Sottas syndrome (DSS), hereditary neuropathy with liability to pressure palsies (HNPP), and congenital hypomyelinating neuropathy (CHN). Charcot-Marie-Tooth disease type 2 (CMT2), listed on the right, is primarily a disorder of the axon. (b) The distribution of the myelin proteins involved in the dysmyelinating peripheral neuropathies is shown by PMP22 (red dots), P_0 (blue dots), and Cx32 (yellow dots). (c) Structure of the myelin proteins.*

Genetics

Charcot-Marie-Tooth disease can be caused by gene alterations at several different loci, but in a single family, only one locus is responsible for the disease phenotype. Although every mendelian inheritance pattern (autosomal dominant, recessive, and X-linked) has been described for CMT, the most commonly observed is autosomal dominant for CMT1, CMT2, and HNPP. Rigorous evaluation of both parents of an affected individual, including electrophysiological studies, is required to firmly establish the inheritance pattern. This is essential because of the remarkable clinical variability in these conditions, which has even been observed in identical twins with the same gene mutation.[17] CMT1 is inherited in some families as an X-linked trait in which males are more severely affected than females, but more females are affected than males. The presence of two X chromosomes in females doubles their risk compared with their XY male relatives). It can often be difficult to establish an X-linked dominant pattern in a given family because females may be only mildly affected or even asymptomatic. Autosomal recessive inheritance is rarely observed. Sporadic neuropathy, perhaps the most frequently observed neuropathy by the practicing clinician, can be a diagnostic challenge. These cases may represent either a dominant new mutation, an autosomal recessive inheritance, or an acquired (nongenetic) neuropathy.

Genetic linkage studies in large families have permitted the identification of a number of loci associated with CMT and related neuropathies[1-6] [*see Figure 2*]. Autosomal dominant forms of CMT1 have been mapped to chromosome 17 (CMT1A) and chromosome 1 (CMT1B), and there are other autosomal dominant pedigrees that do not map to these regions (CMT1C). Autosomal recessive forms of CMT1 have been mapped to chromosomes 5, 8, and 11, and an X-linked form (CMT1X) maps to the long arm of the X chromosome. CMT2, which is less common than CMT1, has been mapped to chromosomes 1, 3, and 7, although no clinical distinctions separate these subcategories. The genes for only three of these 10 loci have been identified. Isolation of the other seven genes, and those from loci yet to be identified, will likely provide substantial insights regarding structure and function of the peripheral nerve.

Molecular Mechanisms

Mutations in the MPZ Gene

Although not the most common cause for inherited peripheral neuropathy, the identification of *MPZ* gene mutations in neuropathy patients has provided insights into the disease process at the molecular level. The gene for the myelin protein P_0 was originally cloned from rats.[18] P_0 is a 28 kd protein exclusively expressed by myelinating Schwann cells; it accounts for more than 50 percent of the protein found in the peripheral nerve myelin sheath[19] [*see Figure 1*]. The protein structure consists of a single membrane-spanning domain, a large glycosylated immunoglobulin-like extracellular domain, and a basic intracellular domain. P_0 functions in the formation and compaction of peripheral nerve myelin, as clearly documented by P_0-knockout mice.[20-22] These animal models show early onset decompaction of myelin and severe hypomyelination in homozygotes ($P_0^{-/-}$) and later-onset segmental demyelination, interruption of compact myelin, and onion bulb formation when heterozygous ($P_0^{+/-}$), resembling Dejerine-Sottas syndrome and CMT1B, respectively. Electron microscopy reveals a uniform structure of myelin, with the homophilic adhesion of the extracellular P_0 domain forming the intraperiod line and electrostatic interactions of the intracellular domain forming the major dense line. The x-ray crystal structure of the extracellular domain of P_0 at 1.9Å resolution shows it is folded as a typical immunoglobulin variable-like domain with a flexible linkage to the membrane.[23] The P_0 extracellular domains emanate from the membranes as tetramers that link to tetramers on the apposing membrane surface and result in the formation of a network of molecules.[23] This network of interacting tetramers, which I call molecular Velcro, enables a powerful adhesive force to tightly compact the multilaminated myelin into a supramolecular structure.

The human gene encoding P_0, *MPZ*, was mapped to chromosome 1, where the CMT1B linkage had already been established[24]; subsequently, P_0 mutations were identified in CMT1B families[25,26] as well as in patients with Dejerine-Sottas syndrome.[27] P_0 point mutations have now been described in several CMT1B and Dejerine-Sottas syndrome patients and in selected pa-

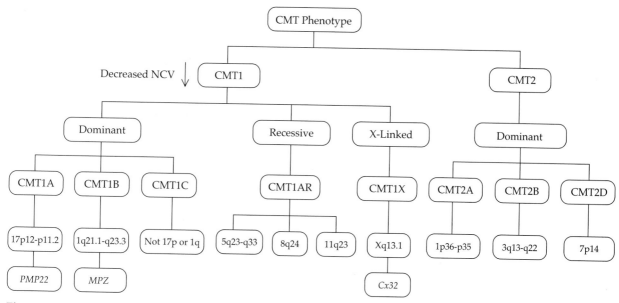

Figure 2 *Genetic heterogeneity underlying the Charcot-Marie-Tooth (CMT) phenotype. Markedly decreased or slow motor nerve conduction velocity (NCV) distinguishes CMT1 from CMT2. For CMT1, there are autosomal dominant, autosomal recessive, and X-linked forms. To date, 10 loci have been associated with a CMT phenotype. Three genes have been identified, and alterations in dosage (PMP22) or structure (Cx32 > MPZ > PMP22) of these gene products account for the majority of patients with inherited neuropathy.*

tients with a clinical diagnosis of congenital hypomyelinating neuropathy.[15,28] In fact, in some instances, different amino acid substitutions at the identical position within P_0 can produce distinct clinical phenotypes (CMT1B versus Dejerine-Sottas syndrome).[3,15,29]

These data document that *MPZ* mutations result in a spectrum of clinical severity and raise the question of how mutations in a single gene can cause such disparate symptoms. It is hypothesized that variation in clinical severity caused by different mutations in *MPZ* can be explained by the alterations in protein function that arise from each mutation. Heterozygous loss-of-function mutations, with one abnormal and one normal gene copy, reduce the total amount of normal protein present, thus producing a less severe CMT1B phenotype. Heterozygous dominant-negative mutations, which disrupt the formation or proper functioning of the P_0 complex, or homozygous loss-of-function mutations with two abnormal copies that completely lack the P_0 protein, result in more severe Dejerine-Sottas syndrome or congenital hypomyelinating neuropathy phenotype.[15] Consistent with the loss-of-function hypothesis is the finding that point mutations resulting in termination or nonsense codons occurring early in the *MPZ* gene yield truncated proteins that probably never reach the membrane, resulting in milder CMT1B phenotypes. A dominant-negative mechanism implies a multimeric P_0 protein complex with disordered protein-protein interactions. Crystalline structural data, which suggest the formation and interaction of homotetramers,[23] are consistent with this proposed model. Also consistent with a dominant-negative model is the observation that ultrastructural changes observed by electron microscopy (to examine myelin membrane packing in sural nerve biopsies from selected patients with *MPZ* mutations) suggest that the altered P_0 is incorporated into the myelin sheath.[30] Furthermore, in vitro cell adhesion assays demonstrate a dominant-negative effect on adhesion by P_0 proteins truncated in their cytoplasmic domain when coexpressed with the wild-type protein.[31] Of course, there is the potential for weak dominant-negative mutations resulting in a more severe CMT1B phenotype than that observed with null alleles, depending on whether or not they are incorporated into the multimeric P_0 complex, and such observations would not be inconsistent with the general hypothesis.

MUTATIONS IN THE *PMP22* GENE

Point mutations in the *PMP22* gene have also been associated with peripheral neuropathies, although these occur less frequently than *MPZ* mutations.[1-14] Peripheral myelin protein-22 (PMP22), encoded by the *PMP22* gene, is a membrane glycoprotein that is highly expressed by myelinating Schwann cells.[32] *PMP22* was isolated by differential complimentary DNA colony screening to identify genes differentially expressed following nerve injury.[33,34] A surprising finding is that *PMP22* is identical to *gas-3*, a gene that encodes a molecule induced in growth-arrested NIH3T3 fibroblasts, thereby leading some researchers to speculate that *PMP22* may also have a potential regulatory role in cell growth cycles.[32] PMP22 consists of 160 amino acids, including four putative transmembrane domains and an N-linked carbohydrate that contains the L2/HNK-1 epitope, which is implicated in adhesive processes[32] [*see Figure 1c*]. This membrane glycoprotein represents up to five percent of the total protein found in the peripheral nerve myelin, and its expression is largely restricted to compact myelin [*see Figure 1b*], although it can be found in nonmyelinating Schwann cells and other nonperipheral nervous system tissues.[32,35]

The mouse *Pmp22* gene was found to be mutated in the allelic *Trembler* (*Tr*) and (*Tr*[J]) murine models for human inherited neuropathies.[36,37] The human *PMP22* gene is 86 percent identical to the mouse gene.[38] *PMP22* point mutations have been identified in some CMT1A patients[39,40] and a larger number of Dejerine-Sottas syndrome[41] patients. These mutations are dominant as the phenotype is expressed in the heterozygous state. One of the CMT1A mutations is identical in position to that of the *Tr*[J],[39] whereas a Dejerine-Sottas syndrome mutation is identical in position to that of *Tr*.[42] *PMP22* loss-of-function mutations result in HNPP.[43] The *PMP22* knockout mice recapitulate the tomaculous neuropathy observed in HNPP[44] and establish an important model for following the disease progression.[45] Interestingly, *PMP22* missense amino acid substitutions associated with CMT1A and Dejerine-Sottas syndrome are all located within one of the four putative transmembrane domains, whereas most of the P_0 mutations are within the immunoglobulin-like extracellular domain.

MUTATIONS IN THE *Cx32* GENE

Mutations in the *Cx32* gene encoding connexin-32 are responsible for X-linked CMT (CMTX).[46-48] Connexins are membrane-spanning proteins that assemble to form gap junctions that facilitate the transfer of ions and small molecules between cells. *Cx32* is highly expressed in the peripheral nervous system but is also expressed in liver and other tissues. However, the clinical manifestations of CMTX appear to be limited to the peripheral nerves, demonstrating that *Cx32* is particularly important to peripheral nerves structure and function and suggesting that other redundant connexins may substitute for Cx32 function in nonperipheral nervous system tissue.[46] The Cx32 protein has two extracellular loops, four transmembrane segments, and three cytoplasmic domains. It localizes by immunofluorescent staining to the nodes of Ranvier and the Schmidt-Lanterman incisures [*see Figure 1b and c*] and may form intracellular gap junctions that connect the folds of Schwann cell cytoplasm.[46] It has been proposed that this would allow transfer of ions, nutrients, and other small molecules around and across the compact myelin to the innermost myelin layers and provide sustenance to the axons. This potentially explains the combined myelin disruption and axonal degeneration observed in CMTX.[46]

DNA REARRANGEMENTS

The largest number of patients with inherited peripheral neuropathy have as the molecular basis of their disease DNA rearrangements—duplication causing CMT1A or deletion causing HNPP. The identification of the CMT1A duplication in 1991 focused research into the molecular causes of inherited peripheral neuropathy[49,50] [*see Figure 3*]. Patients with the CMT1A duplication exhibit the characteristic electrophysiologic and neuropathic features of CMT1.[51,52] The CMT1A duplication is the most common abnormality,[53,54] a fact that is remarkable given that at least seven loci [*see Figure 2*] are responsible for the CMT1 phenotype. Duplication, as a mutational mechanism, may be quite common, with examples having been documented across species.[55] Important questions to be addressed regarding the CMT1A duplication are (1) What is the mechanism of formation?, (2) Why is it so frequent?, and (3) What is the molecular pathophysiology of the disease?

The genomic region duplicated in CMT1A encompasses 1.5 megabases (Mb) and is flanked by a large 24 kilobase (kb) direct repeat termed CMT1A-REP.[56] This finding immediately suggested that the mechanism for the duplication was unequal crossing-over mediated by misaligned flanking repeats.[56] Substantial evidence supporting this mechanism was obtained,[56] [*see Figure 4*] and the predicted reciprocal recombination event that would result in deletion was later identified and associated with a different neuropathy, HNPP.[57] CMT1A and HNPP have distinct clinical features [*see Table 1*]; however, more severe cases of HNPP, particularly late in life, can overlap clinically with CMT1A. Molecular studies firmly established that the CMT1A duplication and HNPP deletion indeed represent products of a reciprocal recombination involving CMT1A-REP.[58]

The flanking 24 kb CMT1A-REP is over 98 percent identical in sequence to CMT1A, suggesting that crossing over could potentially occur anywhere within this region of homology. However, a recombination hot spot[59] has been associated with the crossing-over event. This recombination hot spot was confirmed in multiple independent populations of CMT1A and HNPP patients.[60-62] It was clearly documented by the identification of patient-specific junction fragments that this recombination hot spot did not occur in a region with any greater DNA sequence identity than the surrounding

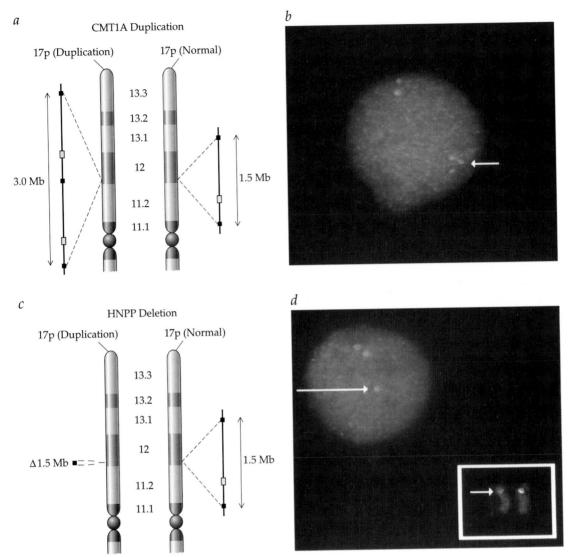

Figure 3 *Structures of CMT1A duplication and HNPP deletion. (a) The G-banded chromosome 17p homologues with the centromeres depicted with a dark gray circle. The normal chromosome is on the right, with the submicroscopic 1.5 Mb region flanked by the CMT1A-REP (black boxes) and containing PMP22 (gray box). On the left is the abnormal chromosome containing the 3.0 Mb CMT1A duplication with three copies of CMT1A-REP and two copies of PMP22. (b) The results of a fluorescent in situ hybridization (FISH) diagnostic test on an interphase nucleus from a peripheral white blood cell of a CMT1 patient. The red signal represents the hybridization to PMP22; green represents a control probe from 17p11.2. Note three signals for the PMP22 gene. (c) The HNPP deletion, with the normal chromosome deleted for 1.5 Mb (Δ 1.5 Mb) on the left. The deletion chromosome retains one copy of CMT1A-REP. (d) Both interphase and metaphase (insert) chromosomes from a patient with HNPP. Note only one red signal for PMP22. The white arrows point to the duplicated (b) or deleted (d) probe.*

sequence.[59] In fact, there were greater stretches of DNA sequence identity outside the hot spot, suggesting that homology alone could not account for the observed hot spot. Interestingly, the analysis of DNA sequence adjacent to the hot spot revealed a structure akin to insect transposable elements. We termed this element MITE for *Mariner* insect transposon-like element.[59] The identification of MITE led to a proposed model[59] wherein a *Mariner* transposase encoded elsewhere in the genome makes a double-stranded DNA

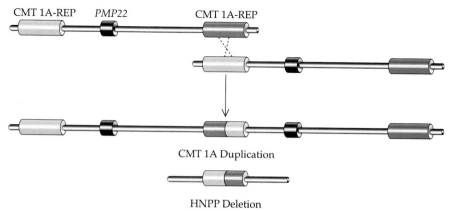

CMT 1A-REP   *PMP22*   CMT 1A-REP

CMT 1A Duplication

HNPP Deletion

Figure 4   *Reciprocal recombination resulting in CMT1A duplication and HNPP deletion. The proximal CMT1A-REP (light blue) and distal CMT1A-REP (dark blue) flank the involved region, which contains the PMP22 gene (black). These CMT1A-REP repeats provide the homology for recombination. Misalignment of the distal repeat from one homologue with the proximal repeat from the other homologue, followed by a recombination between the CMT1A-REPs, results in recombinant chromosomes. Note that the CMT1A duplication contains one recombinant CMT1A-REP plus a proximal and distal copy, resulting in three CMT1A-REPs and two copies of PMP22. The HNPP deletion retains one recombinant CMT1A-REP (the reciprocal copy) and is deleted for PMP22.*

break at the inverted repeat of the MITE element. This break is repaired with the use of homologous sequences between misaligned proximal CMT1A-REP and distal CMT1A-REP. The unequal crossing-over occurs, and the hot spot reflects a preference for resolution of the Holliday junction in a defined region of the CMT1A-REP.

When the CMT1A duplication was initially identified, several hypotheses were formulated to potentially explain how the DNA duplication could affect a specific gene(s).[49] One hypothesis was that gene dosage, having three copies of a CMT1A gene instead of the usual two, would be the ge-

Table 1 Contrasting Features of CMT1A and HNPP

Feature	CMT1A	HNPP
Clinical	Symmetric, slowly progressive	Asymmetric, episodic
Antecedent	None	Minor nerve compression or trauma
Potential early signs	Mild delay in achieving motor milestones, idiopathic toe walking in childhood, absent deep tendon reflexes	None
Presentation	Distal muscle weakness and atrophy, dropped foot, abnormal gait, foot deformity (*pes cavus > pes planus*)	Focal neuropathy, carpal tunnel
Electrophysiology	Slow nerve conduction velocities	Conduction block
Neuropathology	Onion bulbs	Tomacula
Molecular	Duplication	Deletion

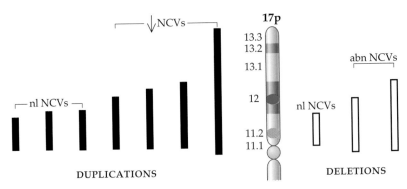

Figure 5 *Cytogenetically visible chromosome 17p rearrangements and neuropathy. The short arm of chromosome 17 (17p) is shown with fluorescent in situ hybridization probes to PMP22 (dark blue) in p12 and to FLI (light blue) in 17p11.2. The left part of the figure depicts the region duplicated (vertical black bars). Note that patients with duplications encompassing the PMP22 gene have decreased nerve conduction velocities (↓NCVs), as observed with CMT1, whereas if the duplication does not include PMP22, such patients have normal motor NCVs (nl NCVs). The right part of the figure depicts patients with deletion of 17p (vertical open bars). Note that if the deletion includes PMP22, then patients have abnormal NCV (abn NCVs), but distinct from CMT1 and consistent with that observed for patients with HNPP.*

netic mechanism responsible. The identification of a partial trisomy 17p patient with the electrophysiological phenotype of decreased NCV consistent with CMT1 showed that gene dosage is a mechanism for CMT1A.[63] Subsequently, it was shown that other patients with cytogenetically visible duplications of 17p displayed a complex phenotype, which included slowed motor NCV and CMT1 only if the *PMP22* gene was duplicated.[63-66] In addition, patients with cytogenetically visible deletions of proximal 17p that included *PMP22* had a complex phenotype, which had electrophysiological abnormalities consistent with HNPP[67,68] [*see Figure 5*]. Such studies have blurred the artificial boundaries that separate two major classes of genetic disorders: the single-gene disorders that segregate in a mendelian fashion and the chromosome syndromes resulting from the incorrect number of chromosomes (aneuploidies, including trisomy and monosomy) or chromosome structural abnormalities (segmental aneuploidies), including contiguous gene syndromes resulting from microduplication or microdeletion.

Substantial evidence supports the contention that the *PMP22* gene is the dosage-sensitive gene responsible for CMT1A when present in three copies and results in HNPP when present in only one copy [*see Table 2*].[39-45,49,63-77] CMT1A and HNPP result from an altered copy number of *PMP22* and constitute an unprecedented model of the phenotypic consequences resulting from a range of possible gene dosages in humans. Furthermore, these studies delineate the concept of a so-called gene expression window for a dosage-sensitive gene. The fact that the genomic region duplicated in CMT1A and deleted in HNPP encompasses 1.5 Mb (potentially 30 to 50 genes), yet only one appears to be dosage sensitive, suggests that dosage-sensitive genes may constitute a small fraction of the genes within the human genome.

Molecular Diagnostics

The CMT1A duplication and HNPP deletion are frequent causes of neuropathy and are examples in humans of consistent large DNA rearrangements that can act as biologic markers for their respective diseases.[78] These disorders are excellent models for investigating methods for detecting large DNA rearrangements responsible for human disease—those too big to be observed by conventional molecular techniques and too small to be seen by conventional cytogenetic techniques. For optimal identification of the CMT1A duplication and HNPP deletion, a test based on detecting the presence or absence of a band or signal is preferable to an assay based on a quantitative measurement of dosage.[78]

The most informative and reliable test depends on directly or indirectly determining the number of copies of the *PMP22* gene. Detection of rearrangement-specific junction fragments utilizing pulsed-field gel electrophoresis to separate large molecular weight DNA is a well-documented method.[79] Molecular cytogenetic testing with a *PMP22* probe by fluorescent in situ hybridization (FISH) also readily detects the CMT1A duplication and HNPP deletion[80] [*see Figure 3*]. A distinct clinical advantage to both of these methods is that either the CMT1A duplication or the HNPP deletion can be detected by the same test.

For patients without the CMT1A duplication or HNPP deletion, a test for direct mutation detection by DNA sequencing of *Cx32* is also commercially

Table 2 Evidence That *PMP22* Gene Dosage Causes CMT1A and HNPP

CMT1A = Trisomic *PMP22*	Homozygous CMT1A duplication patients have more severe phenotype
	Patients with dup(17p) that includes *PMP22* have slow NCV-like CMT1
	PMP22 is contained within the CMT1A duplication
	No *PMP22* coding region mutations are identified in CMT1A duplication patients
	PMP22 point mutations have been found in a small proportion of nonduplication CMT1A patients
	Increased *PMP22* mRNA is found in nerve biopsy samples fromCMT1A duplication patients
	Quantitative increase of PMP22 protein is observed by immunogold labeling in the compact myelin of peripheral nerves from CMT1A duplication patients
	Transgenic animals overexpressing *PMP22*, including rat and two separate mouse models, display clinical, electrophysiological, and neuropathologic features of CMT1
HNPP = Monosomic *PMP22*	HNPP deletion is found in most HNPP patients
	Patients with del(17p) that includes *PMP22* have abnormal NCV consistent with HNPP
	PMP22 is contained within the HNPP deletion
	PMP22 frameshift mutations resulting in loss-of-function alleles are found in a small proportion of nondeletion HNPP patients
	Quantitative decrease of PMP22 protein is observed by immunogold labeling in the compact myelin of peripheral nerves from HNPP deletion patients
	PMP22 knockout mice display clinical, electrophysiologic, and neuropathologic features (including tomacula) of HNPP

available. If a clear X-linked dominant pattern of inheritance is demonstrated in the family, then this is the only molecular test that needs to be considered. *PMP22* and *MPZ* point mutation detection should be considered for patients who do not have the CMT1A duplications. Although point mutations in these genes are relatively infrequent, they are found more often in patients with more severe neuropathy.

Clinical Application of Molecular Diagnostics

DNA diagnostic testing should be considered for patients with CMT1 and related peripheral neuropathies.[78] The inherited disorders of the peripheral nerves can present with wide variation in clinical severity. They can also mimic conditions generally considered to be acquired, such as chronic immune demyelinating polyneuropathy, multifocal neuropathy, and even carpal tunnel syndrome. The detection of the CMT1A duplication, HNPP deletion, or point mutation in *Cx32*, *MPZ*, or *PMP22* in a patient's DNA sample establishes the exact molecular form of the disease in a family. The detection of a specific molecular lesion makes it possible to diagnose or exclude the diagnosis in other family members who are at risk for the disease; establish the inheritance pattern in the family, thus allowing genetic counseling and estimates of recurrence risk; offer prenatal diagnosis; and possibly provide prognostic information.

CMT1A duplication testing should be considered even in the absence of family history of the disease, given the high spontaneous mutation rate. In one study, nine out of 10 patients with sporadic disease had the CMT1A duplication.[81] A study of a larger series of patients identified a new duplication in 71 percent of sporadic cases,[54] and it has been estimated that the de novo duplication index case can be identified in more than 10 percent of autosomal dominant CMT1A families.[82] Reversions of the CMT1A duplication may also occur frequently and may be the mechanism for mosaicism in a patient described with a milder neuropathy phenotype.[83] CMT1 patients who do not have the CMT1A duplication should be screened for *Cx32* mutations, which are the next most common cause of CMT1 and account for five to 10 percent of patients.[54] *Cx32* mutations have also been identified in female patients with a CMT2 diagnosis,[84] likely reflecting the axonal component observed by electrophysiology in females with CMTX who may have nearly normal NCVs. Of course, if an autosomal dominant inheritance pattern is clearly established in the family, determined by observing male-to-male transmission, then no *Cx32* mutation analysis is warranted. CMT patients without CMT1A duplication or *Cx32* mutations can be screened for point mutations in *MPZ* or *PMP22*. However, these account for a small proportion of patients. The *PMP22* and *MPZ* point mutations usually occur in patients with more severe neuropathy, and such analysis should certainly be considered in patients with Dejerine-Sottas syndrome or congenital hypomyelinating neuropathy.

The HNPP deletion should be excluded for all patients with a recurrent demyelinating neuropathy or polyneuropathy of undetermined etiology.[1] It should also be considered for all patients with multifocal neuropathy; in a recent study, almost 50 percent of such patients were shown to have the HNPP deletion.[85] A significant percentage (37 percent) of the index patients with

Congenital Hypomyelination / *MPZ* Point Mutations
Dejerine-Sottas Syndrome / *PMP22, MPZ* Point Mutation
Severe CMT1A / homozygous duplication
CMT1A and CMT1B / *PMP22, MPZ* Point Mutations
X-linked CMT (♂ > ♀) / *Cx32* Point Mutations
CMT1A / 1.5 Mb Duplication
HNPP / 1.5 Mb Deletion

Figure 6 Clinical phenotype/genotype correlations. Generally, hereditary neuropathy with liability to pressure palsies (HNPP) is the mildest phenotype, associated with a 1.5 Mb deletion, and Dejerine-Sottas syndrome and congenital hypomyelination are more severe conditions, associated in some cases with PMP22 or MPZ point mutations. However, the different clinical entities can overlap in severity.

deletion had no affected relatives.[85] The HNPP deletion should also be considered for families with autosomal dominant carpal tunnel syndrome.

Generally speaking, neuropathies resulting from the CMT1A duplication or HNPP deletion are less severe than those attributable to point mutations in *Cx32, MPZ,* or *PMP22* [*see Figure 6*]. Longitudinal studies of CMT1A duplication patients in one family over 22 years showed only a slight reduction (2.2 m/sec) of median motor NCV, and patients reported subjective symptoms of gradually diminishing leg strength.[86] However, there can be variation in motor NCVs and clinical severity between families with the duplication,[53] within the same family,[51] and even in identical twins.[17] A de novo duplication has also been reported in a case of sporadic Dejerine-Sottas syndrome,[87] illustrating an extreme case of variability of expression. These observations suggest that modifier genes and/or environmental factors may modulate disease severity. The phenotypic consequences resulting from a gene-dosage effect may be even more susceptible to variation than those resulting from an altered protein.[17]

A 20-year study of the original CMT1B family, with an *MPZ* mutation altering the P_0 extracellular domain, indicated that affected patients generally show an early age of onset, often indicated by delayed ability to walk.[88] Proximal muscle weakness of the lower extremities is common and often marked, but patients remain ambulatory, and life span does not decrease.[88] As with the CMT1A duplication families, variability of disability between family members suggests that genetic and environmental factors, in addition to the *MPZ* mutation, play a role in the final phenotype.[88]

HNPP usually manifests itself as a milder clinical neuropathy phenotype, and many persons with the HNPP deletion remain undiagnosed. There can be tremendous variability in the clinical expression of the HNPP deletion, with 41 percent of individuals being unaware of their disease and 25 percent almost or totally free of symptoms.[89] However, later in life, and in its extreme form, HNPP can have features that overlap with those of CMT1. In fact, four percent of patients with a clinical diagnosis of CMT1 were found by molecular testing to have the HNPP deletion.[79]

Therapy

An accurate diagnosis of an inherited neuropathy is essential to planning a therapeutic regimen.[90] Genetic counseling should be based on the inheritance pattern or the specific molecular lesion detected. Patients should be counseled on avoiding neurotoxic drugs[91] and trauma to the nerves. Maintaining ideal body weight reduces the strain on weight-bearing muscles and joints. Physical and occupational therapy should be directed toward maintaining function and comfort, ensuring safety, and protecting joints.[90] Ankle-foot orthoses can be used to minimize an abnormal gait. Surgical procedures should be considered for severe foot or hand deformities. Knowledge of the molecular mechanisms may eventually lead to new therapies, perhaps by normalization of the dosage imbalance of *PMP22* in CMT1A duplication and HNPP deletion patients.

Conclusion

The investigation of inherited peripheral neuropathies has uncovered a novel mechanism involving DNA rearrangements and a gene-dosage effect. Studies have yielded some intriguing findings, including the observations that two very different mutational mechanisms (duplication and point mutation) can result in CMT1A, and that different mutations in the same gene can give rise to different neuropathic disease phenotypes.

The CMT1A duplication and HNPP deletion represent the first examples in humans of mendelian syndromes resulting from the reciprocal products of unequal exchange involving relatively large intrachromosomal segments. They also provide the first examples of two human diseases associated with the reciprocal products of the same mutational event. Knowledge of the molecular mechanism has led to useful clinical diagnostic tests. These objective means of establishing a specific diagnosis have enhanced appreciation for the overlapping clinical features of these disorders, which once were thought to be distinct clinical entities. They have shown that many patients exhibiting disorders that were thought to be acquired actually have conditions attributable to genetic alteration. The studies of HNPP families clearly document inherited predisposition to disease; yet these, as well as studies of CMT families and identical twins, demonstrate environmental influences on disease expression. The recent elucidation of the molecular mechanisms for CMT and related neuropathies has important implications for the diagnosis, prognosis, genetic counseling, and rational approaches to therapy for patients with peripheral neuropathies.

References

1. Lupski JR, Chance PF, Garcia CA: Inherited primary peripheral neuropathies: molecular genetics and clinical implications of CMT1A and HNPP. *JAMA* 270:2326, 1993

2. Lupski JR: Charcot-Marie-Tooth disease: a gene-dosage effect. *Hosp Pract* 32:83, 1997

3. Patel PI, Lupski JR: Charcot-Marie-Tooth disease: a new paradigm for the mechanism of inherited disease. *Trends Genet* 10:128, 1994

4. Roa BB, Lupski JR: Molecular genetics of Charcot-Marie-Tooth neuropathy. *Advances in Human Genetics*, Vol 22. Harris H, Hirschhorn K, Eds. Plenum Press, New York, 1994, p 117.

5. Murakami T, Garcia C, Reiter LT, et al: Reviews in molecular medicine: Charcot-Marie-Tooth disease and related neuropathies. *Medicine* 75:233, 1996

6. Warner LE, Reiter LT, Murakami T, et al: Molecular mechanisms for Charcot-Marie-Tooth disease and related demyelinating peripheral neuropathies. *Cold Spring Harb Symp Quant Biol* 61:659, 1996

7. Suter U, Welcher AA, Snipes GJ: Progress in the molecular understanding of hereditary peripheral neuropathies reveals new insights into the biology of the peripheral nervous system. *Trends Neurosci* 16:50, 1993

8. Suter U, Snipes GJ: Biology and genetics of hereditary motor and sensory neuropathies. *Annu Rev Neurosci* 18:45, 1995

9. Snipes GJ, Suter U: Molecular anatomy and genetics of myelin proteins in the peripheral nervous system. *J Anat* 186:483, 1995

10. Snipes GJ, Suter U: Molecular basis of common hereditary motor and sensory neuropathies in humans and in mouse models. *Brain Pathol* 5:233, 1995

11. Harding AE: From the syndrome of Charcot, Marie and Tooth to disorders of peripheral myelin proteins. *Brain* 118:809, 1995

12. Ouvrier R: Correlation between the histopathologic, genotypic, and phenotypic features of hereditary peripheral neuropathies in childhood. *J Child Neurol* 11:133, 1996

13. Garcia CA: The clinical features of Charcot-Marie-Tooth disorders. *Charcot-Marie-Tooth Disorders: A Handbook for Primary Care Physicans*. Parry GJ, Ed. The Charcot-Marie-Tooth Association, Upland, PA, 1995, p 5

14. Dyck PJ, Chance P, Lebo R, et al: Hereditary motor and sensory neuropathies. *Peripheral Neuropathy*, 3rd ed. Dyck PJ, Thomas PK, Griffin JW, et al, Eds. WB Saunders Co, Philadelphia, 1993, p 1094

15. Warner LE, Hilz M, Appel S, et al: Clinical phenotype of different *MPZ* mutations may include Charcot-Marie-Tooth type 1B, Dejerine-Sottas, and congenital hypomyelination. *Neuron* 17:451, 1996

16. Yoshikawa H, Dyck PJ: Uncompacted inner myelin lamellae in inherited tendency to pressure palsy. *J Neuropathol Exp Neurol* 50:649, 1991

17. Garcia CA, Malamut RE, England JD, et al: Clinical variability in two pairs of identical twins with the Charcot-Marie-Tooth disease type 1A duplication. *Neurology* 45:2090, 1995

18. Lemke G, Axel R: Isolation and sequence of a cDNA encoding the major structural protein of peripheral myelin. *Cell* 40:501, 1985

19. Lemke G: The molecular genetics of myelination: an update. *Glia* 7:263, 1993

20. Giese KP, Martini R, Lemke G, et al: Mouse P_0 gene disruption leads to hypomyelination, abnormal expression of recognition molecules, and degeneration of myelin and axons. *Cell* 71:565, 1992

21. Martini R, Mohajeri MH, Kasper S, et al: Mice doubly deficient in the genes for P0 and myelin basic protein show that both proteins contribute to the formation of the major dense line in peripheral nerve myelin. *J Neurosci* 15:4488, 1995

22. Martini R, Zielasek J, Toyka KV, et al: Protein zero (P_0)-deficient mice show myelin degeneration in peripheral nerves characteristic of inherited human neuropathies. *Nat Genet* 11:281, 1995

23. Shapiro L, Doyle JP, Hensley P, et al: Crystal structure of the extracellular domain from P_0, the major structural protein of peripheral nerve myelin. *Neuron* 17:435, 1996

24. Hayasaka K, Himoro M, Wang Y, et al: Structure and chromosomal localization of the gene encoding the human myelin protein zero (MPZ). *Genomics* 17:755, 1993

25. Hayasaka K, Himoro M, Sato W, et al: Charcot-Marie-Tooth neuropathy type 1B is associated with mutations of the myelin P_0 gene. *Nat Genet* 5:31, 1993

26. Kulkens T, Bolhuis PA, Wolterman RA, et al: Deletion of the serine 34 codon from the major peripheral myelin protein P_0 gene in Charcot-Marie-Tooth disease type 1B. *Nat Genet* 5:35, 1993

27. Hayasaka K, Himoro M, Sawaishi Y, et al: *De novo* mutation of the myelin P_0 gene in Dejerine-Sottas disease (hereditary motor and sensory neuropathy type III). *Nat Genet* 5:266, 1993

28. Kozuka N, Tachi N, Ohya K, et al: A de novo insertional mutation of the P_0 gene in a patient with congenital hypomyelination neuropathy. *Am Soc Hum Genet* 59:A398, 1996

29. Meijerink PHS, Hoogendijk JE, Gabreels-Festen AAWM, et al: Clinically distinct codon 69 mutations in major myelin protein zero in demyelinating neuropathies. *Ann Neurol* 40:672, 1996

30. Kirschner DA, Szumowski K, Gabreels-Festen A, et al: Inherited demyelinating peripheral neuropathies: relating myelin packing abnormalities to P_0 molecular defects. *J Neurosci Res* 46:502, 1996

31. Wong M-H, Filbin MT: Dominant-negative effect on adhesion by myelin P_0 protein truncated in its cytoplasmic domain. *J Cell Biol* 134:1531, 1996

32. Suter U, Snipes GJ: Peripheral myelin protein 22: facts and hypotheses. *J Neurosci Res* 40:145, 1995

33. Welcher AA, Suter U, De Leon M, et al: A myelin protein is encoded by the homologue of a growth arrest-specific gene. *Proc Natl Acad Sci USA* 88:7195, 1991

34. Spreyer P, Kuhn G, Hanemann CO, et al: Axon-regulated expression of a Schwann cell transcript that is homologous to a "growth arrest-specific" gene. *EMBO J* 10:3661, 1991

35. Haney C, Snipes GJ, Shooter EM, et al: Ultrastructural distribution of *PMP22* in Charcot-Marie-Tooth disease type 1A. *J Neuropathol Exp Neurol* 55:290, 1996

36. Suter U, Welcher AA, Özcelik T, et al: *Trembler* mouse carries a point mutation in a myelin gene. *Nature* 356:241, 1992

37. Suter U, Moskow JJ, Welcher AA, et al: A leucine-to-proline mutation in the putative first transmembrane domain of the 22-kDa peripheral myelin protein in the *trembler-J* mouse. *Proc Natl Acad Sci U S A* 89:4382, 1992

38. Patel PI, Roa BB, Welcher AA, et al: The gene for the peripheral myelin protein PMP-22 is a candidate for Charcot-Marie-Tooth disease type 1A. *Nat Genet* 1:159, 1992

39. Valentijn LJ, Baas F, Wolterman RA, et al: Identical point mutations of *PMP22* in Trembler-J mouse and Charcot-Marie-Tooth disease type 1A. *Nat Genet* 2:288, 1992

40. Roa BB, Garcia CA, Suter U, et al: Charcot-Marie-Tooth disease type 1A: association with a spontaneous point mutation in the *PMP22* gene. *N Engl J Med* 329:96, 1993

41. Roa BB, Dyck PJ, Marks HG, et al: Dejerine-Sottas syndrome associated with point mutation in the peripheral myelin protein 22 (*PMP22*) gene. *Nat Genet* 5:269, 1993

42. Ionasescu VV, Searby CC, Ionasescu R, et al: Dejerine-Sottas neuropathy in mother and son with same point mutation of *PMP22* gene. *Muscle Nerve* 20:97, 1997

43. Nicholson GA, Valentijn LJ, Cherryson AK, et al: A frameshift mutation in the PMP22 gene in hereditary neuropathy with liability to pressure palsies. *Nat Genet* 6:263, 1994

44. Adlkofer K, Martini R, Aguzzi A, et al: Hypermyelination and demyelinating peripheral neuropathy in *Pmp22*-deficient mice. *Nat Genet* 11:274, 1995

45. Adlkofer K, Frei R, Neuberg DH-H, et al: Heterozygous peripheral myelin protein 2 deficient mice are affected by a progressive demyelinating tomaculous neuropathy. *J Neurosci*, 17:4662, 1997

46. Bergoffen J, Scherer SS, Wang S, et al: Connexin mutations in X-linked Charcot-Marie-Tooth disease. *Science* 262:2039, 1993

47. Fairweather N, Bell C, Cochrane S, et al: Mutations in the connexin 32 gene in X-linked dominant Charcot-Marie-Tooth disease (CMTX1). *Hum Mol Genet* 3:29, 1994

48. Ionasescu V, Searby C, Ionasescu R: Point mutations of the connexin32 (GJB1) gene in X-linked dominant Charcot-Marie-Tooth neuropathy. *Hum Mol Genet* 3:355, 1994

49. Lupski JR, Montes de Oca-Luna R, Slaugenhaupt S, et al: DNA duplication associated with Charcot-Marie-Tooth disease type 1A. *Cell* 66:219, 1991

50. Raeymaekers P, Timmerman V, Nelis E, et al: Duplication in chromosome 17p11.2 in Charcot-Marie-Tooth neuropathy type 1a (CMT1a). *Neuromuscul Disord* 1:93, 1991

51. Kaku DA, Parry GJ, Malamut R, et al: Nerve conduction studies in Charcot-Marie-Tooth polyneuropathy associated with a segmental duplication of chromsome 17. *Neurology* 43:1806, 1993

52. Lupski JR, Garcia CA: Molecular genetics and neuropathology of Charcot-Marie-Tooth disease type 1A. *Brain Pathol* 2:337, 1992

53. Wise CA, Garcia CA, Davis SN, et al: Molecular analyses of unrelated Charcot-Marie-Tooth (CMT) disease patients suggest a high frequency of the CMT1A duplication. *Am J Hum Genet* 53:853, 1993

54. Nelis E, Van Broeckhoven C: Estimation of the mutation frequencies in Charcot-Marie-Tooth disease type 1 and hereditary neuropathy with liability to pressure palsies: a European collaborative study. *Eur J Hum Genet* 4:25, 1996

55. Lupski JR, Roth JR, Weinstock GM: Chromosomal duplications in bacteria, fruit flies, and humans. *Am J Hum Genet* 58:21, 1996

56. Pentao L, Wise CA, Chinault AC, et al: Charcot-Marie-Tooth type 1A duplication appears to arise from recombination at repeat sequences flanking the 1.5 Mb monomer unit. *Nat Genet* 2:292, 1992

57. Chance PF, Alderson MK, Leppig KA, et al: DNA deletion associated with hereditary neuropathy with liability to pressure palsies. *Cell* 72:143, 1993

58. Chance PF, Abbas N, Lensch MW, et al: Two autosomal dominant neuropathies result from reciprocal DNA duplication/deletion of a region on chromosome 17. *Hum Mol Genet* 3:223, 1994

59. Reiter LT, Murakami T, Koeuth T, et al: A recombination hotspot responsible for two inherited peripheral neuropathies is located near a *mariner* transposon-like element. *Nat Genet* 12:288, 1996

60. Kiyosawa H, Lensch MW, Chance PF: Analysis of the CMT1A-REP repeat: mapping crossover breakpoints in CMT1A and HNPP. *Hum Mol Genet* 4:2327, 1995

61. Lopes J, LeGuern E, Gouider R, et al: Recombination hot spot in a 3.2-kb region of the Charcot-Marie-Tooth type 1A repeat sequences: new tools for molecular diagnosis of

hereditary neuropathy with liability to pressure palsies and of Charcot-Marie-Tooth type 1A. *Am J Hum Genet* 58:1223, 1996

62. Yamamoto M, Yasuda T, Hayasaka K, et al: Locations of crossover breakpoints within the CMT1A-REP repeat in Japanese patients with CMT1A and HNPP. *Hum Genet* 99:151, 1997

63. Lupski JR, Wise CA, Kuwano A, et al: Gene dosage is a mechanism for Charcot-Marie-Tooth disease type 1A. *Nat Genet* 1:29, 1992

64. Chance PF, Bird TD, Matsunami N, et al: Trisomy 17p associated with Charcot-Marie-Tooth neuropathy type 1A phenotype: evidence for gene dosage as a mechanism in CMT1A. *Neurology* 42:2295, 1992

65. Upadhyaya M, Roberts SH, Farnham J, et al: Charcot-Marie-Tooth disease 1A (CMT1A) associated with a maternal duplication of chromosome 17p11.2 →12. *Hum Genet* 91:392, 1993

66. Roa BB, Greenberg F, Gunaratne P, et al: Duplication of the *PMP22* gene in 17p partial trisomy patients with Charcot-Marie-Tooth type 1A neuropathy. *Hum Genet* 97:642, 1996

67. Zori RT, Lupski JR, Heju Z, et al: Clinical, cytogenetic, and molecular evidence for an infant with Smith-Magenis syndrome born from a mother having mosaic 17p11.2p12 deletion. *Am J Med Genet* 47:504, 1993

68. Trask BJ, Mefford H, van den Engh G, et al: Quantification by flow cytometry of chromosome-17 deletions in Smith-Magenis syndrome patients. *Hum Genet* 98:710, 1996

69. LeGuern E, Gouider R, Mabin D, et al: Patients homozygous for the 17p11.2 duplication in Charcot-Marie-Tooth type 1A disease. *Ann Neurol* 41:104, 1997

70. Yoshikawa H, Nishimura T, Nakatsuji Y, et al: Elevated expression of messenger RNA for peripheral myelin protein 22 in biopsied peripheral nerves of patients with Charcot-Marie-Tooth disease type 1A. *Ann Neurol* 35:445, 1994

71. Warner LE, Roa BB, Lupski JR: Absence of PMP22 coding region mutations in CMT1A duplication patients: further evidence supporting gene dosage as a mechanism for Charcot-Marie-Tooth disease type 1A. *Hum Mutat* 8:362, 1996

72. Sereda M, Griffiths I, Pühlhofer A, et al: A transgenic rat model of Charcot-Marie-Tooth disease. *Neuron* 16:1049, 1996

73. Huxley C, Passage E, Manson A, et al: Construction of a mouse model of Charcot-Marie-Tooth disease type 1A by pronuclear injection of human YAC DNA. *Hum Mol Genet* 5:563, 1996

74. Magyar JP, Martini R, Ruelicke T, et al: Impaired differentiation of Schwann cells in transgenic mice with increased *PMP22* gene dosage. *J Neurosci* 16:5351, 1996

75. Vallat J-M, Sindou P, Preux P-M, et al: Ultrastructural *PMP22* expression in inherited demyelinating neuropathies. *Ann Neurol* 39:813, 1996

76. Schenone A, Nobbio L, Mandich P, et al: Underexpression of messenger RNA for peripheral myelin protein 22 in hereditary neuropathy with liability to pressure palsies. *Neurology* 48:445, 1997

77. Young P, Wiebusch H, Stogbauer F, et al: A novel frameshift mutation in *PMP22* accounts for hereditary neuropathy with liability to pressure palsies. *Neurology* 48:450, 1997

78. Lupski JR: DNA diagnostics for Charcot-Marie-Tooth disease and related inherited neuropathies. *Clin Chem* 42:995, 1996

79. Roa BB, Ananth U, Garcia CA, et al: Molecular diagnosis of CMT1A and HNPP. *Lab Med Int* 12:22, 1995

80. Shaffer LG, Kennedy GM, Spikes AS, et al: Diagnosis of CMT1A duplications and HNPP deletions by interphase FISH: implications for testing in the cytogenetics laboratory. *Am J Med Genet* 69:325, 1997

81. Hoogendijk JE, Hensels GW, Gabreëls-Festen AAWM, et al: De-novo mutation in hereditary motor and sensory neuropathy type 1. *Lancet* 339:1081, 1992

82. Blair IP, Nash J, Gordon MJ, et al: Prevalence and origin of de novo duplications in Charcot-Marie-Tooth disease type 1A: first report of a de novo duplication with a maternal origin. *Am J Hum Genet* 58:472, 1996

83. Liehr T, Rautenstrauss B, Grehl H, et al: Mosaicism for the Charcot-Marie-Tooth disease type 1A duplication suggests somatic reversion. *Hum Genet* 98:22, 1996

84. Timmerman V, De Jonghe P, Spoelders P, et al: Linkage and mutation analysis of Charcot-Marie-Tooth neuropathy type 2 families with chromosomes 1p35-p36 and Xq13. *Neurology* 46:1311, 1996

85. Tyson J, Malcolm S, Thomas PK, et al: Deletions of chromosome 17p11.2 in multifocal neuropathies. *Ann Neurol* 39:180, 1996

86. Killian JM, Tiwari PS, Jacobson S, et al: Longitudinal studies of the duplication form of Charcot-Marie-Tooth polyneuropathy. *Muscle Nerve* 19:74, 1996

87. Silander K, Meretoja P, Nelis E, et al: A *de novo* duplication in 17p11.2 and a novel mutation in the P_0 gene in two Dejerine-Sottas syndrome patients. *Hum Mut* 8:304, 1996

88. Bird TD, Kraft GH, Lipe H, et al: Clinical and pathological phenotype of the original family with Charcot-Marie-Tooth type 1B: a 20 year study. *Ann Neurol* 41:463, 1997

89. Pareyson D, Scaioli V, Taroni F, et al: Phenotypic heterogeneity in hereditary neuropathy with liability to pressure palsies associated with chromosome 17p11.2-12 deletion. *Neurology* 46:1133, 1996

90. Garcia CA: Familial neuropathies. *Current Therapy in Neurologic Disease*, 5th ed. Johnson RT, Griffin JW, Eds. Mosby-Year Book, St. Louis, 1997, p 386

91. Graf WD, Chance PF, Lensch MW, et al: Severe vincristine neuropathy in Charcot-Marie-Tooth disease type 1A. *Cancer* 77:1356, 1996

Acknowledgment

This research has been supported in part by the National Institute of Neurological Disorders and Stroke (National Institutes of Health) and the Muscular Dystrophy Association. The author wishes to thank the patients, the families, and their physicians and health care providers without whom these studies would not have been possible.

Figures 1, 3a, 3c, and 5   Tom Moore.

Figures 3b and d   Photographs courtesy Dr. Lisa G. Shaffer.

Figures 2, 4, and 6   Marcia Kammerer.

CHAPTER 14

Ion Channel Defects in the Hereditary Myotonias and Periodic Paralyses

Stephen C. Cannon, M.D., Ph.D.

The myotonias and periodic paralyses are rare inherited disorders of skeletal muscle in which the primary defect is an alteration in the electrical excitability of the muscle fiber.[1-5] Electrical excitability is critical for normal muscle function. To trigger synchronous contraction over an entire muscle, action potentials initiated at the neuromuscular junction (NMJ) must propagate both longitudinally along the fiber and radially into the core of the fiber. The inward spread of depolarization is propagated along an intricate system of invaginations of the surface membrane, the transverse tubule, where dihydropyridine-sensitive calcium channels trigger the release of Ca^{2+} from the sarcoplasmic reticulum (SR). This coupling between the arrival of input at the NMJ and release of Ca^{2+} from the SR is essential for the rapidity and fidelity of motor responses. Derangements of this electrical coupling occur in both myotonia and periodic paralysis, which are now known to be caused by mutations of voltage-gated ion channels.

Myotonia

Myotonia is not a single disease entity but rather a clinical term used to describe an impairment in the relaxation of muscle force. The inability of myotonic muscle to relax normally produces a sensation of muscle stiffness. The severity of myotonic stiffness fluctuates substantially in an affected individual and is dependent on the level of past muscular activity. Myotonia is greatest during the first three or four contractions following a period of inactivity for 15 minutes or longer. Myotonia diminishes with repeated muscle activity, a phenomenon called warm-up. Excitability is enhanced in myotonic muscle. A single brief stimulus elicits a train of action potentials, the myotonic run, lasting seconds to a minute or longer. The aberrant excitability originates in the muscle itself because myotonia is not prevented by nondepolarizing block of neuromuscular transmission by curare.[6] The myotonic run of afterdischarges triggers Ca^{2+} release from the SR, which produces the delayed relaxation in force that patients experience as use-dependent muscle stiffness.

Periodic Paralysis

Periodic paralysis is characterized by transient attacks of muscle weakness lasting from minutes to hours or even days. Attacks are episodic rather than periodic and may render a patient bedridden, unable to sit, or even unable to raise a limb against gravity. Respiration, deglutition, and eye movements are generally spared even with severe attacks, and no impairment of consciousness or sensation occurs. In contrast to myotonia, periodic paralysis is caused by a loss of membrane excitability. During attacks of weakness, the electromyogram is silent, muscle fibers are depolarized at -50 to -40 mV from a normal resting potential of -95 mV, and action potentials cannot be generated, even by direct galvanic shock. Strength is normal between attacks, and dystrophy is not a prominent feature of periodic paralysis, although mild permanent proximal weakness with vacuolar myopathy may develop in the fifth or sixth decades of life.

Patients with some forms of periodic paralysis may also have myotonia, which may manifest itself either symptomatically or electromyographically.

Table 1 Classification of the Hereditary Myotonias and Periodic Paralyses

Disorder	Myotonia congenita (dominant)	Myotonia congenita (recessive)	Potassium-aggravated myotonia*	Paramyotonia congenita	Hyperkalemic periodic paralysis	Hypokalemic periodic paralysis
Eponym	Thomsen's disease	Becker's generalized myotonia		Eulenburg's disease	Gamstorp's disease	None
Myotonia	Moderate, constant	Severe, constant	Variable, fluctuating, K-aggravated	Moderate, paradoxical, cold-aggravated	Mild, fluctuating, K-aggravated	None
Weakness	None	Mild, transient	None	Moderate, cold-aggravated	Moderate-severe, ↑[K+] worsens	Severe, ↓[K+] worsens
Inheritance	AD	AR	AD	AD	AD	AD
Penetrance	High	High	High	High	High	↓Females
Functional defect	↓Cl⁻ conductance	↓Cl⁻ conductance	Mildly impaired fast inactivation	Impaired fast inactivation	Impaired fast and slow inactivation	Unknown
Locus	7q35	7q35	17q23-q25	17q23-q25	17q23-q25	1q31-q32
Gene	CLCN1	CLCN1	SCNA4	SCNA4	SCNA4	CACNL1A3
Gene product	Cl channel, ClC-1	Cl channel, ClC-1	Na channel, α-subunit	Na channel, α-subunit	Na channel, α-subunit	Ca channel, α₁-subunit

*Includes myotonia fluctuans, myotonia permanens, and acetazolamide-responsive myotonia.

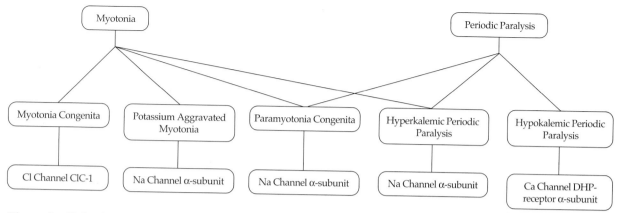

Figure 1 *Skeletal muscle disorders along the spectrum from myotonia to periodic paralysis. Muscle stiffness predominates in myotonia congenita and potassium-aggravated myotonia. Both stiffness and weakness are present in paramyotonia congenita and hyperkalemic periodic paralysis. Severe weakness without stiffness occurs in hypokalemic periodic paralysis. Chloride channel defects lead to myotonia, calcium channel mutations cause periodic paralysis, and sodium channel mutations may cause both stiffness and weakness.*

This overlap has led to the view that the myotonias and periodic paralyses lie along a spectrum of altered muscle excitability ranging from enhanced to impaired, respectively. Several distinct clinical disorders with distinguishing features have been delineated along this spectrum [*see Table 1*]. At either extreme [*see Figure 1*], patients experience only myotonia (myotonia congenita) or only weakness (hypokalemic periodic paralysis), whereas the syndromes near the middle (paramyotonia congenita and hyperkalemic periodic paralysis) have elements of both symptoms. One common cause of inherited myotonia, myotonic dystrophy, has been omitted from this classification. It is a multisystem disease affecting brain, heart, eye, and endocrine organs as well as skeletal muscle that is associated with an expanded trinucleotide repeat on chromosome 19 [*see Chapter 4*] and not a mutation of an ion channel gene.

Candidate Disease Genes Implicated by Electrophysiological Studies

The first insights into the etiology of the myotonias and periodic paralyses were gained from in vitro electrophysiological studies of diseased muscle fibers. In a series of classic studies, Bryant and colleagues discovered that the membrane resistance was increased threefold in muscle from myotonic goats.[7] Additional experiments demonstrated that this high resistance was attributable to a severe reduction in the resting membrane conductance to chloride. (Conductance is the reciprocal of resistance and, in the context of excitable cells, equals the ionic current flow across the membrane divided by the driving voltage). These investigators provided further support for the role of chloride by showing that normal muscle fibers became myotonic when the chloride conductance was reduced by blocking agents or a chloride-free bath.[8] In comparison to the human disorders, the myotonia in goats is most similar to autosomal dominant myotonia congenita (MC). In vitro recordings of fibers obtained from MC patients also revealed a severe reduction in the resting chloride conductance.[9] These studies firmly established that a loss of the resting chloride conductance is sufficient to cause myotonia and that a heritable form of myotonia in humans may arise from a chloride channel defect.

The prevailing view in the 1970s and early 1980s was that all myotonias were caused by a defect in the resting chloride conductance. This view was revised by a series of landmark studies by Lehmann-Horn and colleagues.[10-12] These investigators measured the electrical behavior of muscle fibers obtained from patients with myotonia that occurred in conjunction with periodic paralysis. Quite unexpectedly, the resting chloride conductance was normal. Other responses, however, were not normal. An aberrant inward current and depolarized resting potentials were found in diseased, but not control, muscle fibers. Both abnormalities were prevented by application of tetrodotoxin, a potent and specific blocker of voltage-gated Na channels. These data implied that a Na channel defect might be the cause of hyperkalemic periodic paralysis (HyperPP) and paramyotonia congenita (PMC).

The next major advance in elucidating the etiology of various myotonias and periodic paralyses was the cloning of full-length complementary DNAs (cDNAs) for several voltage-dependent ion channels of skeletal muscle. With the knowledge of these sequences, hybridization probes were designed to test for genetic linkage between candidate disease genes implicated by electrophysiological studies and specific syndromes in the myotonia–periodic paralysis spectrum. Initial attempts were met with rapid success. Linkage analysis supported the hypothesis that the major chloride channel of skeletal muscle, ClC-1 encoded by the *CLCN1* gene on chromosome 7q35, was the site of the defect in MC.[13] HyperPP and PMC were linked to the adult skeletal muscle isoform of the Na channel α-subunit gene, *SCNA4*, at 17q23-25.[14,15] The molecular defect in hypokalemic periodic paralysis (HypoPP) eluded detection by electrophysiological testing. In the absence of a candidate disease gene, a genome-wide screen using highly polymorphic dinucleotide repeats was required to map the HypoPP locus to chromosome 1q31-32.[16] The α₁-subunit gene of the dihydropyridine-sensitive L-type Ca channel (*CACNL1A3*) had previously been mapped to this region and was found to cosegregate with HypoPP. Subsequent genetic analysis has revealed mutations of the L-type Ca channel (*CACNL1A3*) in HypoPP, of the Na channel (*SCN4A*) in families with HyperPP and PMC, and of the chloride channel (*CLCN1*) in MC. Thus, the underlying molecular defects have now been identified for all the nondystrophic myotonias and periodic paralyses [*see Figure 1*].

Chloride Channel Defects Cause Myotonia

Mutations in *CLCN1*, the gene coding for the major skeletal muscle Cl channel, cause myotonia that may be inherited as either an autosomal dominant or an autosomal recessive trait. The dominant form was first reported in 1876 by Thomsen,[17] who had the disorder himself and wished to establish "tonic cramps in voluntary muscles in consequence of inherited psychic disposition" as a disease so that his affected son might avoid military service. The major features of the disease described in this initial report were generalized myotonic stiffness at birth or within the first two decades, autosomal dominant inheritance, and a nonprogressive course with no dystrophy. This disorder became known as myotonia congenita or Thomsen's disease. Half a century elapsed before Becker recognized an autosomal recessive form of myotonia congenita.[18] In comparison to the Thomsen type, the Becker form has a later onset and more severe myotonia, which may be mildly progressive over the first decade and accompanied by brief weakness during the first seconds after a forceful contraction is initiated. Heterozygous carriers are asymptomatic but may have latent myotonia revealed by electromyography. The recessive variant of myotonia congenita has been termed Becker myotonia or recessive generalized myotonia. (In this chapter, MC without further clarification refers to the dominant and recessive forms collectively; the inheritance pattern is specified for the different subtypes as required for clarity.) An important diagnostic feature of MC is that muscle cooling or ingestion of potassium salts does not aggravate the myotonia, in contrast to myotonic syndromes associated with Na channel defects.

Chloride Channel Mutations

The identification of chloride channel mutations associated with MC became possible after the initial member of this gene superfamily was cloned. Twenty years after the electrophysiological studies by Bryant and colleagues identified a defect in the chloride conductance of MC fibers,[7] a chloride channel cDNA was isolated from the electric fish *Torpedo marmorata* by expression cloning.[19] This breakthrough paved the way toward identifying a human skeletal muscle–specific chloride channel, hClC-1. The full-length cDNA is about three kilobases (kb) long and encodes a 988 amino acid protein [*see Figure 2*]. The corresponding gene on chromosome 7q35, *CLCN1*, contains 22 introns and spans at least 40 kb.[20] Both the dominant and the recessive forms of MC are tightly linked to this locus.[13]

A combination of screening strategies has been used to identify over 30 mutations in *CLCN1*. Six missense mutations and a nonsense codon leading to a premature stop near the carboxy tail have been found in association with dominant MC. The mutation in Thomsen's own family has been identified as a Pro480Leu substitution.[21] Myotonia levior, a disease of previously uncertain relation to MC, with mild dominantly inherited myotonia affecting bulbar muscles and sparing the limbs, was found to also be caused by a missense mutation, Gln552Arg.[22] Reported mutations in recessive MC are more varied and include 14 missense substitutions, four nonsense codons, five deletions, two splice site errors, and one insertion. Mutations have been identified in about 50 percent of all chromosomes screened from patients with recessive MC. Three mutations—Phe413Cys, Arg894x, and a 14 base pair (bp) deletion Δ1437-1450—occur commonly and account for nearly one third of all known defects.[23] Most cases of recessive MC are compound heterozygotes in which the two *CLCN1* alleles harbor different mutations. A recent review of all genotyped families with recessive MC reported that only 11 independent kindreds were available in which both mutant alleles had been identified.[23] Three were homozygous mutations in families with consanguinous marriages, and eight were compound heterozygotes.

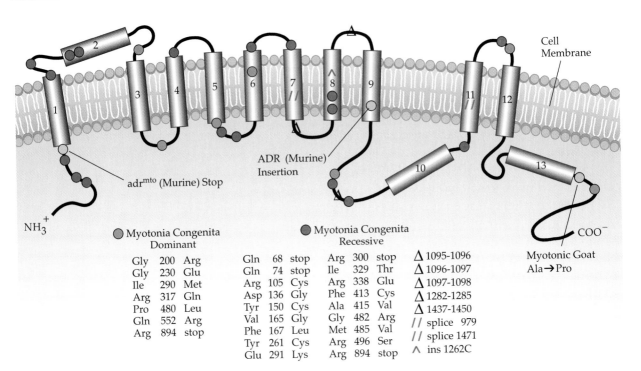

Figure 2 *Proposed structure of the skeletal muscle chloride channel and location of mutations associated with myotonia congenita. The membrane-folding topology of the predicted α-helical segments (cylinders) has not been confirmed experimentally. Missense and nonsense mutations are listed as the normal amino acid, followed by its location in the primary sequence, and then the substituted residue.*

Neither the location nor the nature of the mutations in *CLCN1* allows an a priori prediction of which ones are expected to be associated with dominant versus recessive phenotypes. Indeed, Ile290Met causes dominant MC, whereas a substitution at an adjacent amino acid, Glu291Lys, is associated with recessive MC. One common nonsense mutation, Arg894X, may cause either dominant or recessive MC. The most likely explanation is that the functional deficit for this nearly full-length channel straddles the boundary between more severe dominant-negative mutants and the milder defects produced by recessive mutations.[23] The literature contains conflicting reports on whether other mutations, such as Gly230Glu,[24,25] give rise to dominant or recessive MC. There is agreement that these changes are mutations and not benign polymorphisms. The assignment of dominant or recessive is defined solely by the inheritance pattern, not the mutation. Hence, a likely contributor to the discrepancy is the difficulty in ascertaining affected individuals. With dominant MC, myotonia may be very mild, and penetrance is not 100 percent, both of which features might give the impression of a recessive inheritance pattern.

A myotonia congenita–like syndrome in goats and mice is also caused by mutations in ClC-1. The congenital myotonia in goats is severe and is expressed in an autosomal dominant pattern. The initial recognition of a reduced chloride conductance in myotonic muscle was made in the myotonic goat,[7] and affected descendants from this same colony contain an alanine to proline missense mutation at the equivalent of residue 885 in human ClC-1.[26] A murine equivalent of recessive myotonia congenita has arisen from two separate mutational events. The arrested development of righting (ADR) mouse has a transposon insertion near the coding region for the ninth α-helical segment [*see Figure 2*], which destroys the coding potential for the ClC-1 gene.[27] Another murine allele associated with recessive myotonia, adr^mto, has a nonsense mutation at Arg47 causing a severely truncated ClC-1. The ADR mouse has become an important model system for studying the pathophysiology and treatment of myotonia congenita.

Functional Consequences of Chloride Channel Defects and the Pathogenesis of Myotonia

The chloride conductance must be severely compromised to cause myotonia. Recordings from MC fibers showed a threefold decrease in the total membrane conductance.[9] This implies a fivefold reduction in the chloride conductance because about 70 percent of the resting membrane conductance is attributable to chloride ion flow. In agreement with this estimate, pharmacological models have shown that 80 percent or more of the chloride current must be blocked to produce myotonia.[28]

Chloride ions are passively distributed across the muscle cell and therefore play no active role in adjusting the set point of the resting membrane potential. More important, the unusually high Cl conductance of skeletal muscle helps maintain the membrane potential near its previously established resting value. In response to any deviation in membrane voltage, very large Cl currents flow to quickly reestablish the resting potential. This electrical buffering by the chloride conductance is particularly important for skeletal muscle because of the transverse tubules. The T tubules provide an effective means of transmitting action potentials from the surface into the core of the fiber. With each action potential, however, the T tubular potassium rises transiently because of an egress of intracellular K^+ into this long, narrow, diffusion-limited space. Elevated K^+ in the T tubule (an extracellular compartment) depolarizes the fiber. This afterdepolarization is blunted by a repolarizing current flowing through the Cl conductance. If the conductance is substantially decreased by a Cl channel mutation, then the afterdepolarization may trigger another action potential and thereby produce a self-sustained myotonic run. In support of this pathomechansim, disruption of the T tubules by osmotic shock abolishes the self-sustained train of discharges in myotonic muscle.[8]

How might the mutations in ClC-1 cause such a profound disruption in channel function? The answers are just now coming into focus. Unlike voltage-activated Na, K, or Ca channels, the regions of the ClC-1 protein that give rise to the ion-conducting pore or voltage-sensing gates have not been identified. Nevertheless, it has been established that dominant MC does not occur simply from a 50 percent reduction in the number of functional Cl channels (haploinsufficiency). First, a 50 percent reduction in Cl current density is not sufficient to cause myotonia. Second, drastic truncations of ClC-1 near the amino terminus that certainly lead to a nonfunctioning channel (e.g., Gln68X or Gln74X) are associated with recessive, not dominant, MC. Rather than gene dosage, the mechanism of dominant inheritance is a dominant-negative suppression of channel function. Specifically, coexpression of wild-type (WT) ClC-1 and dominant MC mutant messenger RNA (mRNA) in frog oocytes results in substantially less chloride current than if the WT mRNA alone were expressed.[21] The magnitude of the reduction in chloride current density is comparable to the 80 percent loss observed in MC fibers. These results also imply that chloride channels in skeletal muscle are multimeric complexes consisting of several (probably four) ClC-1 subunits. Because the dominant-negative effect in the oocyte expression system is so robust, Jentsch's group has suggested this test could be used to determine whether a novel mutation identified in a pedigree with an ambiguous inheritance pattern should be classified as dominant or recessive.[21,23]

In contrast to the effects of dominant MC mutations, coexpression of WT and recessive MC mutant mRNAs results in normal chloride current densities, as if the mutant mRNA were an innocuous bystander. Thus, the gene product in recessive MC either is unable to coassemble with WT ClC-1 or does not significantly alter function in heteromeric WT/mutant channel complexes.

Sodium Channel Defects Cause Myotonia or Periodic Paralysis or Both

Mutations in *SCN4A*, the gene for the adult skeletal muscle isoform of the Na channel α-subunit, cause HyperPP, PMC, and potassium-aggravated myotonia (PAM). Each of these disorders has an autosomal dominant mode of inheritance with high penetrance.

HyperPP presents as episodic attacks of weakness lasting hours to a day or more.[5] Serum K^+ is often high during an attack (more than 4.5 mEq/L), but it may be normal. A more consistent finding is provocation of weakness by K salt administration. Foods with high K content, rest after exercise, stress, or ethanol ingestion may induce an attack. Mild muscle stiffness or latent myotonia revealed by electromyography occurs in some families with HyperPP, but weakness is always the predominant symptom.

Patients with PMC complain of muscle stiffness, especially in the hands and face. The myotonia is aggravated by cold weather, and a cold-water immersion test is often used to elicit limb myotonia in the clinic. A more specific sign of PMC is myotonia that paradoxically worsens with repeated muscle activity (paramyotonia), in contrast to the warm-up phenomenon in MC. Episodic weakness occurs in PMC, either sporadically or from cold provocation.

PAM is the most recently recognized disorder of skeletal muscle Na channels.[29] The clinical presentation is myotonia, without paradoxical behavior, that may be either mild and fluctuating or severe and constant. As implied by the name PAM, myotonia is aggravated by K administration. By definition, attacks of episodic weakness do not occur. Aside from the K sensitivity, milder cases of PAM are clinically indistinguishable from dominant MC. In fact, before the advent of molecular genetic testing, many families with PAM had been misdiagnosed as having dominant MC, a chloride channel disorder. Distinct diseases within the general category of PAM have been delineated to reflect the range from severe (myotonia permanens)[29] to mild (myotonia fluctuans)[30] impairment by myotonia.

Sodium Channel Mutations

At least 19 mutations in *SCN4A* have been found in association with Hyper-

PP, PMC, and PAM. After linkage to 17q23-q25 was established by use of a candidate gene approach,[14] mutations in the α-subunit were identified by direct sequencing of cDNA[31] or by identifying abnormal conformers of single-strand DNA amplified from genomic material.[32] *SCN4A* consists of 24 exons spanning 35 kb[33] and codes for a 1836 amino acid protein.[34] Sodium channels in adult innervated skeletal muscle are heterodimers of the large pore-forming α-subunit (SkM1) and a smaller accessory β₁-subunit.[35] The β₁-subunit gene maps to chromosome 19q13.1 and has not been linked to any human disease. SkM1 is expressed at significant levels only in skeletal muscle,[34] which explains the absence of cardiac, peripheral nerve, or central nervous system symptoms in these disorders.

Each of the 19 mutations of *SCN4A* results in a missense substitution at a highly conserved amino acid [*see Figure 3*]. In general, each specific mutation is consistently associated with a particular disorder (HyperPP, PMC, or PAM).[36] Some mutations are more prevalent than others. In HyperPP families for which a mutation has been identified, the Thr704Met mutation occurs with a frequency of 60 percent and Met1592Val about 32 percent. The remaining HyperPP mutations are rare, having occurred in one or two families. One amino acid substitution, Val781Ile, was previously suspected to be a novel mutation causing HyperPP with cardiac dysrhythmia[37] but was subsequently proved to be a benign polymorphism by functional studies and further screening of normal chromosomes.[38] Haplotype analysis suggests that common mutations probably arose independently in unrelated families.[39] Two mutations predominate in PMC: Thr1313Met, which has been detected in 45 percent of affected families, and Arg1448Cys, in 35 percent.

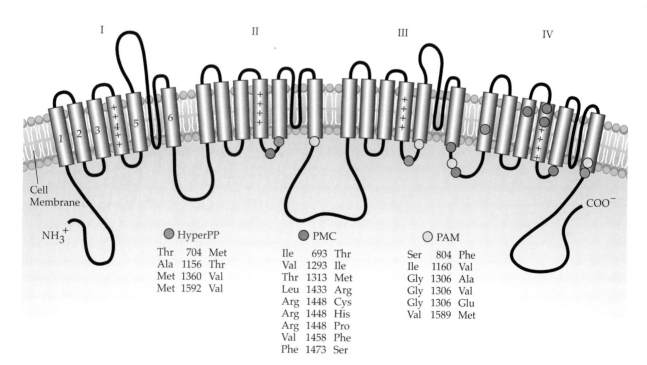

Figure 3 *Mutations in sodium channel α-subunit. Model of the Na channel α-subunit and locations of mutations associated with hyperkalemic periodic paralysis (HyperPP), paramyotonia congenita (PMC), and potassium-aggravated myotonia (PAM).*

Functional Defects in Mutant Sodium Channels

The functional consequences of missense mutations in the α-subunit have been studied by recording Na currents in muscle samples obtained from patients[29,40] or in mammalian cells transfected with mutant cDNA.[41-46] The primary defect, shared by virtually every mutation tested to date, is an impairment of channel inactivation. Normally, Na channels open quickly (activate) in response to membrane depolarization and then shut within a millisecond to a special closed conformation, the fast-inactivated state.[47] Once inactivated, Na channels cannot reopen until the membrane is hyperpolarized, which allows channels to recover from inactivation to a resting closed conformation. Fast inactivation limits the duration of the action potential and initiates repolarization of the muscle fiber. The slow recovery from inactivation (several milliseconds at a resting potential of –90 mV) causes the refractory period and limits the maximal firing rate of the cell.

A defect of Na channel inactivation in HyperPP muscle was first identified from patch-clamp recordings of single-channel behavior.[40] Fast inactivation was impaired in HyperPP myotubes as evidenced by aberrant bursts of reopenings and prolonged durations of individual openings [*see Figure 4*]. The defect in fast inactivation was shown by the HyperPP Myotube to be intermittent [*see Figure 4*]. For some trials, such as the first or the last, inactiva-

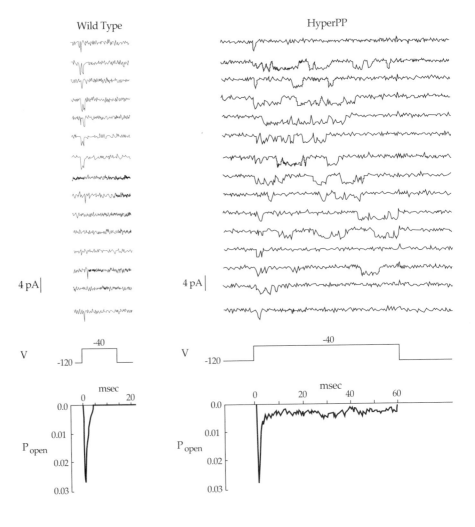

Figure 4 Defective fast inactivation of Na channels in HyperPP (Met1592Val). Tracings show currents recorded from single Na channels in cultured human myotubes. In response to depolarization, both wild-type and mutant Na channels quickly open (pulse-like downward deflection) within a few milliseconds, indicating that activation was not impaired. In selected consecutive trials, failure of fast inactivation produced bursts of reopenings and prolonged open durations in channels with the Met1592Val mutation. Ensemble averages (below tracings) show that mutant channels inactivate for the majority of trials (rapid decline in P_{open}) but that intermittent bursts cause a small persistent current. pA, picoamps; V, volts; P_{open} probability.

tion was normal. During other trials, bursts of openings indicated that inactivation was severely disrupted for the same channel containing the Met152Val mutation. The averaged response shown below demonstrates that for most trials the mutant channel inactivated normally. Intermittent failure of fast inactivation caused only a small persistent Na current (open probability after 50 milliseconds of 0.02 to 0.05 compared with approximately 0.001 in normal channels).

For most of the *SCN4A* mutations, the alteration in channel function has been studied in heterologous expression systems. cDNA coding for a specific α-subunit mutation is constructed and transfected into fibroblast-like (HEK) cells, which then synthesize mutant Na channels. Patch recordings from cells expressing the engineered Met1592Val mutation revealed the same inactivation defect that was observed in HyperPP muscle from a family in which Met1592Val was detected.[41] These heterologous expression studies demonstrate that the missense changes identified by molecular genetic screening are not benign polymorphisms. A single missense mutation out of the 1,836 residues in the α-subunit is sufficient to recreate the functional defect observed for mutant Na channels expressed endogenously in HyperPP muscle.

Additional defects of fast inactivation have been identified as more missense mutations have been screened in the HEK cell expression system.[48] Some mutant channels will fully inactivate, but the time course is slowed by two to 10 times. Another defect is a depolarized shift in the voltage dependence of fast inactivation. In other words, a stronger depolarization is required for mutant channels to achieve the same degree of fast inactivation that a mild depolarization causes in WT channels. For other mutants, destabilization of the inactivated state causes a faster recovery from inactivation. When this occurs, the cell is able to fire at higher discharge rates. For two mutations, Thr704Met and Gly1306Glu, the voltage dependence of activation is shifted to more negative voltages by about 10 mV. This shift allows mutant Na channels to open in response to milder depolarizing stimuli. Approximately 15 of the 20 *SCN4A* mutations have been studied, and none has caused an alteration in the ability of Na^+ to pass through the open channel.

The observation that the primary defect for all the *SCN4A* mutations is a disruption of fast inactivation is consistent with contemporary structure-function models of the Na channel and has been used to identify new functional domains of the protein. A large body of evidence drawn from a comparison of Na channel isoforms, site-directed mutagenesis, antipeptide antibodies, and protein-modifying reagents supports the following conceptual model.[49] The α-subunit consists of four homologous repeats, each containing six predicted membrane-spanning segments [*see Figure 3*]. The fourth segment, S4, in each repeat contains basic residues (positively charged lysine or arginine) at every third position and is thought to act as the voltage-sensing mechanism. The four short segments linking S5 and S6 come together to form the ion-conducting pore of the channel. The 50 amino acid cytoplasmic loop that links the III and IV homologous domains is critical for fast inactivation and is thought to act as a gate or hinged lid that occludes the inner mouth of the pore. In the context of this structural model, 13 of the 19 mutations in PMC-PAM-

HyperPP are at locations previously identified as being critical for fast inactivation. Four mutations (three PAM at Gly1306 and one PMC at Thr1313) are in the III-IV loop. Nine other missense mutations occur at the cytoplasmic ends of S5 or S6 segments at the inner mouth of the pore, where the III-IV loop probably binds to inactivate the channel. Functional studies of disease mutants have identified an additional domain of the α-subunit that is critical for fast inactivation. A cluster of PMC mutations occurs at or near an arginine in the S4 segment of domain IV (Arg1448). These mutations dramatically slow the time course of fast inactivation, a finding that led to the realization that the IVS4 segment plays a special role in coupling voltage-dependent activation to inactivation.[46]

Pathogenesis of Myotonia and Periodic Paralysis

If the Na channel mutations are to be accepted as the cause of HyperPP-PMC-PAM, then it is essential to show that the observed defects in channel behavior are sufficient to cause myotonia or paralysis. Both animal[50] and computer[51] models have demonstrated that the fast-inactivation defects of mutant Na channels, although quite mild, can produce muscle stiffness and weakness. A sea anemone toxin, ATXII, was applied to rat fast-twitch muscle in vitro to mimic the inactivation defect of mutant Na channels.[50] Micromolar ATXII partially disrupted inactivation such that about two percent of Na channels remained open during prolonged membrane depolarizations. This mild defect was sufficient to cause a train of myotonic discharges in toxin-exposed muscle [*see Figure 5*]. The mechanical behavior of toxin-exposed muscle was also myotonic, with a 10-fold slowing of twitch force relaxation. Each successive action potential during a myotonic run ends at a progressively more depolarized potential. This phenomenon triggers the discharges after the stimulus pulse and is thought to arise from activity-dependent trapping of K+ in the T tubule. This hypothesis was tested by disrupting the T tubules mechanically by shrinking and reswelling fibers in a hyperosmolar bath. The surface membrane (sarcolemma) remained intact and continued to exhibit enhanced excitability in the presence of the Na channel inactivation defect, as shown by the repetitive discharges. In detubulated fibers, however, no progressive depolarized shift in the membrane potential and no discharges after the current stimulus occurred. These experiments show that even a mild impairment of fast inactivation (approximately two percent of Na channels) is sufficient to cause myotonia. Furthermore, an intact T tubule system is required for the generation of self-sustained myotonic discharges regardless of whether the enhanced excitability arises from a large reduction in the Cl conductance (MC) or a mild impairment of Na channel inactivation (HyperPP, PMC, PAM).

The mechanism by which disrupted inactivation causes myotonia and paralysis has been explored further with a simulated fiber.[51] The model contains two electrically coupled compartments to simulate the surface and T tubule membranes. In agreement with the toxin studies, a small fraction of noninactivating Na channels (approximately 0.8 to two percent) is sufficient to cause myotonic runs of discharges [*see Figure 6*]. A slightly larger defect of inactivation (three percent of channels or more) results in a train of action potentials of varying amplitude that runs down to a stable depolar-

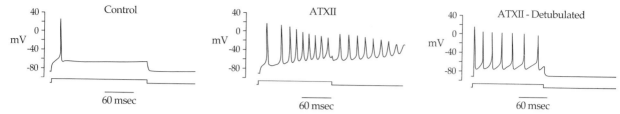

Figure 5 *Myotonia results from a partial disruption of Na channel inactivation and requires intact transverse tubules. Action potentials were recorded in vitro from rat extensor digitorum longus in response to an injected current pulse. A single action potential is elicited in normal muscle, and the fiber repolarizes at the end of the pulse (left). After application of sea anemone toxin, ATXII, sufficient to disrupt fast inactivation for about two percent of Na channels, the same stimulus elicits a myotonic run of discharges that continued after the end of the pulse (middle). Disruption of the transverse tubules by hyperosmotic shock abolishes the afterdischarges (right).*

ized resting potential near –40 mV. The simulated fiber is incapable of generating action potentials from this aberrantly depolarized potential. This is the model equivalent of the depolarization-induced weakness observed in patients with periodic paralysis. The inexcitability occurs because the vast majority of Na channels (97 percent in this simulation) are inactivated at –40 mV. This model provides an explanation for the dominant inheritance pattern in the Na channel disorders. In a molecular genetic view, the inactivation defect causes a gain of function (more Na current) that exerts a dominant-negative effect. The aberrant Na current conducted by a small fraction of mutant channels depolarizes the fiber and inactivates most Na channels, both mutant and WT.

In response to this model, another sodium channel defect was suspected because of the prolonged duration of depolarization-induced weakness in HyperPP.[52] In addition to the fast-inactivation mechanism that operates on a millisecond time scale and shapes the time course of the action potential, a separate voltage-dependent process termed slow inactivation modulates the availability of Na channels on a time scale of seconds to minutes.[53] The fast-inactivation defects described for mutant Na channels can initiate an

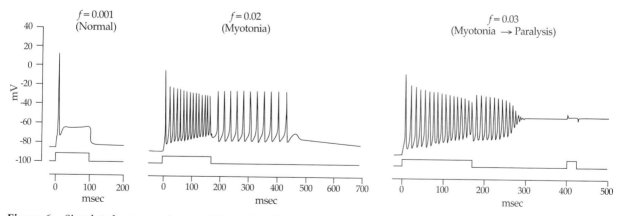

Figure 6 *Simulated responses in a model muscle cell confirm that even a small noninactivating Na current is sufficient to cause myotonia and paralysis. With normal values for the model parameters (f = 0.001), application of a prolonged current pulse elicited a single action potential. When two percent of the Na current failed to inactivate (f = 0.02) the same stimulus elicited a myotonic run of discharges, followed by repolarization of the membrane. Any greater fraction of noninactivating Na current (f = 0.03) caused the myotonic run to settle at an anomalously depolarized resting potential from which subsequent discharges could not be elicited.*

attack of depolarization-induced weakness, but because weakness may persist for hours, slow inactivation should also be impaired. Otherwise, slow inactivation would shut off the aberrant Na current and allow the fiber to repolarize. Patch-clamp studies of HyperPP mutants expressed in HEK cells confirmed that slow inactivation is disrupted for the two most common mutations, Thr704Met and Met1592Val.[54,55] Slow inactivation was normal, however, in a rare mutation causing HyperPP, Met1360Val, and in a PMC mutation, Thr1313Met, associated with prolonged episodes of weakness. Thus, a defect in slow inactivation causes a predilection to, but is not obligatory for, prolonged attacks of weakness.

Computer simulation also provides insights as to why some inactivation defects cause myotonia, whereas others lead to paralysis. The consistent association of a particular phenotype with a subset of *SCNA4* mutations [*see Figure 3*] suggests that a common functional defect should be shared by these mutations. Indeed, subtypes of inactivation defect have been observed. Mutations associated with HyperPP cause an increased persistent current [*see Figure 4*], a disruption of slow inactivation, or a hyperpolarized shift in channel activation. PMC mutations dramatically slow the rate of inactivation, shift its voltage dependence, and accelerate recovery. PAM mutations cause a milder slowing of inactivation and a depolarized shift in the voltage-dependence of inactivation. Each of these specific types of inactivation defect can easily be incorporated into the model cell. Simulations show that aberrant depolarization of the resting potential, and hence paralysis, occurs most readily for abnormal persistent Na currents in combination with disrupted slow inactivation, as observed with HyperPP mutations.[51,55] A sluggish rate of inactivation, as seen in PMC mutants, broadens the action potential duration, which in turn augments the K^+ accumulation in the T tubule and thereby promotes myotonia. The 5 to 10 mV shift in the voltage dependence of fast inactivation in PMC and PAM leads to myotonia by reducing the fraction of Na channels that will be inactivated near the threshold voltage for initiating another action potential.[43]

The computer and animal models provide compelling evidence that the inactivation defects observed in mutant Na channels are sufficient to cause myotonia and paralysis. Despite this success in elucidating the pathomechanisms of these disorders, perplexing questions remain. How do environmental factors, such as K^+ loading, cooling, and exercise, aggravate the symptoms in these disorders? What leads to the late-onset fixed proximal weakness and vacuolar myopathy? The answers to these questions await the availability of suitable animal models—transgenic mice, for example—in which to explore these phenomena.

Calcium Channel Defects Cause Periodic Paralysis

Missense mutations in the α_1-subunit of the skeletal muscle dihydropyridine receptor (DHPR), an L-type Ca channel, cause HypoPP. In common with other forms of periodic paralysis, attacks of weakness in HypoPP last for hours to days. Two distinguishing features of HypoPP are that during an attack, serum K^+ is almost invariably decreased (less than 3.5 mEq/L) and that myotonia never occurs.[5] Stress, rest after exercise, and behaviors that lower serum K+, such as carbohydrate ingestion and insulin use or diuretic use, may trigger an attack. Administration of K salts usually hastens recovery. HypoPP is the most familial periodic paralysis (1:100,000). Inheritance is autosomal dominant, with reduced penetrance in females. Episodes of weakness begin around the age of puberty, and a progressive vacuolar myopathy develops by the fourth or fifth decade.

Physiologic studies of HypoPP fibers in vitro revealed an abnormality of the resting potential but failed to identify the underlying defective channel or pump.[56] Reducing the bath K+ to 1 mM caused HypoPP fibers to *depolarize* by 20 mV, whereas normal fibers *hyperpolarized* by 10 mV. The depolarized fibers were flaccid and inexcitable, showing that hypokalemia alone is sufficient to cause paralysis in HypoPP. The aberrant depolarization was not blocked by tetrodotoxin, which excluded an Na channel defect, and the resisting K and Cl conductances were normal. The mechanism of the resting potential defect remained controversial, and the weakness was attributed to depolarization block of action potentials by inactivation of Na channels.

Calcium Channel Mutations

In the absence of a candidate disease gene, the molecular defect in HypoPP was identified by linkage analysis using a genome-wide search with dinucleotide repeat polymorphisms.[57] A HypoPP locus was mapped to chromosome 1q21-31 and cosegregated with *CACNL1A3*, a previously mapped gene encoding the α_1-subunit of the skeletal muscle dihydropyridine receptor. Analysis of *CACNL1A3* cDNA has identified three missense mutations in HypoPP.[58,59] Interestingly, all three are at positively charged arginine residues in S4 segments [*see Figure 7*], the proposed voltage sensors of the channel. The incomplete penetrance in females is unique to the Arg528His mutation, which occurs in half of the families with HypoPP.[60] The

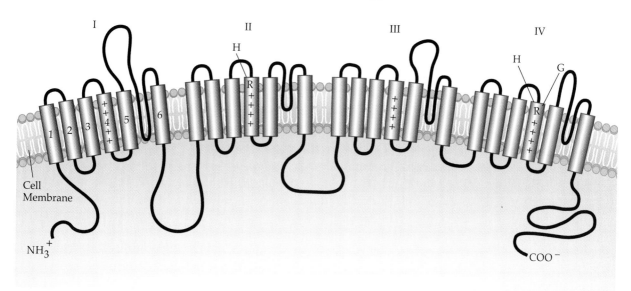

Figure 7 *Mutations in skeletal muscle calcium channel. Membrane-folding model for the α_1-subunit of the skeletal muscle Ca channel and location of missense mutations in HypoPP. Single-letter amino acid abbreviations: R, arginine; H, histidine; G, glycine.*

Arg1239His mutation accounts for most of the remaining genotyped families, whereas the Arg1239Gly mutation has been detected in only a single kindred. A founder effect is not likely to be the cause for the high prevalence of the Arg528His and Arg1239His mutations, as shown by haplotype analysis and the identification of de novo mutations.[60] One large family with HypoPP is not linked to the locus on 1q31-q32.[61] Identification of a second disease gene may provide clues to the pathogenesis of HypoPP, which has remained elusive.

Pathogenesis of HypoPP Remains a Mystery

Despite the unequivocal evidence that HypoPP is caused by missense mutations of *CACNL1A3*, the relevant functional defect in channel behavior and the pathophysiological mechanism of the attacks remain to be elucidated. Calcium channels in skeletal muscle are localized to the T tubule membrane. The channel is a heteropentameric complex composed of a large pore-forming α_1-subunit that contains the dihydropyridine binding site, a disulfide-linked $\alpha_2\delta$-subunit, and a membrane-spanning γ-subunit.[49] This complex functions both as a voltage-activated Ca channel and a voltage-activated sensor for excitation-contraction (E-C) coupling.[62,63] In skeletal muscle, E-C coupling is independent of external Ca^{2+} entry. The α_1-subunit, in particular the interdomain II-III loop, appears to be coupled to the Ca release channel or ryanodine receptor of the SR.[64] Because both functions are highly voltage dependent and each mutation in HypoPP is in a charged residue of the S4 voltage sensor, the pathogenesis of the weakness may arise from defects in Ca channel activation or E-C coupling or both. On the other hand, the aberrant depolarization in low K^+ suggests there must be a defect in regulating the Ca conductance because internal Ca^{2+} release from the SR should not affect the membrane potential. The depolarization could arise directly from an altered Ca current through the channel, or from modulation of another channel or pump by Ca^{2+} acting as a second messenger.

A gain-of-function defect is expected because HypoPP is autosomal dominant, and each channel contains a single α_1-subunit. Reduced channel density from haploinsufficiency is unlikely to be the mechanism, based on an observation from research on mutant mice. A frameshift mutation in the α_1 gene is clinically and electrophysiologically silent in heterozygous mice, whereas homozygotes die at birth from muscular dysgenesis (mdg) with absent E-C coupling and DHP-sensitive L-type Ca currents in skeletal muscle.[63,65] The mdg myotube provides a system to test the integrity of E-C coupling for exogenously expressed α_1-subunits. Intranuclear injection of WT α_1-subunit cDNA restores depolarization-induced Ca^{2+} release and contraction.[65] Injection of mutant cDNAs, coding for either the Arg528His or the Arg1239His mutation, also restored Ca^{2+} release and contraction, which implies that neither HypoPP mutation drastically alters E-C coupling.[66]

Direct measurement of Ca currents in HypoPP mutants has yielded conflicting results. Initially, a defect of channel inactivation was observed in recordings from cultured human myotubes with the Arg528His mutation.[67] The voltage dependence of inactivation was shifted by 40 mV in the hyperpolarized direction. Subsequent experiments on myotubes cultured from additional biopsy samples failed to consistently detect the aberrant shift. Heterologous expression studies, with the equivalent of Arg528His engineered into the rabbit skeletal muscle α_1-subunit, resulted in a decreased current density compared with WT channels, but no shift occurred.[68] When the homologous mutation was made in the cardiac α_1-subunit, which was expressed better than the skeletal muscle isoform, no functional defects were detected.[69] In human myotubes harboring the Arg1239His mutation, no alterations in voltage dependence were detected, but the Ca current density was reduced by 50 percent.[67] None of the reported defects in Ca current, even if genuine, readily leads to a plausible explanation for the aberrant depolarization during attacks of weakness in HypoPP or for the sensitivity to reduced K^+. One strategy for elucidating the pathomechanism of HypoPP is to develop a transgenic animal model. The ability to study fully differentiated HypoPP muscle fibers may permit identification of the mechanism of episodic weakness and may reveal a previously unrecognized role of the L-type Ca channel in maintaining the resting potential of skeletal muscle.

Prospectus and Clinical Implications

The molecular genetic bases for all the familial myotonias and periodic paralyses are now known. Both dominant and recessive forms of myotonia congenita are caused by mutations in the major Cl channel of skeletal muscle, ClC-1. Hyperkalemic periodic paralysis, paramyotonia congenita, and the potassium-aggravated myotonias result from missense mutations in the α-subunit of the skeletal muscle Na channel. Hypokalemic periodic paralysis is caused by missense mutations in the α_1-subunit of the skeletal muscle DHPR/L-type Ca channel. These recent discoveries have shifted the classification of these disorders from a phenotype- to a genotype-based system. Molecular genetic testing has led to the recognition of a new class of dominant myotonia without weakness. Previously, all such cases were diagnosed as dominant (Thomsen) myotonia congenita, a Cl channel disorder. In fact, a large portion of these patients are now known to have potassium-aggravated myotonia, a Na channel disorder in which K loading worsens the muscle stiffness.

In general, the number and diversity of mutations in the nondystrophic myotonias and periodic paralyses precludes the development of a rapid molecular screening test for clinical use. There is one caveat regarding young patients with myotonia: If the possibility of myotonic dystrophy exists, a test for the expanded trinucleotide repeat on chromosome 19q13.2-q13.3 as well as a slit-lamp examination for premature cataract should be performed to exclude myotonic dystrophy, which has a much poorer prognosis than myotonia congenita. Because a few common missense mutations account for the majority of cases of HyperPP and HypoPP, a polymerase chain-reaction–based assay could be developed as a highly specific test with modestly reduced sensitivity, but no such service is commercially available at present.

An accurate diagnosis among the different hereditary myotonias and periodic paralysis is important for optimal patient management. Mexiletine, a class I antiarrhythmic agent and use-dependent blocker of Na channels, is the drug of choice for prophylactic treatment of disabling myotonia.[4] The antimyotonic effects of mexiletine are beneficial in disorders of both Cl (MC) and Na (PAM, PMC) channels. Acetazolamide, a potent inhibitor of carbonic anhydrase, has been beneficial in preventing or reducing myotonia and weakness in PMC and HyperPP.[70] Use of thiazide diuretics to chronically deplete total body K^+ has helped some patients with HyperPP, but side effects of these drugs may limit their usefulness. Many patients prefer to use nonpharmacological measures to reduce attack frequency.[5] Thus, patients with PMC avoid exposure to cold, and those with PAM or HyperPP limit their intake of K-rich foods. A high-carbohydrate snack, which shifts K^+ intracellularly by cotransport with glucose, may avert an impending attack of weakness. Conversely, HypoPP patients should avoid high-carbohydrate meals, and oral K supplements may avert an impending attack. Acetazolamide also helps prevent attacks of weakness in HypoPP patients.[71] A trial of the Ca channel blocker verapamil (80 mg t.i.d.) did not reduce the frequency of paralytic attacks in a double-blind crossover study of HypoPP patients.[72]

In comparison to studies of other heritable disorders of the nervous system, progress has been particularly rapid in elucidating the molecular defects and pathogenic bases for the familial myotonias and periodic paralyses. The search for the molecular defects in the Cl and Na channel disorders was greatly accelerated by the identification of candidate disease genes from electrophysiological studies of diseased muscle. The previous cloning of many voltage-dependent ion channels from skeletal muscle circumvented the necessity of laborious positional cloning. Moreover, in contrast to discoveries of novel disease genes whose function is unknown, more than 40 years of research had established the role of Na and Cl channels in regulating the excitability of skeletal muscle. Finally, animal and computer models have provided powerful confirmation that the functional defects identified in mutant ion channels are sufficient to cause myotonia and paralysis. Two important questions remain: How do environmental factors such as K^+, temperature, and exercise modulate the frequency of attacks? What is the pathogenesis of the weakness in HypoPP? Despite these gaps in our knowledge, the familial myotonias and periodic paralyses stand out as one of the clearest examples of human disease for which the pathophysiological basis can be traced from gene defect to altered protein function to patient symptoms.

References

1. Cannon SC: Ion-channel defects and aberrant excitability in myotonia and periodic paralysis. *Trends Neurosci* 19:3, 1996
2. Lehmann-Horn F, Rüdel R: Hereditary nondystrophic myotonias and periodic paralyses. *Curr Opin Neurol* 8:402, 1995
3. Rüdel R, Lehmann-Horn F: Membrane changes in cells from myotonia patients. *Physiol Rev* 65:310, 1985
4. Rüdel R, Lehmann-Horn F, Ricker K: The nondystrophic myotonias. *Myology*, 2nd ed. Engel AG, Franzini-Armstrong C, Eds. McGraw-Hill, New York, 1994, p 1291
5. Lehmann-Horn F, Engel AG, Ricker K, et al: The periodic paralyses and paramyotonia congenita. *Myology*, 2nd ed. Engel AG, Franzini-Armstrong C, Eds. McGraw-Hill, New York, 1994, p 1303

6. Brown GL, Harvey AM: Congenital myotonia in the goat. *Brain* 62:341, 1939

7. Bryant SH: Cable properties of external intercostal muscle fibres from myotonic and nonmyotonic goats. *J Physiol* 204:539, 1969

8. Adrian RH, Bryant SH: On the repetitive discharge in myotonic muscle fibres. *J Physiol* 240:505, 1974

9. Lipicky RJ, Bryant SH, Salmon JH: Cable parameters, sodium, potassium, chloride, and water content, and potassium efflux in isolated external intercostal muscle of normal volunteers and patients with myotonia congenita. *J Clin Invest* 50:2091, 1971

10. Lehmann-Horn F, Rüdel R, Dengler R, et al: Membrane defects in paramyotonia congenita with and without myotonia in a warm environment. *Muscle Nerve* 4:396, 1981

11. Lehmann-Horn F, Rüdel R, Ricker K, et al: Two cases of adynamia episodica hereditaria: in vitro investigation of muscle cell membrane and contraction parameters. *Muscle Nerve* 6:113, 1983

12. Lehmann-Horn F, Kuther G, Ricker K, et al: Adynamia episodica hereditaria with myotonia: a non-inactivating sodium current and the effect of extracellular pH. *Muscle Nerve* 10:363, 1987

13. Koch MC, Steinmeyer K, Lorenz C, et al: The skeletal muscle chloride channel in dominant and recessive human myotonia. *Science* 257:797, 1992

14. Fontaine B, Khurana TS, Hoffman EP, et al: Hyperkalemic periodic paralysis and the adult muscle sodium channel alpha-subunit gene. *Science* 250:1000, 1990

15. Ebers GC, George AL, Barchi RL, et al: Paramyotonia congenita and hyperkalemic periodic paralysis are linked to the adult muscle sodium channel gene. *Ann Neurol* 30:810, 1991

16. Fontaine B, Vale-Santos J, Jurkat-Rott K, et al: Mapping of the hypokalaemic periodic paralysis (HypoPP) locus to chromosome 1q31-32 in three European families. *Nat Genet* 6:267, 1994

17. Thomsen J: Tonische Krämpfe in willkürlich beweglichen Muskeln in Folge von ererbter psychischer Disposition. *Arch Psychiatr Nervenkr* 6:702, 1876

18. Becker P: Zur Frage der Heterogenie der erblichen Myotonien. *Nervenarzt* 28:455, 1957

19. Jentsch TJ, Steinmeyer K, Schwarz G: Primary structure of *Torpedo marmorata* chloride channel isolated by expression cloning in Xenopus oocytes. *Nature* 348:510, 1990

20. Lorenz C, Meyer-Kleine C, Steinmeyer K, et al: Genomic organization of the human muscle chloride channel ClC-1 and analysis of novel mutations leading to Becker-type myotonia. *Hum Mol Genet* 3:941, 1994

21. Steinmeyer K, Lorenz C, Pusch M, et al: Multimeric structure of ClC-1 chloride channel revealed by mutations in dominant myotonia congenita (Thomsen). *EMBO J* 13:737, 1994

22. Lehmann-Horn F, Mailander V, Heine R, et al: Myotonia levior is a chloride channel disorder. *Hum Mol Genet* 4:1397, 1995

23. Meyer-Kleine C, Steinmeyer K, Ricker K, et al: Spectrum of mutations in the major human skeletal muscle chloride channel gene (CLCN1) leading to myotonia. *Am J Hum Genet* 57:1325, 1995

24. George AL Jr, Crackower MA, Abdalla JA, et al: Molecular basis of Thomsen's disease (autosomal dominant myotonia congenita). *Nat Genet* 3:305, 1993

25. Zhang J, George A, Griggs R, et al: Mutations in the human skeletal muscle chloride channel gene (*CLCN1*) associated with dominant and recessive myotonia congenita. *Neurology* 47:993, 1996

26. Beck CL, Fahlke C, George AL Jr: Molecular basis for decreased muscle chloride conductance in the myotonic goat. *Proc Natl Acad Sci USA* 93:11248, 1996

27. Steinmeyer K, Klocke R, Ortland C, et al: Inactivation of muscle chloride channel by transposon insertion in myotonic mice. *Nature* 354:304, 1991

28. Furman RE, Barchi RL: The pathophysiology of myotonia produced by aromatic carboxylic acids. *Ann Neurol* 4:357, 1978

29. Lerche H, Heine R, Pika U, et al: Human sodium channel myotonia: slowed channel inactivation due to substitutions for a glycine within the III-IV linker. *J Physiol* 470:13, 1993

30. Ricker K, Moxley RT 3rd, Heine R, et al: Myotonia fluctuans: a third type of muscle sodium channel disease. *Arch Neurol* 51:1095, 1994

31. Rojas CV, Wang JZ, Schwartz LS, et al: A Met-to-Val mutation in the skeletal muscle Na+ channel alpha-subunit in hyperkalaemic periodic paralysis. *Nature* 354:387, 1991

32. Ptacek LJ, George AL Jr, Griggs RC, et al: Identification of a mutation in the gene causing hyperkalemic periodic paralysis. *Cell* 67:1021, 1991

33. McClatchey AI, Lin CS, Wang J, et al: The genomic structure of the human skeletal muscle sodium channel gene. *Hum Mol Genet* 1:521, 1992

34. George AL Jr, Komisarof J, Kallen RG, et al: Primary structure of the adult human skeletal muscle voltage-dependent sodium channel. *Ann Neurol* 31:131, 1992

35. Barchi RL: Molecular pathology of the skeletal muscle sodium channel. *Annu Rev Physiol* 57:355, 1995

36. Rüdel R, Ricker K, Lehmann-Horn F: Genotype-phenotype correlations in human skeletal muscle sodium channel diseases. *Arch Neurol* 50:1241, 1993

37. Baquero JL, Ayala RA, Wang JW, et al: Hyperkalemic periodic paralysis with cardiac dysrhythmia: a novel sodium channel mutation? *Ann Neurol* 37:408, 1995

38. Green DS, Hayward LJ, George AL Jr, et al: A proposed mutation, Val781Ile, associated with hyperkalemic periodic paralysis and cardiac dysrhythmia is a benign polymorphism. *Ann Neurol.* 42:253, 1997

39. Plassart E, Reboul J, Rime C, et al: Mutations in the muscle sodium channel gene (SCN4A) in 13 French families with hyperkalemic periodic paralysis and paramyotonia congenita: phenotype to genotype correlations and demonstration of the predominance of two mutations. *Eur J Hum Genet* 2:110, 1994

40. Cannon SC, Brown RH Jr, Corey DP: A sodium channel defect in hyperkalemic periodic paralysis: potassium-induced failure of inactivation. *Neuron* 6:619, 1991

41. Cannon SC, Strittmatter SM: Functional expression of sodium channel mutations identified in families with periodic paralysis. *Neuron* 10:317, 1993

42. Cummins TR, Zhou J, Sigworth FJ, et al: Functional consequences of a Na+ channel mutation causing hyperkalemic periodic paralysis. *Neuron* 10:667, 1993

43. Hayward LJ, Brown RH Jr, Cannon SC: Inactivation defects caused by myotonia-associated mutations in the sodium channel III-IV linker. *J Gen Physiol* 107:559, 1996

44. Mitrovic N, George AL Jr, Lerche H, et al: Different effects on gating of three myotonia-causing mutations in the inactivation gate of the human muscle sodium channel. *J Physiol* 487:107, 1995

45. Yang N, Ji S, Zhou M, et al: Sodium channel mutations in paramyotonia congenita exhibit similar biophysical phenotypes in vitro. *Proc Natl Acad Sci USA* 91:12785, 1994

46. Chahine M, George AL Jr, Zhou M, et al: Sodium channel mutations in paramyotonia congenita uncouple inactivation from activation. *Neuron* 12:281, 1994

47. Aldrich RW, Corey DP, Stevens CF: A reinterpretation of mammalian sodium channel gating based on single channel recording. *Nature* 306:436, 1983

48. Cannon SC: From mutation to myotonia in sodium channel disorders. *Neuromusc Disord* 7:241, 1997

49. Catterall WA: Structure and function of voltage-gated ion channels. *Annu Rev Biochem* 64:493, 1995

50. Cannon SC, Corey DP: Loss of Na+ channel inactivation by anemone toxin (ATX II) mimics the myotonic state in hyperkalaemic periodic paralysis. *J Physiol* 466:501, 1993

51. Cannon SC, Brown RH Jr, Corey DP: Theoretical reconstruction of myotonia and paralysis caused by incomplete inactivation of sodium channels. *Biophys J* 65:270, 1993

52. Ruff RL: Slow Na+ channel inactivation must be disrupted to evoke prolonged depolarization-induced paralysis. *Biophys J* 66:542, 1994

53. Almers W, Stanfield PR, Stühmer W: Slow changes in currents through sodium channels in frog muscle membrane. *J Physiol* 339:253, 1983

54. Cummins TR, Sigworth FJ: Impaired slow inactivation in mutant sodium channels. *Biophys J* 71:227, 1996

55. Hayward LJ, Brown RH Jr, Cannon SC: Slow inactivation differs among mutant Na channels associated with myotonia and periodic paralysis. *Biophys J* 72:1204, 1997

56. Rüdel R, Lehmann-Horn F, Ricker K, et al: Hypokalemic periodic paralysis: in vitro investigation of muscle fiber membrane parameters. *Muscle Nerve* 7:110, 1984

57. Fontaine B, Vale-Santos J, Jurkat-Rott K, et al: Mapping of the hypokalemic periodic paralysis (HypoPP) locus to chromosome 1q31-32 in three European families. *Nat Genet* 6:267, 1994

58. Jurkat-Rott K, Lehmann-Horn F, Albaz A, et al: A calcium channel mutation causing hypokalemic periodic paralysis. *Hum Mol Genet* 3:1415, 1994

59. Ptacek LJ, Tawil R, Griggs RC, et al: Dihydropyridine receptor mutations cause hypokalemic periodic paralysis. *Cell* 77:863, 1994

60. Elbaz A, Vale-Santos J, Jurkat-Rott K, et al: Hypokalemic periodic paralysis and the dihydropyridine receptor (CACNL1A3): genotype/phenotype correlations for two predomi-

nant mutations and evidence for the absence of a founder effect in 16 caucasian families. *Am J Hum Genet* 56:374, 1995

61. Plassart E, Elbaz A, Santos JV, et al: Genetic heterogeneity in hypokalemic periodic paralysis. *Hum Genet* 94:551, 1994

62. Rios E, Brum G: Involvement of dihydropyridine receptors in excitation-contraction coupling in skeletal muscle. *Nature* 325:717, 1987

63. Beam KG, Knudson CM, Powell JA: A lethal mutation in mice eliminates the slow calcium current in skeletal muscle cells. *Nature* 320:168, 1986

64. Tanabe T, Beam KG, Adams BA, et al: Regions of the skeletal muscle dihydropyridine receptor critical for excitation-contraction coupling. *Nature* 346:567, 1990

65. Tanabe T, Beam KG, Powell JA, et al: Restoration of excitation-contraction coupling and slow calcium current in dysgenic muscle by dihydropyridine receptor complementary DNA. *Nature* 336:134, 1988

66. Beam KG: Personal communication, December 1995.

67. Sipos I, Jurkat-Rott K, Harasztosi C, et al: Skeletal muscle DHP receptor mutations alter calcium currents in human hypokalaemic periodic paralysis myotubes. *J Physiol* 483:299, 1995

68. Lapie P, Goudet C, Nargeot J, et al: Electrophysiological properties of the hypokalaemic periodic paralysis mutation (R528H) of the skeletal muscle alpha 1s subunit as expressed in mouse L cells. *FEBS Lett* 382:244, 1996

69. Lerche H, Klugbauer N, Lehmann-Horn F, et al: Expression and functional characterization of the cardiac L-type calcium channel carrying a skeletal muscle DHP-receptor mutation causing hypokalaemic periodic paralysis. *Pflugers Arch* 431:461, 1996

70. Griggs RC, Moxley RT, Riggs JE, et al: Effects of acetazolamide on myotonia. *Ann Neurol* 3:531, 1978

71. Resnick JS, Engle WK, Griggs RC, et al: Acetazolamide prophylaxis in hypokalemic periodic paralysis. *N Engl J Med* 278:582, 1968

72. Links TP: Personal communication, December 1995.

Acknowledgments

Figures 1, 4, 5, and 6   Stephen C. Cannon. Adapted by Marcia Kammerer.

Figures 2, 3, and 7   Stephen C. Cannon. Adapted by Kathy Konkle.

CHAPTER 15

Genetic Mitochondrial Disorders

Donald R. Johns, M.D., Ricardo N. Fadic, M.D.

Mitochondria are the main source of energy in the cell. Most organ systems are at least partially dependent on mitochondrial energy supply, and disorders of oxidative phosphorylation could theoretically cause a wide range of signs and symptoms. Despite this potential, diseases primarily associated with defective mitochondrial function were first described just three decades ago. The 1962 report by Luft and colleagues of a patient with euthyroid hypermetabolism[1] was followed by numerous descriptions of abnormal muscle and mitochondria morphology and abnormal oxidative phosphorylation in a variety of disorders ranging from ophthalmoplegia to encephalopathies. The required association among the clinical phenotypes, the abnormal histologic and ultrastructural abnormalities in the mitochondria, and the biochemical abnormalities in oxidative phosphorylation made the diagnosis difficult in individual patients and was the source of conflict in the classification of these entities.

Mitochondrial genetics became a major topic in neurology after mutations in the mitochondrial DNA (mtDNA) were described in 1988.[2,3] Over the past 10 years, a wide variety of human diseases were found to be caused by or associated with mtDNA mutations. These disorders are relatively uncommon, but they were the first molecularly defined examples of many cardinal neurologic diseases, including seizures, stroke, and optic neuropathy. Mutations in mtDNA may also contribute to normal aging and neurodegenerative disorders. This chapter examines advances in mitochondrial genetics, with particular focus on disorders attributable to defects in oxidative phosphorylation.

Mitochondrial DNA

Mitochondrial DNA is the only extrachromosomal DNA in humans. It is a small genome (16,569 nucleotides) that contains only 37 genes specifying 13 polypeptides (all of which are oxidative phosphorylation chain subunits), 22 transfer RNAs (tRNAs), and two ribosomal RNAs (rRNAs) [*see Figure 1*]. Of the 13 polypeptides coded in mtDNA, seven are subunits of the respiratory chain complex I, one is a subunit of complex III, three are subunits of complex IV, and two are subunits of complex V.[4] Each of these four complexes also contains subunits encoded by nuclear genes. All four subunits of complex II (succinate-ubiquinone oxidoreductase) are encoded in the nuclear DNA, as are all of the other proteins in the mitochondria. Each mitochondrion has an average of five mtDNA molecules. Although less than one percent of the total cellular DNA is typically mtDNA, the copy number is high, ranging from approximately 10^3 to 10^4 genomes per cell. Mitochondria probably evolved from independent organisms that became endosymbiotically incorporated into the cell. Consequently, they have independent replication, transcription, and translation systems that have both prokaryotic and eukaryotic features.[5] mtDNA has a unique genetic code and cannot be translated in the cytoplasmic compartment, and nuclear DNA cannot be translated within the mitochondria. Like bacterial protein synthesis, mitochondrial protein synthesis is initiated with formylmethionine and is sensitive to the bacterial ribosome inhibitor chloramphenicol.

mtDNA Genetics

The cytoplasmic location of the mtDNA results in a unique inheritance pattern called maternal inheritance, the transmission of mtDNA exclusively from a mother to her children.[6] Paternal transmission of mtDNA normally does not occur. Mothers can transmit errors in mtDNA sequence (mutations) to both sons and daughters, but only daughters can pass the mutation to their descendants. mtDNA does not recombine and has a poorly developed repair system[7,8] leading readily to the accumulation of mutations along maternal lines. Because of the variability of the phenotypic manifestation associated with mtDNA mutations, analysis of large pedigrees is necessary to prove maternal transmission.

A mutation may be present in all (homoplasmy) or in a fraction (heteroplasmy) of mtDNA molecules. All mtDNA duplications and deletions are heteroplasmic. Pathogenic mtDNA point mutations can be either homoplasmic or heteroplasmic, whereas neutral non–disease-causing polymorphisms are almost always homoplasmic. A mother with a homoplasmic mutation will transmit only mutant mtDNA to her offspring, and the mutation will be stably maintained in successive generations. In contrast, a mother with a heteroplasmic mutation may pass very different levels of mutant mtDNA to her children, and the genotype may change in a few generations. Moreover, if the mother is heteroplasmic, mutated mtDNA is not always transmitted from mother to child.[9] Experimental evidence for both a random genetic drift and a restriction-amplification event (bottleneck) during oogenesis is cited to explain the widely variant level of mutant mtDNA present in siblings.[10,11]

Mitochondria, and the mtDNAs contained therein, segregate at random into daughter cells during mitosis (mitotic segregation). The proportion of a heteroplasmic mtDNA mutation may therefore vary widely between different tissues of an individual and even between different cells of the same tis-

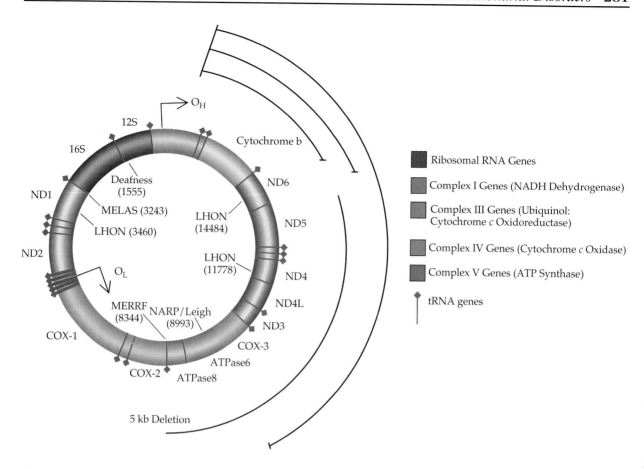

Figure 1 *Human mitochondrial DNA (mtDNA). The most common point mutations and the associated clinical phenotypes are indicated within the circular molecule. The most common single mtDNA deletion, the 5-kb deletion, and representative multiple mtDNA deletions are shown outside the circle. LHON, Leber's hereditary optic neuropathy; MELAS, mitochondrial encephalomyopathy, lactic acidosis, and strokelike symptoms; MERRF, myoclonic epilepsy and ragged red fiber disease; NARP, neurogenic muscle weakness, ataxia, and retinitis pigmentosum; O_H and O_L, replication origins on the H and L mtDNA strands.*

sue.[12,13] To illustrate this concept, consider patients with a single common mtDNA deletion. The onset of blood dyscrasia in Pearson's syndrome (pancytopenia and exocrine pancreas dysfunction) commonly occurs in childhood, and survival into the second decade may be complicated by the development of Kearns-Sayre syndrome. The improvement of the pancytopenia is likely because of the positive selection during cell division of hematopoietic cells without the deletion. The opposite occurs in muscle, a postmitotic tissue, in which the proportion of deleted mtDNA increases in sequential muscle biopsies.[14,15]

The phenotypic consequences of mtDNA mutations depend on the severity of the oxidative phosphorylation defect and the energy requirements of different organs and tissues. Each tissue requires a different level (threshold) of mitochondrial adenosine triphosphate (ATP) production to sustain normal cellular function, and organs that are more dependent on oxidative metabolism (e.g., brain, retina, heart, skeletal muscle) are more frequently affected. In families harboring a heteroplasmic mtDNA mutation, the severity of the oxidative phosphorylation defect varies with the percentage of normal mtDNA. Different family members may inherit various proportions of mutant mtDNA and thus may demonstrate strikingly variable clinical manifestations.[16]

Table 1   Mitochondrial Mutations and Associated Phenotypes

Primary mtDNA mutations
  mtDNA rearrangements (deletions/duplications)
    Chronic progressive external ophthalmoplegia
    Kearns-Sayre syndrome
    Pearson's syndrome
    Diabetes mellitus and deafness
    Mitochondrial myopathy
  mtDNA protein coding gene mutations
    Leber's hereditary optic neuropathy
    Leber's hereditary optic neuropathy plus dystonia
    Maternally inherited Leigh syndrome
    Neurogenic weakness, ataxia, retinitis pigmentosa (NARP) syndrome
  Mitochondrial tRNA gene mutations
    Mitochondrial encephalomyopathy, lactic acidosis, and strokelike (MELAS) episodes
    Myoclonic epilepsy and ragged-red fibers (MERRF) syndrome
    Chronic progressive external ophthalmoplegia

    Diabetes mellitus and deafness
    Maternally inherited cardiomyopathy
    Maternally inherited deafness
    Mitochondrial myopathy
  Mitochondrial rRNA gene mutations
    Maternally inherited deafness with aminoglycoside sensitivity
Nuclear DNA mutations
  Multiple mtDNA deletions
    Autosomal dominant and recessive chronic progressive external ophthalmoplegia
    Myoneurogastrointestinal encephalopathy (MNGIE) syndrome
    Sensory ataxic neuropathy, dysarthria, and ophthalmoplegia (SANDO) syndrome
    Acute rhabdomyolysis
  Succinate dehydrogenase (complex II) mutation
    Leigh syndrome

Because mtDNA is maternally inherited, pathogenetic point mutations will follow a maternal pattern of inheritance. However, a number of nuclear DNA genes are involved in the maintenance and function of mtDNA, and mendelian inheritance patterns are also seen in mitochondrial syndromes. Multiple mtDNA deletions are inherited most commonly in an autosomal dominant pattern. At least three nuclear loci responsible for the integrity of the mtDNA have been identified by linkage studies in different families with dominant multiple mtDNA deletions.[17,18] An autosomal recessive pattern of inheritance was described in two families with chronic progressive ophthalmoplegia and cardiomyopathy in association with multiple mtDNA deletions.[19] The first pathogenetic mutation in a nuclear gene encoding a respiratory chain subunit was identified in a child with Leigh syndrome and severe succinate dehydrogenase deficiency[20] [*see Table 1*].

Large single mtDNA deletions generally occur sporadically. Mitochondrial genomes with deletions are transcribed but not translated, including genes not encompassed by the deletion.[21] The deleted genomes appear to be functionally dominant over the wild type.[22] In patients with single deletions, an identical species of deleted mtDNA molecules is present in all tissues of the body, albeit at different proportions in different tissues. This suggests that the deletions are new mutations occurring in the oocytes, zygote, or during early embryogenesis. Small amounts of partially deleted mtDNA molecules are detectable in oocytes,[23] and a single family was reported with diabetes and deafness associated with a unique maternally inherited mtDNA deletion.[24]

Pathophysiology of mtDNA Mutations

The prevailing view of the pathogenesis of mitochondrial diseases is that the decline in mitochondrial ATP-generating capacity leads to energetic failure and cell death in affected tissues. There are three types of mtDNA mutations: (1) mtDNA rearrangements in which genes are deleted or duplicated; (2) point mutations in tRNA or rRNA genes, resulting in defects in mitochondrial protein synthesis; and (3) missense mutations in peptide subunits of the respiratory chain. tRNA genes harbor a disproportionately high number of pathogenetic mtDNA point mutations, and all well-documented mtDNA deletions include at least one tRNA gene, suggesting a vital tRNA role in pathogenesis.

The precise relation between the different mtDNA mutations, the impairment of oxidative phosphorylation, and the associated clinical phenotypes is not well understood. A human cell line depleted of mtDNA by exposure to ethidium bromide, which inhibits mtDNA replication, was created by King and Attardi.[25] These ρo cell lines, devoid of mtDNA, are fused with donor cytoplasts that contain the mtDNA mutation of interest. Defects in mitochondrial protein synthesis or impairment of oxidative phosphorylation or both can then be studied in these constructs.[26,27] The ρo cybrid cell lines have made important contributions to understanding the pathogenetic mechanism of mtDNA mutations. For instance, the mutation in the tRNA^{Lys} gene associated with the myoclonic epilepsy with ragged red fibers (MERRF) phenotype causes a protein synthesis defect with the formation of peptides of abnormal size.[28] This is attributable to premature termination of translation at or near each lysine codon because of a defect in the specific aminoacylation capacity of the mutant tRNA^{Lys}.[28] Biochemical evidence of a respiratory chain complex I impairment was provided for mutations associated with different forms of Leber's hereditary optic neuropathy (LHON).[29,30]

The detailed data from studies at the molecular and cellular levels do not fully explain the breadth of clinical phenotypic manifestations in patients with mitochondrial diseases. In some mitochondrial syndromes, interactions with environmental factors are necessary for the phenotypic manifestations. A mutation in the 12s-subunit of the rRNA gene can cause deafness when the individual is exposed to aminoglycosides.[31] The biochemical defects associated with the LHON-associated mutations do not explain the male predominance, the acute onset of the vision loss, the limitation of the pathology to the optic nerve, or the long latency preceding onset of the symptoms. Alcohol and tobacco exposure are important epigenetic factors in LHON.[32]

Classic Phenotypes Associated with mtDNA Mutations

Let us first discuss briefly the cardinal, most established clinical mitochondrial disease syndromes [*see Table 1*]. The diagnosis of a mitochondrial disease is not straightforward, but the availability of molecular genetic tests for pathogenetic mtDNA mutations has helped significantly. The same mtDNA mutation can be associated with different clinical syndromes, and a given clinical syndrome may be attributable to different mutations. Mitochondrial diseases are multisystemic syndromes with very protean manifestations and intrafamilial and interfamilial variability. Many patients cannot be neatly classified into discrete subcategories of syndromes and may fall into multiple, overlapping groups. The number of mutations reported in mtDNA, particularly point mutations in tRNA genes, continues to increase [*see Table 2*]. More extensive and detailed descriptions of these mutations and the associated clinical phenotypes can be found in comprehensive reviews.[33,34]

Chronic Progressive External Ophthalmoplegia

Chronic progressive external ophthalmoplegia (CPEO) is characterized by ptosis and symmetric weakness of the extraocular muscles, usually in association with proximal limb myopathy. CPEO is insidious in onset; ptosis is usually the initial symptom and may, although rarely, be the only manifestation. A later age of onset is correlated with limited and milder symptoms. The term CPEO-plus designates the presence of other neurologic or somatic manifestations. Kearns-Sayre syndrome is a subset of CPEO-plus defined as ophthalmoplegia with age of onset younger than 20 years and pigmentary retinopathy with the ancillary features of elevated cerebrospinal fluid protein, heart block, or ataxia. Some patients demonstrate in infancy the variant Pearson's syndrome. CPEO is typically associated with mtDNA deletions, and approximately 80 to 90 percent of patients with Kearns-Sayre syndrome, 70 percent of those with CPEO-plus, and 40 percent of those with CPEO alone harbor mtDNA rearrangements.[35,36] Genetic testing in blood samples is not useful for the diagnosis of mtDNA deletions, because they tend to segregate to very low levels in leukocytes; skeletal muscle is the best tissue to use.

CPEO displays several different modes of inheritance. Most cases are clinically sporadic, but maternally inherited CPEO occurs in association with point mutations in the tRNA$^{Leu (UUR)}$ and tRNA^{Asn} genes. PEO is the cardinal clinical manifestation associated with autosomally inherited (dominant and recessive) multiple mtDNA deletions.

Table 2 Representative Pathogenetic mtDNA Point Mutations

Nucleotide Position	mtDNA Gene	Phenotypes	Nucleotide Change	Amino Acid Substitution
1555	12S rRNA	Deafness	A to G	NA
3243	tRNA$^{Leu(UUR)}$	MELAS Diabetes PEO	A to G	NA
3260	tRNA$^{Leu(UUR)}$	Cardiomyopathy	A to G	NA
3271	tRNA$^{Leu(UUR)}$	MELAS	T to C	NA
3460	ND-1	LHON	G to A	Ala/Thr
5549	tRNA^{Trp}	Dementia + Chorea	G to A	NA
7445	tRNA$^{Ser(UCN)}$	Deafness	A to G	NA
8344	tRNA^{Lys}	MERRF	A to G	NA
8356	tRNA^{Lys}	MERRF	T to C	NA
8363	tRNA^{Lys}	Cardiomyopathy/ Deafness	G to A	NA
8993	ATPase 6	NARP/MILS	T to G T to C	Leu/Arg Leu/Pro
9804	COX III	LHON	A to G	Ala/Thr
11778	ND-4	LHON	G to A	Arg/His
14459	ND-6	LHON + Dystonia	G to A	Ala/Val
14484	ND-6	LHON	T to C	Met/Val
15257	Cytochrome *b*	LHON	G to A	Asp/Asn

Leber's Hereditary Optic Neuropathy

LHON is a maternally inherited form of subacute, painless loss of central vision and dyschromatopsia that predominantly affects young men. The visual field loss consists initially of an enlarged blind spot and progresses to involve central vision as a large centrocecal scotoma. Symptoms generally start in one eye and occur in the other within weeks or month. Although in most cases only the optic nerve is involved, associated features seen in rare pedigrees include dystonia, myelopathy, ataxia, athetosis, spasticity, peripheral neuropathy, and cardiac conduction abnormalities. No major differences between men and women are noted in the clinical presentations of LHON other than a later onset in women in the same pedigree and the association with a multiple sclerosislike illness in women from families with the 11778 mtDNA mutation.[37]

LHON may serve as a paradigm for homoplasmic missense mutation-mediated mitochondrial diseases. Although complex I is predominantly involved, LHON-associated mtDNA mutations exhibit a great deal of genetic heterogeneity, with mutations in at least eight genes that encode subunits of three respiratory chain complexes. Four mutations, at nucleotide positions 3460, 11778, 14484, and 15257, are considered to have primary pathogenetic importance. Several other mtDNA mutations have been associated with LHON, and their interaction with the primary mutations may be important in the predisposition to the disease.[38,39] Each of the primary LHON-associated mtDNA mutations has a genotype-specific phenotype, with the prognosis for recovery of vision being the most notable differential feature.[40] Most LHON-associated mtDNA mutations are homoplasmic, and all are readily detectable in blood samples.

Mitochondrial Encephalomyopathy, Lactic Acidosis, and Strokelike Episodes

The hallmarks of mitochondrial encephalomyopathy, lactic acidosis, and stroke-like episodes (MELAS syndrome) are acute recurrent focal neurologic deficits that resemble strokes. The lesions have a posterior predominance and thus produce cortical blindness or hemianopia more than hemiparesis. Strokes may be followed by complete recovery, but residual symptoms can occur, as can progressive encephalopathy. Other common symptoms include focal or generalized seizures, growth retardation, dementia, recurrent headaches, and vomiting. Most patients with MELAS harbor a point mutation at nucleotide position (np) 3243 in the tRNA$^{Leu(UUR)}$ gene, although other mutations in the same gene are also described. The 3243 mtDNA mutation is the most pleiotropic of all pathogenetic mtDNA mutations and is associated with a broad spectrum of clinical syndromes, including CPEO, myopathy, dystonia, cardiomyopathy, diabetes mellitus, and deafness.

Myoclonic Epilepsy with Ragged Red Fibers

MERRF is characterized by myoclonus, cerebellar ataxia, and seizures. Associated features include myopathy, deafness, dementia, vascular headache, peripheral neuropathy, cervical lipomas, and pyramidal signs. There is marked variability in the clinical course and prognosis. Some patients have onset of symptoms in early childhood, with a severe course leading to death during the second decade, whereas other family members are asymptomatic or have a milder, later onset. There is some correlation between the proportion of mutant mtDNA and the severity of the clinical manifestations. Two mtDNA point mutations at np 8344 and 8356 in the tRNA^{Lys} gene are associated with MERRF and are readily detectable in blood.

Systemic Manifestations of mtDNA Mutations

Multisystemic involvement is a characteristic feature of mitochondrial diseases. Isolated somatic organ system involvement can also occur and has been increasingly recognized as a phenotypic manifestation of mtDNA mutations. This is now recognized in several organ systems, but the most informative paradigm is diabetes mellitus. Clues for the contribution of mtDNA mutations to the development of diabetes mellitus arose from two observations: (1) diabetes mellitus is frequently associated with a variety of mitochondrial encephalomyopathies and mtDNA mutations, and (2) epidemiological studies show a predominance of maternal transmission of non–insulin-dependent diabetes mellitus (NIDDM). Two large families whose primary clinical manifestations were diabetes mellitus and deafness were reported in 1992. One of the pedigrees was associated with a 10.4 kilobase deletion in the mtDNA[24] and the other with the np 3243 point mutation in the tRNA$^{Leu(UUR)}$ gene.[41] Several other tRNA mutations are associated with multisystemic diseases that have diabetes mellitus as an associated manifestation. Overall, the prevalence of the 3243 mtDNA mutation has been estimated to be 1.5 percent of all diabetics and is about two to five times higher in those with a family history of diabetes mellitus.[42] The prevalence of diabetes mellitus in association with other mtDNA mutations is unknown. Among the other manifestations of mitochondrial disease, an increasing number of mtDNA mutations have been described in association with cardiomyopathy[43-48] and sensorineural hearing loss.[48-51]

Somatic mtDNA Mutations in Aging and Neurodegenerative Diseases

The mutation rate of mtDNA is at least 10 times that of the nuclear genome. mtDNA is particularly susceptible to the mutagenic effects of free radicals because mtDNA is close to the site of free radical formation in the respiratory chain, has few noncoding sequences, lacks protective histones, and has limited DNA repair mechanisms. Aging in humans is associated with an increase in the number of cytochrome c oxidase–deficient fibers in skeletal muscle and a decline in both respiratory rate and oxidative phosphorylation enzyme activities. Evidence for cumulative mtDNA oxidative damage comes from the increased frequency of deletions, point mutations, and oxidized forms of mtDNA with aging. The age-related increase in somatic tissue mtDNA mutations, particularly in postmitotic tissues, may be an important variable in aging and a contributor to cell death in the neurodegenerative diseases.[52]

Mitochondrial dysfunction has been postulated to play a role in a number of common neurodegenerative diseases, including Parkinson's disease, Alzheimer's disease, and Huntington's disease. The selective vulnerability of certain neuronal subpopulations is a shared feature of these neurodegenerative diseases and the classic mitochondrial diseases such as MELAS and

MERRF. Several studies have demonstrated a specific defect in complex I in different tissues in Parkinson's disease, and levels of brain lactate are elevated in Huntington's disease. Preliminary evidence for putative mtDNA mutations in cytochrome *c* oxidase genes in association with Alzheimer's disease has recently been reported.[53] However, these data may be artifactual.

The production of ATP within the mitochondria is accompanied by the production of reactive oxygen species that can cause oxidative damage to mtDNA. In a destructive pattern, impairment of oxidative phosphorylation accelerates the production of free radicals, which in turn cause further impairment of oxidative phosphorylation. Highly oxidative cells within the central nervous system would be particularly sensitive to this oxidative damage and could undergo programmed cell death.

Mitochondria may also play a role in the pathophysiology of neurodegenerative diseases caused by defined nuclear DNA mutations. Friedreich's ataxia is an autosomal recessive disorder that has many clinical features reminiscent of a mitochondrial disorder: ataxia, dysarthria, areflexia, posterior column sensory loss, cardiomyopathy, and diabetes mellitus. The defective protein, frataxin, has recently been shown to localize to the mitochondria and to be involved in the control of oxidative phosphorylation, potentially via an effect on mtDNA stability.[54,55]

Diagnosis and Treatment

A detailed history and physical examination of the patient and relatives may provide evidence of a characteristic pattern of signs and symptoms that is suggestive of a mitochondrial disorder [*see Table 3*]. Discernment of a maternal inheritance pattern may be obscured by incomplete, oligosymptomatic manifestations in maternal relatives. Evidence of multisystemic involvement in the context of a cardinal clinical manifestation of mitochondrial disease heightens clinical suspicion of a mitochondrial disease. Elevated lactate levels in blood or cerebrospinal fluid aid in the diagnosis of a mitochondrial disorder; normal values or inconsistent variable elevations do not rule out mitochondrial dysfunction. Clinical neurophysiological and neuroimaging procedures can provide supportive evidence of mitochondrial disease[56]; for example, electromyography may demonstrate an asymptomatic myopathy, or magnetic resonance imaging may indicate that characteristic signal abnormalities are present in the basal ganglia or elsewhere.

Muscle biopsy is the most helpful diagnostic procedure for the comprehensive evaluation of a mitochondrial encephalomyopathy. An open biopsy is preferable to obtain adequate tissue for histologic, biochemical, and molecular genetic studies. The classic histochemical findings are ragged red fibers demonstrated on Gomori-Wheatley trichrome staining, but the presence of frequent cytochrome *c* oxidase–negative fibers is a more sensitive histochemical marker. Molecular genetic determination is commercially available for the most common mtDNA mutations, and point mutations are usually readily detectable in blood. The absence of a known point mutation in leukocyte DNA does not, however, preclude the diagnosis, and the diagnostic yield can be improved by testing muscle mtDNA. Indeed, muscle mtDNA is required for molecular genetic detection of single and multiple mtDNA deletions.

Table 3 Neurologic and Systemic Manifestations
of Mitochondrial Encephalomyopathies

Neurologic
    Myopathy
    Exercise intolerance
    Recurrent myoglobinuria
    Neuropathy
    Ophthalmoplegia
    Optic neuropathy
    Pigmentary retinopathy
    Nystagmus
    Seizures
    Myoclonus
    Strokelike episodes (posterior predominant)
    Dementia
    Depression
    Vascular headache
    Central hypoventilation
    Sensorineural hearing loss
    Dystonia
    Ataxia
    Myelopathy
Systemic
    Lactic acidosis
    Diabetes mellitus
    Hypothyroidism
    Hypoparathyroidism
    Hypogonadism
    Hyperaldosteronism
    Primary ovarian failure
    Adrenocorticotrophin deficiency
    Short stature
    Cardiomyopathy
    Cardiac conduction defects
    Exocrine pancreas dysfunction
    Hepatopathy
    Intestinal pseudo-obstruction
    Renal tubulopathies
    Pancytopenia
    Sideroblastic anemia
    Corneal opacities and cataracts

Effective treatment of mitochondrial syndromes continues to be limited.[57] Vigilant, prospective monitoring for complications such as diabetes mellitus, deafness, and heart block is important. Current pharmacological treatment options include ubiquinone (coenzyme Q_{10}), idebenone, dichloroacetate, corticosteroids, and antioxidant vitamins.[57] Most of the reported beneficial effects are anecdotal and have not been confirmed by the few limited larger trials. Aerobic conditioning to improve muscular strength and stamina is under investigation.

An intriguing possibility for future treatment is mitochondrial gene therapy, which would involve overcoming an additional set of obstacles compared with nuclear gene therapy, such as targeting to the correct mitochondrial compartment. Mitochondrial gene therapy could focus on the replacement of a defective gene, the enhancement of normal mtDNA function, or the inhibition of mutant mtDNA. The latter strategy was recently employed in vitro in the selective inhibition of replication of mutant, but not wild-type, mtDNA by peptide nucleic acids (nucleic acids that contain a peptide bond and thus are not catabolized within the cell).[58]

References

1. Luft R, Ikkos D, Palmieri G, et al: A case of severe hypermetabolism of nonthyroid origin with a defect in the maintenance of mitochondrial respiratory control: a correlated clinical, biochemical, and morphological study. *J Clin Invest* 41:1776, 1962

2. Holt IJ, Harding AE, Morgan-Hughes JA: Deletions of muscle mitochondrial DNA in patients with mitochondrial myopathies. *Nature* 331:717, 1988

3. Wallace DC, Singh G, Lott MT, et al: Mitochondrial DNA mutation associated with Leber's hereditary optic neuropathy. *Science* 242:1427, 1988

4. Anderson S, Bankier AT, Barrell BG, et al: Sequence and organization of the human mitochondrial genome. *Nature* 290:457, 1981
5. Johns DR: Mitochondrial DNA and disease. *N Engl J Med* 333:638, 1995
6. Giles RE, Blanc H, Cann R, et al: Maternal inheritance of human mitochondrial DNA. *Proc Natl Acad Sci USA* 77:6715, 1980
7. Clayton DA, Doda JN, Friedberg EC: The absence of a pyrimidine dimer repair mechanism in mammalian mitochondria. *Proc Natl Acad Sci U S A* 71:2777, 1975
8. Pettepher CC, LeDoux SP, Bohr VA, et al: Repair of alkali-labile sites within the mitochondrial DNA of RINr 38 cells after exposure to the nitrosourea streptozotocin. *J Biol Chem* 266:3113, 1991
9. Larsson NG, Tulinius MH, Holme E, et al: Segregation and manifestations of the mtDNA tRNA^{Lys} A→G $^{(8344)}$ mutation of myoclonus epilepsy and ragged-red fibers (MERRF) syndrome. *Am J Hum Genet* 51:1201, 1992
10. Jenuth JP, Peterson AC, Fu K, et al: Random genetic drift in the female germline explains the rapid segregation of mammalian mitochondrial DNA. *Nat Genet* 14:146, 1996
11. Marchington DR, Hartshorne GM, Barlow D, et al: Homopolymeric tract heteroplasmy in mtDNA from tissues and single oocytes: support for a genetic bottleneck. *Am J Hum Genet* 60:408, 1997
12. Ciafaloni E, Ricci E, Shanske S, et al: MELAS: clinical features, biochemistry, and molecular genetics. *Ann Neurol* 31:391, 1992
13. MacMillan C, Lach B, Shoubridge EA: Variable distribution of mutant mitochondrial DNAs (tRNALeu[3243]) in tissues of symptomatic relatives with MELAS: the role of mitotic segregation. *Neurology* 43:1686, 1993
14. Larsson NG, Holme B, Kristiansson B: Progressive increase of the mutated mitochondrial DNA fraction in Kearns-Sayre syndrome. *Pediatr Res* 28:131, 1990
15. McShane MA, Hammans SR, Sweeney M, et al: Pearson syndrome and mitochondrial encephalopathy in a patient with a deletion of mtDNA. *Am J Hum Genet* 48:39, 1991
16. Wallace DC, Zheng X, Lott MT, et al: Familial mitochondrial encephalomyopathy (MERRF): genetic, pathophysiological and biochemical characterization of a mitochondrial DNA disease. *Cell* 61:601, 1988
17. Bourgeron T, Rustin P, Chretien D, et al: Mutation of a nuclear succinate dehydrogenase gene results in mitochondrial respiratory chain deficiency. *Nat Genet* 11:114, 1995
18. Suomalainen A, Kaukonen J, Amati P, et al: An autosomal locus predisposing to deletions of mitochondrial DNA. *Nat Genet* 9:146,1995
19. Kaukonen JA, Amati P, Suomalainen A, et al: An autosomal locus predisposing to multiple deletions of mtDNA on chromosome 3p. *Am J Hum Genet* 58:763, 1996
20. Bohlega S, Tanji K, Santorelli SM, et al: Multiple mitochondrial DNA deletions associated with autosomal recessive ophthalmoplegia and severe cardiomyopathy. *Neurology* 46:1329, 1996
21. Nakase H, Moraes CT, Rizzuto R, et al: Transcription and translation of deleted mitochondrial genomes in Kearns-Sayre syndrome: implications for pathogenesis. *Am J Hum Genet* 46:418, 1990
22. Shoubridge EA, Karpati G, Hastings KEM: Deletion mutants are functionally dominant over wild-type mitochondrial genomes in skeletal muscle fiber segments in mitochondrial disease. *Cell* 62:43,1990
23. Chen X, Prosser R, Somonetti S, et al: Rearranged mitochondrial genomes are present in human oocytes. *Am J Hum Genet* 57:239, 1995
24. Ballinger SW, Shoffner JM, Hedaya EV, et al: Maternally transmitted diabetes and deafness associated with a 10.4 kb mitochondrial DNA deletion. *Nat Genet* 1:11, 1992
25. King MP, Attardi G: Human cells lacking mtDNA: repopulation with exogenous mitochondria by complementation. *Science* 246:500, 1989
26. Chomyn A, Meola G, Bresolin N, et al: In vitro genetic transfer of protein synthesis and respiration defects to mitochondrial DNA-less cells with myopathy-patient mitochondria. *Mol Cell Biol* 11:2236, 1991
27. Hayashi J-I, Ohta S, Kikuchi A, et al: Introduction of disease-related mitochondrial DNA deletions into HeLa cells lacking mitochondrial DNA results in mitochondrial dysfunction. *Proc Natl Acad Sci USA* 88:10614, 1991
28. Enriquez JA, Chomyn A, Attardi G: mtDNA mutation in MERRF syndrome causes defective aminoacylation of tRNA^{Lys} and premature translation termination. *Nat Genet* 10:47, 1995
29. Jun AS, Trounce IA, Brown MD, et al: Use of transmitochondrial cybrids to assign a complex I defect to the mitochondrial DNA encoded NADH dehydrogenase subunit 6 gene mutation at nucleotide pair 14459 that causes Leber hereditary optic neuropathy and dystonia. *Mol Cell Biol* 16:771, 1996
30. Hofhaus G, Johns DR, Hurko O, et al: Respiration and growth defects in transmitochondrial lines carrying the 11778 mutation associated with Leber's hereditary optic neuropathy. *J Biol Chem* 271:13155, 1996
31. Prezant TR, Agapian JV, Bohlman MC, et al: Mitochondrial ribosomal RNA mutation associated with both antibiotic-induced and non-syndromic deafness. *Nat Genet* 4:289, 1993

32. Cullom ME, Heher KL, Miller NR, et al: Leber's hereditary optic neuropathy masquerading as tobacco-alcohol amblyopia. *Arch Ophthalmol* 111:1482, 1993

33. Ernster L, Luff R, Orrenius S, Eds: Nobel symposium 90: mitochondrial diseases. *Biochim Biophys Acta* 1271:1, 1995

34. Shoffner JM, Wallace DC: Oxidative phosphorylation diseases. *The Metabolic and Molecular Bases of Inherited Disease,* 7th ed. Scriber CR, Beaudet AL, Sly WS, et al, Eds. McGraw-Hill Book Co, New York, 1995, p 1535

35. Holt IJ, Harding AE, Cooper JM, et al: Mitochondrial myopathies: clinical and biochemical features of 30 patients with major deletions of muscle mitochondrial DNA. *Ann Neurol* 26:699, 1989

36. Moraes CT, DiMauro S, Zeviani M, et al: Mitochondrial DNA deletions in progressive external ophthalmoplegia and Kearns-Sayre syndrome. *N Engl J Med* 320:1293, 1989

37. Harding AE, Sweeney MG, Miller DH, et al: Occurrence of a multiple sclerosis-like illness in women who have a Leber's hereditary optic neuropathy mitochondrial DNA mutation. *Brain* 115:979, 1992

38. Johns DR, Neufeld MJ, Park RD: An ND-6 mitochondrial DNA mutation associated with Leber hereditary optic neuropathy. *Biochem Biophys Res Comm* 187:1551, 1992

39. Brown MD, Sun F, Wallace DC: Clustering of caucasian Leber hereditary optic neuropathy patients containing the 11778 or 14484 mutations on a mtDNA lineage. *Am J Hum Genet* 60:381, 1997

40. Johns DR: Genotype-specific phenotypes in Leber hereditary optic neuropathy. *Clin Neurosci* 2:146, 1994

41. Van den Ouweland JMW, Lemkes HHPJ, Ruitenbeek W, et al: Mutation in the mitochondrial tRNA$^{Leu(UUR)}$ gene in a large pedigree with maternally transmitted type II diabetes and deafness. *Nat Genet* 1:368,1992

42. Gerbitz K-D, van den Ouweland JMW, Maassen JA, et al: Mitochondrial diabetes mellitus: a review. *Biochim Biophys Acta* 1271:253, 1995

43. Casali C, Santorelli FM, D'Amati G, et al: A novel mtDNA point mutation in maternally inherited cardiomyopathy. *Biochem Biophys Res Commun* 213:588,1995

44. Santorelli FM, Mak SH, Vazquez-Acevedo M, et al: A novel mitochondrial DNA point mutation associated with mitochondrial encephalocardiomyopathy. *Biochem Biophys Res Commun* 216:835, 1995

45. Silvestri G, Santorelli FM, Shanske S, et al: A new mtDNA mutation in the tRNALeu (UUR) gene associated with maternally inherited cardiomyopathy. *Hum Mutat* 3:37, 1994

46. Merante F, Tein I, Benson L, Robinson BH: Maternally inherited hypertrophic cardiomyopathy due to a novel T-to-C transition at nucleotide 9997 in the mitochondrial tRNA glycine gene. *Am J Hum Genet* 55:437,1994

47. Merante F, Myint T, Tein I, et al: An additional mitochondrial tRNA^{Ile} point mutation (A-to-G at nucleotide 4295) causing hypertrophic cardiomyopathy. *Hum Mutat* 8:216, 1996

48. Santorelli FM, Mak SC, El-Schahawi M, et al: Maternally inherited cardiomyopathy and hearing loss associated with a novel mutation in the tRNA^{Lys} gene (G8363). *Am J Hum Genet* 58:993, 1996

49. Reid FM, Vernham GA, Jacobs HT: A novel mitochondrial point mutation in a maternal pedigree with sensorineural deafness. *Hum Mutat* 3:243, 1994

50. Vernham GA, Reid FM, Rundle PA, Jacobs HT: Bilateral sensorineural hearing loss in members of a maternal lineage with a mitochondrial point mutation. *Clin Otolaryngol* 19:314, 1994

51. Tiranti V, Chariot P, Carella F, et al: Maternally inherited hearing loss, ataxia and myoclonus associated with a novel point mutation in mitochondrial tRNA Ser(UCN) gene. *Hum Mol Genet* 4:1421, 1995

52. Beal F: Aging, energy, and oxidative stress in neurodegenerative diseases. *Ann Neurol* 38:357, 1995

53. Davis RE, Miller S, Herrnstadt C, et al: Mutations in mitochondrial cytochrome c oxidase genes segregate with late-onset Alzheimer disease. *Proc Natl Acad Sci U S A* 94:526, 1997

54. Koutnikova H, Campuzano V, Foury F, et al: Studies of human, mouse and yeast homologues indicate a mitochondrial function for frataxin. *Nat Genet* 16:345, 1997

55. Wilson RB, Roof DM: Respiratory deficiency due to loss of mitochondrial DNA in yeast lacking the frataxin homologue. *Nat Genet* 16:352, 1997

56. Fadic R, Johns DR: Clinical spectrum of mitochondrial diseases. *Semin Neurol* 16:11,1996

57. Fadic R, Johns DR: Treatment of mitochondrial encephalomyopathies. *Mitochondria and Free Radicals in Neurodegenerative Diseases.* Beal MF, Howell N, Bodis-Wollmer I, Eds. Wiley-Liss, New York, (in press)

58. Taylor RW, Chinnery PF, Turnbull DM, et al: Selective inhibition of mutant human mitochondrial DNA replication *in vitro* by peptide nucleic acids. *Nat Genet* 15:212, 1997

Acknowledgment

Figure 1   Dimitry Schidlovsky.

CHAPTER 16

Epilogue

Joseph B. Martin, M.D., Ph.D.

It seemed appropriate to summarize briefly some of the important advances in neurogenetics that have occurred since the chapters for this volume were submitted. The most seminal of these advances are reviewed by consideration of selected diseases in which genetic analysis continues to contribute new concepts and hypotheses, adding almost weekly to our understanding of the pathogenesis of these disorders.

Alzheimer's Disease

Three genes for early-onset familial Alzheimer's disease (FAD)—amyloid precursor protein (APP), presenilin 1, and presenilin 2—and genetic risk factors associated with late-onset disease, particularly apolipoprotein E4 (APOE4) appear to account for about 50 percent of total genetic susceptibility to AD. Borchelt and coworkers[1] report that transgenic mice bearing mutations in both APP and presenilin 1 develop accelerated deposition of β-amyloid (Aβ). The cellular interaction of these two proteins remains unclear, however.

An appreciation of the full impact of the APOE4 isoform in AD is provided by a meta-analyis of 40 studies,[2] which examined the association between APOE genotype and incidence of AD by age and sex in various ethnic and racial groups. The results show definitively that the E4 allele is a major risk factor for AD in all groups and has an effect across ages between 40 and 90 years. Modest racial differences were noted, with African Americans and Hispanics showing less strongly associated risk. This analysis also shows the striking impact on incidence of AD of the rare E4/E4 homozygous genotype in both men and women.

The authors succinctly summarize current speculation regarding neurotoxicity in AD. Cell death "may result from a concatenation of factors such as enhanced cellular processing of APP, releasing amyloidogenic Aβ peptides; interaction of Aβ with cell membranes, both at specific cell-binding sites and, presumably, by directing action on the membrane itself; non-enzymatic glycation of macromolecules whose turnover is delayed in AD brain, producing modified structures that can form cross-links and generate reactive oxygen intermediates; and abnormality in the protective response to oxidant stress and/or susceptibility to apoptotic stimuli."[2]

Researchers continue to search the genome for other risk factors. In a large interinstitutional study, investigators examined two subsets of families with increased incidence of AD[3] (subset 1: 16 families, 135 total members, 52 of whom had AD; subset 2: 38 families with 216 members, 89 with AD). A complete genomic screen using 280 markers pointed to a new locus on chromosome 12, with the suggestion of additional loci on chromosomes 4, 6, and 20. No clues exist at present to suggest what the gene or genes are. Another report suggests synergy between the K isoform for butyryl cholinesterase on chromosome 3 and APOE4 in late-onset AD.[4]

A new entry in the contest for primary factors involved in AD is a protein identified by the powerful yeast two-hybrid system. This technique selects for proteins of unknown type that interact with a known protein, in this case APP. The new protein, labeled ERAB (for endoplasmic reticulum-associated binding protein),[5] was identified in four positive clones (one from human brain and three from a HeLa cell DNA library). ERAB, a 262 amino acid protein, shares structural homology to a family of alcohol dehydrogenases, including hydroxysteroid dehydrogenase and acetoactyl coenzyme A reductase, enzymes involved in cholesterol biosynthesis. ERAB, as the name indicates, is found in the endoplasmic reticulum. Lacking a signal sequence, it normally stays there, but in concert with Aβ, it can be translocated to the inner aspect of the plasma membrane.[6] Intense ERAB reactivity is found associated with senile plaques in AD brains. The authors hypothesize that excessive production of Aβ (as occurs in all the genetic lesions identified to date) may contribute to aberrant translocation of ERAB to the cell membrane, where deleterious actions may lead to neural cell death.

How might these effects occur? A surprising discovery by Tienari and coworkers[7] points to an action of Aβ in axonal sorting. Is it possible that Aβ, through interaction with a sorting receptor, is passaged into axonal transport vesicles? ERAB, acting as an intracellular Aβ binding protein, may, in the presence of a mutation in Aβ or the presenilins, lead to aberrant intracellular trafficking of APP or of both AP and ERAB, with cytotoxic effects.[5,6]

These observations in AD research point to the fundamental fact in cell biology that no protein acts alone. Through an appreciation of the cascade of events involving the culprits identified, we will eventually understand the interactions of APP, APOE4, presenilins, ERAB, and doubtless many others in the pathogenesis of AD. The development of effective treatments may need to await clarification of these interactions.

Trinucleotide Repeat Disorders

The search continues for the link between the elongated polyglutamine tract in the CAG repeat disorders and the selective neurotoxicity characteristic of each. Recent studies in Huntington's (HD) disease have focused attention on intranuclear inclusions of the ubiquinated huntingtin protein.[8-10] Observations in transgenic mice described by Young [*see Chapter 3*] have been extended to humans in elegant studies by DiFiglia and associates.[10] Using antibodies to the N-terminal region of mutant huntingtin, they demonstrated positive staining, neuronal intranuclear inclusions, and labeling of dystrophic neurites in HD cortex and striatum. There was a significant correlation between polyglutamine length and extent of huntingtin accumulation in these structures. Ubiquitin staining was found in each location, leading to the suggestion that abnormal huntingtin is targeted for proteolysis (by ubiquitinization) but is resistant to removal. Because these observations conform closely to the sites of neuronal loss in HD (striatum and cortex), the authors suggest this defect may be part of the pathogenic mechanism of cell death in HD. These observations in mice and humans are the first to provide evidence for regional selective neuronal abnormalities that correspond to the neuropathologic selectivity found in HD.

In spinocerebellar ataxia type 1 (SCA1), Skinner and coworkers[11] examined the subcellular distribution of human ataxin-1 (encoded by the *SCA1* gene). Ataxin-1 localizes to Purkinje cell nuclei in a rather uniform distribution pattern, whereas in mutant ataxin-1 (with polyglutamine tract expanded), localization is restricted to a single nuclear structure. Only affected neurons show this abnormality; other regions of the brain are normal. COS-1 cells transfected with normal or mutant ataxin-1 also showed a nuclear localization. The ataxins associate with a nuclear matrix-associated protein. Ataxin-1 interacts with a specific cerebellar leucine-rich nuclear protein thatforms part of the nuclear matrix-associated subnuclear structure.[12] It is hypothesized that mutant ataxin-1 protein, as a result of the elongated polyglutamine tract, interacts more avidly with this protein complex and that the effects are specific to those neurons that degenerate.

Intranuclear inclusions have also been identified in SCA3 (Machado-Joseph disease).[13] How do these intranuclear bodies form, and how do they lead to neuronal degeneration? Max Perutz has predicted that they may form β sheets similar to those found with Aβ in AD or in the prion diseases.[14] Lunkes and Mandel[15] point out that the CAG repeat disorders have "... joined a group of degenerative diseases caused by aberrant deposition of protein aggregates in the brain."

A recent report adds another likely candidate to the list of CAG repeat disorders: autosomal dominant pure spastic paraplegia, linked to chromosome 2p21-24.[16] The list has now grown to nine disorders.[15]

Motor Neuron Disease: Amyotrophic Lateral Sclerosis and Spinal Muscular Atrophy

In the continuing search for the role of mutations in the copper/zinc superoxide dismutase (SOD1) gene in causing neural cell death, Kostic and coworkers[17] examined transgenic mice bearing such mutations for factors that affect motor neuron survival. Mice with overexpression of the proto-oncogene *bcl-2* demonstrated delayed onset of motor neuron disease accompanied by prolonged survival, although the manipulation did not affect the duration of the disease. Transgenic mice with *bcl-2* overexpression showed strong immunoreactivity in all motor neurons of the spinal cord. At death, heterozygous mice with the mutant SOD1 gene (glycine substituted by alanine at position 93) showed a 50 percent loss of motor neurons in the spinal cord, accompanied by comparable reduction in myelinated axons in the phrenic nerve, and severe muscle atrophy. In contrast, the transgenic mice with SOD1 mutations crossed with *bcl-2* overexpression showed reduction in each of these three effects. Bcl-2 is a potent inhibitor of apoptotic cell death. These observations in transgenic mice are consistent with the notion that cells in SOD1 mutant mice, and presumably in humans with amyotrophic lateral sclerosis (ALS), die via apoptosis and are partially protected by Bcl-2. It is known that Bcl-2 has potent antioxidant properties, which may explain its beneficial effects.

Another set of observations also suggests that this hypothesized apoptotic mechanism of cell death can be modified by genetic manipulations.[18] Interleukin-1β converting enzyme (ICE) is an important cell death gene. Inhibition of ICE activity in SOD1 mutant transgenic mice more than doubled survival time, from 11 to 27 days, although the onset of the disease was similar to that of control subjects. The authors speculate that inhibition of ICE might be of value in the treatment of ALS.

Parkinson's Disease

The importance of genetic factors in Parkinson's disease (PD) has only recently been fully appreciated. A few large pedigrees of autosomal dominant PD have been delineated and linkage analysis applied. The most exciting development is the discovery of a mutation in the gene that encodes a synaptic protein, α-synuclein, in four families: one of Italian extraction, and three (unrelated) of Greek origin.[19]

After linkage to chromosome 4q21-23, candidate genes in the region were examined. α-Synuclein is a member of a diverse family of synaptic proteins derived by alternative splicing of a single gene. Synuclein is concentrated in presynaptic regions in close proximity to synaptic vesicles. An alanine-to-threonine mutation in position 53 of the 140 amino acid protein was identified. Whether the mutation causes a loss or a gain of function is unknown. Synuclein is not distributed uniquely to cells that die in PD (i.e., substantia nigra). Hence, another factor or explanation for selective cell loss must be found to account for the regional selectivity of neuronal death in PD. Synuclein is found in the Lewy bodies, the characteristic cellular signature of PD. Spillantini and coworkers[20] speculate that α-synuclein may be the main component of the Lewy body; however, no quantitative data have been reported in support of this hypothesis.

Torsion Dystonia

The hereditary dystonias consist of at least six different gene loci: (1) early-onset torsion dystonia at 9q34, (2) dopa-responsive dystonia at 14q21-22, (3) paroxysmal dystonia on 2q, (4) a mixed phenotype on chromosome 8, (5) late-onset focal dystonia on 18p, and (6) Lubag dystonia-parkinsonism on Xq13.1.[21] Ozelius and coworkers[22] recently identified the likely mutation in early-onset torsion dystonia after nearly a decade of work involving many centers and supported by a substantial amount of federal and private funding. The disease is autosomal dominant with 30 to 40 percent penetrance. Ashkenazi Jews are particularly susceptible because of the so-called founder effect, but the disease also occurs in other populations.

The gene encodes a protein called torsinA that by structural homology appears to be an adenosine triphosphate binding protein. A GAG deletion was found in 64 Ashkenazi pedigrees and in 88 other affected individuals from four non-Ashkenazi families. The GAG deletion causes full phenotypic expression in the heterozygote state. The finding of a single genetic alteration in all subjects tested will greatly facilitate genetic testing in susceptible individuals.

Prion Diseases

The announcement that Stanley Prusiner was awarded the 1997 Nobel Prize in Physiology or Medicine in recognition of his outstanding contributions has not entirely settled the continuing controversy surrounding his hypothesis of a prion-protein–only pathogenesis for the prion diseases.[23] Presciently, the October 10 issue of *Science*, which reported the persisting controversy following the awarding of the Nobel Prize, also contains a superb review by Prusiner of the prion diseases and bovine spongiform encephalopathy.[24] The review complements Chapter 10 by Prusiner in this volume.

Those of us who have followed this emerging work over the past 20 years eagerly await a final proof for the hypothesis. As reviewed by Aguzzi and Weissman,[25] "... the holy grail of the protein-only faction is to produce prions in vitro from components that have never been in contact with the infectious agent—that is, from PrP made in healthy eukaryotic cell lines, *Escherichia coli*, or, better still, chemically from amino acids." The identity and function of protein X [*see Chapter 10*], speculated by Prusiner to have a critical role in spongiform encephalopathy, remains unknown.

References

1. Borchelt D, Ratovitski T, van Lare J, et al: Accelerated amyloid deposition in the brains of transgenic mice coexpressing mutant presenilin 1 and amyloid precursor proteins. *Neuron* 19:939, 1997

2. Farrer L, Cupples L, Haines J, et al: Effects of age, sex, and ethnicity on the association between apolipoprotein E genotype and Alzheimer disease. *JAMA* 278:1349, 1997

3. Pericak-Vance M, Bass M, Yamaoka L, et al: Complete genomic screen in late-onset familial Alzheimer disease. *JAMA* 278:1237, 1997

4. Lehmann D, Johnston C, Smith A: Synergy between the genes for butyrylcholinesterase K variant and apolipoprotein E4 in late-onset confirmed Alzheimer's disease. *Hum Mol Genet* 6:1933, 1997

5. Yan S, Fu J, Soto C, et al: An intracellular protein that binds amyloid-β peptide and mediates neurotoxicity in Alzheimer's disease. *Nature* 389:689, 1997.

6. Beyreuther K, Masters C: The ins and outs of amyloid-β. *Nature* 289:677, 1997.

7. Tienari P, De Strooper B, Ikonen E, et al: The β-amyloid domain is essential for axonal sorting of amyloid precursor protein. *EMBO J* 15:5218, 1996

8. Davies S, Turmaine M, Cozens B, et al: Formation of neuronal intranuclear inclusions underlies the neurological dysfunction in mice transgenic for the HD mutation. *Cell* 90:537, 1997

9. Scherzinger E, Lurz R, Turmaine M, et al: Huntingtin-encoded polyglutamine expansions form amyloid-like protein aggregates in vitro and in vivo. *Cell* 90:549, 1997

10. DiFiglia M, Sapp E, Chase K, et al: Aggregation of huntingtin in neuronal intranuclear inclusions and dystrophic neurites in brain. *Science* 277:1990, 1997

11. Skinner P, Koshy B, Cummings C, et al: Ataxin-1 with an expanded glutamine tract alters nuclear matrix-associated structures. *Nature* 389:971, 1997

12. Matilla A, Koshy B, Cummings C, et al: The cerebellar leucine-rich acidic nuclear protein interacts with ataxin-1. *Nature* 389:974, 1997

13. Paulson H, Perez M, Trottier Y, et al: Intranuclear inclusions of expanded polyglutamine protein in spinocerebellar ataxia type 3. *Neuron* 19:333, 1997

14. Stott K, Blackburn J, Butler P, et al: Incorporation of glutamine repeats makes protein oligomerize: implications for neurodegenerative diseases. *Proc Natl Acad Sci USA* 92:6509, 1995

15. Lunkes A, Mandel J-L: Polyglutamines, nuclear inclusions and neurodegeneration. *Nature Med* 3:1201, 1997

16. Nielsen J, Koefoed P, Abell K, et al: CAG repeat expansion in autosomal dominant pure spastic paraplegia linked to chromosome 2p21-p24. *Hum Mol Genet* 6:1811, 1997

17. Kostic V, Jackson-Lewis V, de Bilbao F, et al: Bcl-2: prolonging life in a transgenic mouse model of familial amyotrophic lateral sclerosis. *Science* 277:559, 1997

18. Friedlander R, Brown R, Gagliardini V, et al: Inhibition of ICE slows ALS in mice. *Nature* 388:31, 1997

19. Polymeropoulos M, Lavedan C, Leroy E, et al: Mutation in the α-synuclein gene identified in families with Parkinson's disease. *Science* 276:2045, 1997

20. Spillantini M, Schmidt M, Lee V, et al: α-synuclein in Lewy bodies. *Nature* 388:839, 1997

21. Trifiletti R: Identification of the early-onset torsion dystonia gene. *Neurol Alert* 16:22, 1997

22. Ozelius L, Hewett J, Page C, et al: The early onset torsion dystonia gene (DYT1) encodes an ATP-binding protein. *Nature Genet* 17:40, 1997

23. Vogel G: Prusiner recognized for once-heretical prion theory. *Science* 278:214, 1997

24. Prusiner S: Prion diseases and the BSE crisis. *Science* 278:245, 1997

25. Aguzzi A, Weissmann C: Prion research: the next frontiers. *Nature* 389:795, 1997

Appendix

Molecular Definition of Defects in Neurogenetic Disorders

Name of Disorder	Inheritance Mode	Chromosomal Location	Locus Name*	OMIM Number†	Neurologic/ Neuromuscular Phenotype‡	Nature of Genetic Defect
Adrenoleukodystrophy X-linked	XR	Xq28	ALD	300100	Degenerative neurologic disorder, spastic paraplegia, peripheral neuropathy	Inactivation of a peroxisomal membrane transporter-like gene
Adrenoleukodystrophy, neonatal	AR	12p13	NALD (PXR1)	202370	Mental retardation, seizures	Inactivation of peroxisome receptor (PTS1 receptor)
Alzheimer's disease	AD	21q21.3-q22.05	AD1 (APP)	104760	Dementia, memory loss, early onset	Altered amyloid protein precursor
Alzheimer's disease	Complex	19q13.2	AD2 (APOE)	107741	Dementia, memory loss, late onset	Allele 4 of the apolipoprotein E gene
Alzheimer's disease	AD	14q24.3	AD3	104311	Dementia, memory loss, early onset	Altered presenilin 1 gene
Alzheimer's disease	AD	1q31-q42	AD4	600759	Dementia, memory loss, early onset	Altered presenilin 2 gene
Amyotrophic lateral sclerosis	AD	21q22.1	ALS1 (SOD1)	105400	Progressive motor function loss, lower motor neuron manifestations	Altered superoxide dismutase 1
Aniridia	AD	11p13	AN2 (PAX6)	106210	Optic nerve hypoplasia	Mutation in paired homeo box gene 6

This table presents representative disorders that have been investigated using positional cloning or candidate gene approaches and does not list all neurogenetic disorders for which a primary defect is known.

AD—autosomal dominant, AR—autosomal recessive, XR—X-linked recessive, XD—X-linked dominant, M—mitochondrial.

*When a gene has been assigned a locus symbol by the Human Genome Organization based on its protein product, that preferred symbol is given in brackets after any symbol based on the disease name.

†The number refers to the reference entry in OMIM (Online Mendelian Inheritance in Man), on the World Wide Web at http://www.ncbi.nlm.nih.gov/Omim/.

‡Although many disorders include other phenotypes that are more characteristic, only neurologic/neuromuscular phenotypes are listed here.

Molecular Definition of Defects in Neurogenetic Disorders *continued*

Name of Disorder	Inheritance Mode	Chromosomal Location	Locus Name*	OMIM Number†	Neurologic/ Neuromuscular Phenotype‡	Nature of Genetic Defect
Ataxia telangiectasia	AR	11q22.3	ATM	208900	Cerebral ataxia, oculomotor apraxia, dystonia	Mutation in a gene for a protein with domains similar to phosphatidylinositol 3′ kinase, a signal transduction mediator, and rad3, a DNA repair/cell cycle regulator
Ataxia with vitamin E deficiency	AR	8q13.1-q13.3	AVED (ITPA)	277460	Spinocerebellar ataxia, areflexia, proprioception loss	Mutation in the gene for α-tocopherol transfer protein
Basal cell nevus syndrome	AD	9q22.3	BCNS (PTCH)	109400	Mental retardation, medulloblastoma	Mutation in the human homologue of the *Drosophila* patched gene
Dementia, hereditary multi-infarct	AD	19p13.1	CADASIL (NOTCH3)	125310	Relapsing strokes, dementia, seizures, depression, manic episodes, motor disability	Mutation in the human homologue of a *Drosophila* notch receptor
Central core disease (allelic with MHS3)	AD	19q13.1	CCD (RYR1)	117000	Slowly progressive muscle weakness, muscle cramps after exercise, myopathy, neonatal hypotonia	Mutation in gene for ryanodine receptor 1, the calcium-release channel of skeletal muscle sarcoplasmic reticulum
Charcot-Marie-Tooth disease 1A (allelic with DSDA)	AD	17p11.2	CMT1A (PMP22)	118220	Muscle atrophy and weakness, weak or absent deep tendon reflexes, sensory defects, chronic sensorineural polyneuropathy	Mutations in the gene for peripheral myelin protein 22
Charcot-Marie-Tooth disease 1B (allelic with DSDB)	AD	1q22	CMT1B (MPZ)	118200	Muscle atrophy and weakness, weak or absent deep tendon reflexes, sensory defects, chronic sensorineural polyneuropathy	Mutations in the myelin protein zero gene
Charcot-Marie-Tooth disease 1, X-linked	XR	Xq13.1	CMTX1 (GJB1)	302800	Muscle atrophy and weakness, neuropathy, areflexia, sensory loss	Mutations in the gene for gap junction protein connexin 32
Choroideremia	XR	Xq21.2	CHM	303100	Choroidoretinal degeneration, progressive vision loss	Inactivation of Rab escort protein 1
Congenital stationary night blindness (allelic with autosomal recessive retinitis pigmentosa)	AD	4p16.3	PDE6B	165300	Night blindness	Mutation in the human homologue of the mouse *rd* (retinal degeneration) gene encoding a cGMP phosphodiesterase

Molecular Definition of Defects in Neurogenetic Disorders   *continued*

Name of Disorder	Inheritance Mode	Chromosomal Location	Locus Name*	OMIM Number†	Neurologic/ Neuromuscular Phenotype‡	Nature of Genetic Defect
Creutzfeldt-Jakob disease (allelic with *GSS* and *FFI*)	AD	20pter-p12	*CJD (PRNP)*	123400	Chronic dementia, trembling hand movements, unsteady gait, myoclonus, extrapyramidal muscular rigidity	Mutation in the prion protein gene
Dentatorubropallido-luysian atrophy	AD	2p13.31	*DRPLA*	125370	Myoclonus epilepsy, dementia, ataxia, choreoathetosis	Expanded CAG repeat in atrophin 1 gene
Dystonia, early onset, generalized	AD	9q34	*DYT1*	128100	Involuntary trunk or neck posturing, torsion dystonia	Loss of a glutamate codon in a novel gene
Dystonia, dopa responsive, with diurnal variation	AD	14q22.1-q22.2	*DYT5 (GCH1)*	128230	Dystonic posture/ movement, parkinson-ism, dopa responsive, alleviated by sleep	Mutation in the gene for GTP cyclohydrolase I
Dystonia, dopa responsive	AR	11p15.5	*TH*	191290	Dopa-responsive dystonia	Mutation in tyrosine hydroxylase gene
Epilepsy, nocturnal frontal lobe type	AD	20q13.2-q13.3	*CHRNA4*	121200	Nocturnal frontal lobe epilepsy	Altered α-polypep-tide-4 of the nico-tinic cholinergic receptor
Epilepsy, progressive myoclonus 1	AR	21q22.3	*EPM1 (STFB)*	254800	Generalized tonic-clonic seizures, proxi-mal limb muscle myoclonus, dementia	Inactivation of cysteine protease cystatin B (stefin B)
Episodic ataxia 1	AD	12p13 (KCNA1)	*EA1*	160120	Myokymia, continu-ous muscle move-ment, periodic ataxia	Altered potassium channel
Episodic ataxia 2 (allelic with *MHP* and *SCA6*)	AD	19p13	*EA2 (CACNL1A4)*	108500	Episodic ataxia, cere-bellar ataxia, vertigo	Altered calcium channel
Fatal familial insomnia (allelic with *GSS* and *CJD*)	AD	20pter-p12	*FFI (PRNP)*	600072	Progressive insomnia, dysautonomia, pyrexia	Mutation in the prion protein gene
Fragile X syndrome	XR	Xq27.3	*FMR1*	309550	Severe mental retardation	Inactivation of the gene for a novel RNA binding pro-tein, usually by an expanded CGG repeat in the 5′ untranslated region
Friedreich's ataxia	AR	9q13-q21.1	*FRDA*	229300	Cerebellar ataxia	Inactivation of the gene for frataxin, a novel mitochondrial protein, sometimes by expanded GAA repeat in an intron
Gerstmann-Sträussler disease (allelic with *CJD* and *FFI*)	AD	20pter-p12	*GSD (PRNP)*	137440	Cerebellar ataxia, progressive dementia, absent leg reflexes	Mutation in the prion protein gene

Molecular Definition of Defects in Neurogenetic Disorders   *continued*

Name of Disorder	Inheritance Mode	Chromosomal Location	Locus Name*	OMIM Number†	Neurologic/ Neuromuscular Phenotype‡	Nature of Genetic Defect
Glycerol kinase deficiency	XR	Xp21.3-p21.2	GK	307030	Mental retardation, acidotic stupor and coma	Inactivation of gene for glycerol kinase
Haw River syndrome (allelic with DRPLA)	AD	2p13.31	DRPLA	140340	Ataxia, seizures, chorea, progressive dementia	Expanded CAG repeat in atrophin 1 gene
Hemiplegic migraine (allelic with EA2 and SCA6)	AD	19p13	MHP (CACNL1A4)	141500	Hemiplegic migraine	Mutation in calcium channel
Hereditary hemorrhagic telangiectasia (Osler-Weber-Rendu disease 1)	AD	9q34.1	ORW1 (ENG)	187300	Cerebral arteriovenous fistula, migraine headaches	Altered endoglin
Hereditary hemorrhagic telangiectasia (Osler-Weber-Rendu disease 2)	AD	12q11-q14	ORW2 (ACVRL1)	600376	Cerebral arteriovenous fistula, migraine headaches	Altered activin receptor–like kinase 1
Huntington's disease	AD	4p16.3	HD	143100	Chorea, dementia, depression	Expanded CAG repeat in the gene for huntingtin
Hydrocephalus due to congenital stenosis of aqueduct of Sylvius	XR	Xq28	L1CAM	307000	Mental retardation, spastic paraplegia, aqueductal stenosis, hydrocephalus	Mutation in the gene for L1 neural cell adhesion protein
Hyperekplexia	AD	5q33.2-q33.3	GLRA1	149400	Flexor hypertonia, seizures, exaggerated startle response	Mutation in gene for α_1-subunit of the glycine receptor
Hyperkalemic periodic paralysis (allelic with PMC)	AD	17q23.1-q25.3	HYPP (SCN4A)	170500	Periodic paralysis associated with increased potassium	Alteration of muscle sodium channel α-subunit
Hypokalemic periodic paralysis	AD	1q32	HOPP (CACNL1A3)	170400	Periodic paralysis associated with low potassium	Alteration of muscle dihydropyridine (DHP)-sensitive calcium channel α_1-subunit
Hypertrophic neuropathy of Dejerine-Sottas (allelic with CMT1B)	AD	1q22	DSDB (MPZ)	145900	Hypertrophic neuropathy, distal sensory change and muscle weakness, hyporeflexia	Mutation in myelin protein zero gene
Hypertrophic neuropathy of Dejerine-Sottas (allelic with CMT1A)	AD	17p11.2	DSDA (PMP22)	145900	Hypertrophic neuropathy, distal sensory change and muscle weakness, hyporeflexia	Mutation in peripheral myelin protein 22 gene
Kallmann's syndrome	XR	Xp22.3	KAL	308700	Olfactory lobe agenesis, mirror hand movements, ataxia	Mutation in gene for novel protein with fibronectin-like domains
Leber's optic atrophy	M DNA	mitochondrial	LHON	535000	Optic atrophy, sudden central visual loss, headaches at onset, early onset dystonia, bilateral basal ganglia lesions	Missense mutation in complex I, III, and IV polypeptides

Molecular Definition of Defects in Neurogenetic Disorders *continued*

Name of Disorder	Inheritance Mode	Chromosomal Location	Locus Name*	OMIM Number†	Neurologic/ Neuromuscular Phenotype‡	Nature of Genetic Defect
Limb girdle muscular dystrophy 2A	AR	15q15.1-q21.1	LGMD2A (CAPN3)	253600	Pelvic and shoulder girdle muscle wasting	Mutations in the gene for the protease calpain
Limb girdle muscular dystrophy 2C	AR	13q12	LGMD2C	253700	Calf muscle pseudo-hypertrophy, limb girdle and truncal muscle wasting	Mutations in gene for γ-sarcoglycan
Limb girdle muscular dystrophy 2D	AR	17q12-q21.33	LGMD2D (ADL)	600119	Pelvic and shoulder girdle muscle wasting (adhalin)	Mutations in gene for α-sarcoglycan
Limb girdle muscular dystrophy 2E	AR	4q12	LGMD2E	600900	Pelvic and shoulder girdle muscle wasting	Mutations in gene for β-sarcoglycan
Limb girdle muscular dystrophy 2F	AR	5q33	LGMD2F	601287	Pelvic and shoulder girdle muscle wasting	Mutations in gene for δ-sarcoglycan
Lowe's oculocerebrorenal syndrome	XR	Xq26.1	OCRL	309000	Hypotonia, reduced deep tendon reflexes, mental retardation	Mutations in gene for a novel protein related to inositol polyphosphate 5-phosphatase II
MASA syndrome (allelic with SPG1 and hydrocephalus due to congenital stenosis of aqueduct of Sylvius)	XR	Xq28	L1CAM	303350	Mental retardation, aphasia, hyperactive leg deep tendon reflexes	Mutation in the gene for L1 neural cell adhesion protein
Machado-Joseph disease	AD	14q24.3-q32	MJD1 (SCA3)	109150	Ataxia, parkinsonian features, dystonia	Expanded CAG repeat in gene for novel protein
Malignant hyperthermia 1 (allelic with CCD)	AD	19q13.1	MHS1 (RYR1)	145600	Hyperthermia, myopathy	Mutation in gene for ryanodine receptor 1, the calcium-release channel of skeletal muscle sarcoplasmic reticulum
Malignant hyperthermia 3	AD	7q21-q22	MHS3 (CACNL2A)	154276	Hyperthermia, myopathy	Mutation in gene for skeletal muscle L-type voltage-dependent calcium channel α_2-/δ-subunit
MELAS syndrome	M	Mitochondrial DNA	MTTL1A	540000	Episodic sudden headache, intermittent migraine, grand mal seizures, hemiparesis, strokelike episodes dementia, encephalopathy, sensorineural deafness, blindness, myopathy	Mutation in the mitochondrial leucine tRNA (Leu-UUR) gene
Menkes' disease	XR	Xq12-q13 (ATP7A)	MNK	309400	Mental retardation, spastic diplegia	Mutation in the gene encoding Cu^{2+}-transporting ATPase, α-polypeptide

Molecular Definition of Defects in Neurogenetic Disorders   *continued*

Name of Disorder	Inheritance Mode†	Chromosomal Location	Locus Name*	OMIM Number†	Neurologic/ Neuromuscular Phenotype‡	Nature of Genetic Defect
Muscular dystrophy, Duchenne and Becker types	XR	Xp21.2	DMD	310200	Hypotonia, waddling gait, hyporeflexia, mild mental retardation	Mutation in the gene for dystrophin
Muscular dystrophy, Emery-Dreifuss type	XR	Xq28	STA	310300	Muscular dystrophy, waddling gait, toe walking, mental retardation	Mutations in the gene for emerin
Myoclonus epilepsy with ragged-red fibers	M	Mitochondrial DNA	MTTK	545000	Myoclonus epilepsy, ataxia, spasticity, muscle weakness, myopathy, sensori-neural hearing loss	Mutation in the mitochondrial lysine tRNA
Myopathy, benign congenital with contractures	AD	21q22.3	COL6A1 or COL6A2	158810	Benign congenital myopathy, proximal muscle weakness and atrophy	Mutation in the collagen 6A1 or 6A2 gene
Myotonic dystrophy	AD	19q13.2-q13.3	DM	160900	Myotonia, muscle wasting, swallowing and speech disabilities, polyneuropathy	Expanded CTG repeat in the 3′ untranslated region of the gene for myotonin-protein kinase
Myotubular myopathy 1, X-linked	XR	Xq28	MTM1	310400	Neonatal hypotonia, facial diplegia, ptosis, strabismus, external ophthalmoplegia, hyporeflexia, seizures	Mutations in the gene for myotubularin, a probable tyrosine phosphatase
Nemaline myopathy	AD	1q22-q23 (TPM3)	NEM1	161800	Nemaline myopathy, neonatal hypotonia	Mutation in the tropomyosin-3 gene
Neurofibromatosis type 1	AD	17q11.2	NF1	162200	Neurofibroma, neurofibrosarcoma, optic glioma, learning disability	Inactivation of the gene for neurofibro-min, a modulator of Ras-mediated signal transduction
Neurofibromatosis type 2	AD	22q12.2	NF2	101000	Bilateral vestibular schwannoma, meningioma, spinal schwannoma	Inactivation of the gene for merlin, a novel member of the ERM family of cytoskeleton-membrane linker proteins
Neuronal ceroid lipofuscinosis 1	AR	1p32	CLN1 (PPT)	256730	Mental retardation, loss of speech, infantile onset of psychomotor deterioration	Mutation in the gene for palmitoyl-protein thioesterase
Neuronal ceroid lipofuscinosis 3 (Batten disease)	AR	16p12.1	CLN3	204200	Slow loss of intellect, seizures, psychosis, visual loss, juvenile onset	Inactivation of a novel gene of unknown function
Neurosensory nonsyndromic recessive deafness 1	AR	13q12	GJB2	220290	Deafness	Mutation in gene for gap junction pro-tein connexin 26

Molecular Definition of Defects in Neurogenetic Disorders *continued*

Name of Disorder	Inheritance Mode	Chromosomal Location	Locus Name*	OMIM Number†	Neurologic/ Neuromuscular Phenotype‡	Nature of Genetic Defect
Norrie's disease	XR	Xp11.4	NDP	310600	Retinopathy, mental retardation	Inactivation of a novel gene for a possible growth factor
Paramyotonia congenita (allelic with HYPP)	AD	17q23.1-q25.3	PMC (SCN4A)	168300	Myotonia, increased by cold exposure	Alteration of muscle sodium channel α-subunit
Pelizaeus-Merzbacher disease (allelic with SPG2)	XR	Xq22	PLP	312080	Neonatal hypotonia, rotary head movements, ataxia, spasticity, dementia, rigidity, tremor	Mutation in the myelin proteolipid protein gene
Retinitis pigmentosa, autosomal recessive	AR	4p16.3	PDE6B	268000	Retinitis pigmentosa	Mutation in the human homologue of the mouse rd (retinal degeneration) gene encoding a cGMP phosphodiesterase
Retinitis pigmentosa	AD and AR	3q21-q24	RP4 (RHO)	180380	Retinitis pigmentosa	Mutation in the rhodopsin gene
Retinitis pigmentosa, peripherin-related	AD	6p21.1-cen	RP7 (RDS)	179605	Retinal dystrophy, adult-onset retinitis pigmentosa	Mutation in the human homologue of the mouse rds (retinal degeneration slow) gene
Retinitis pigmentosa, X-linked	XR	Xp21.1	RP3	312610	Choroidoretinal degeneration, pigmentary retinopathy	Mutation in gene for novel protein similar to the guanine nucleotide exchange factor of the Ras-like GTPase Ran
Spastic paraplegia 1, X-linked (allelic with MASA syndrome and hydrocephalus due to congenital stenosis of aqueduct of Sylvius)	XR	Xq28	SPG1 (L1CAM)	312900	Spastic paraplegia, ataxia	Mutation in the gene for L1 neural cell adhesion protein
Spastic paraplegia 2, X-linked (allelic with Pelizaeus-Merzbacher disease)	XR	Xq22	SPG2 (PLP)	312920	Spastic paraplegia, hyperreflexia, spastic gait	Mutation in the myelin proteolipid protein gene
Spinal muscular atrophy	AR	5q12.2-q13.3 (SMN1, NAIP)	SMA1	253300	Muscle atrophy and weakness, hypotonia, decreased spontaneous activity, and deep tendon reflexes	Mutation in the survival of motor neuron gene and often in an adjacent gene encoding neuronal apoptosis inhibitor
Spinobulbar muscular atrophy (Kennedy's syndrome)	XD	Xq12	SMAX1 (AR)	313200	Bulbar signs, facial fasciculations, dysphagia, lower-motor and primary sensory neuronopathy, muscular atrophy	Expanded CAG repeat in the gene for the androgen receptor

Molecular Definition of Defects in Neurogenetic Disorders   *continued*

Name of Disorder	Inheritance Mode	Chromosomal Location	Locus Name*	OMIM Number†	Neurologic/ Neuromuscular Phenotype‡	Nature of Genetic Defect
Spinocerebellar ataxia 1	AD	6p23	SCA1	164400	Spinocerebellar ataxia	Expanded CAG in the gene for a novel nuclear protein, ataxin 1
Spinocerebellar ataxia 2	AD	12q24	SCA2	183090	Spinocerebellar ataxia	Expanded CAG in the gene for a novel protein, ataxin 2
Spinocerebellar ataxia 3 (allelic with Machado-Joseph disease)	AD	14q24.3-q32	SCA3	183085	Spinocerebellar ataxia	Expanded CAG repeat in the gene for a novel protein
Spinocerebellar ataxia 6 (allelic with *MHP* and *EA2*)	AD	19p13	SCA6 (CACNL1A4)	183086	Spinocerebellar ataxia, dementia	Expanded CAG repeat in a calcium channel gene
Thomsen's disease	AD	7q35	CLCN1	160800	Myotonia, especially eyelids, hands, legs	Mutation in skeletal muscle chloride channel gene
Tuberous sclerosis 2	AD	16p13.3	TSC	191092	Infantile spasms, seizures, mental retardation, hydrocephaly secondary to cerebral glial nodules	Mutation in the gene for tuberin, a novel protein with a Rap-GAP homology domain
von Hippel-Lindau disease	AD	3p26-p25	VHL	193300	Retinal angiomata, cerebellar hemangioblastoma, spinal cord hemangioma	Mutation in the gene for a novel protein
Wilson's disease	AR	13q14.3-q21.1 (ATP7B)	WND	277900	Tremor, speech difficulty, dysphagia, dementia, poor motor coordination, dystonia	Mutation in a gene encoding a novel copper-transporting ATPase (related to Menkes' disease gene)

Index

A

Acetazolamide, antimyotonic effects of, 273–274
Acetylcholine transporter, in brain, 164
N-Acetylcysteine, palliative effect, in amyotrophic lateral sclerosis, 230
Adrenal medulla, transplantation, in Parkinson's disease treatment, 159
Adrenoleukodystrophy
  neonatal
    defect in, molecular definition of, 297
    NALD (PXR1) gene in, 297
  X-linked
    ALD gene in, 297
    defect in, molecular definition of, 297
Aging
  midbrain dopamine cell loss in, 156, 156*f*, 159, 160*f*
  neuropathologic lesions with, 55–56
  somatic mtDNA mutations in, 286–287
ALD gene. *See* Adrenoleukodystrophy
Aldolase, response to hypoxia, 119
ALS. *See* Amyotrophic lateral sclerosis
Alzheimer's disease. *See also* Familial Alzheimer's disease
  AD1 (APP) gene in, 297
  AD2 (APOE) gene in, 58, 297
  AD3 gene in, 297
  AD4 gene in, 297
  animal model for, 58
  clinical features of, 55
  defect in, molecular definition of, 291–292, 297
  deterministic genetic component of, 56
  diagnosis of, 55
  early-onset, 291–292
    myoclonus in, 56
  epidemiology of, 55–56
  genes, for age of onset older than 70 years, 59–60, 291–292
  inheritance patterns in, 78
  late-onset, 13, 291–292
    familial clustering in, 56
  molecular genetics of, 55–73, 291–292
  multifactorial etiology of, 142–143
  neurogenetics, recent advances in, 291–292
  neuropathogenesis of, 67
  neuropathologic features of, 55
  risk factors for, 56, 291–292
  somatic mtDNA mutations and, 286–287
  therapeutic options in, 67–68
Amines, endogenous, neurotoxicity, 165–166
Amphetamine, and dopamine toxicity in Parkinson's disease, 166–167
β-Amyloid, 55, 291–292
Amyloid plaques, in Alzheimer's disease, 4, 291–292
β-Amyloid precursor protein, 55, 57*f*
  Aβ domain, 56, 57*f*
  APP gene for, 4
  cloning, 57
  mapping, 56–57
  mutations, 57
    in Alzheimer's disease, 78
    in familial Alzheimer's disease, 60, 62
  induction, by cerebral ischemia, 119–120, 120*f*
  processing of, 61–62, 62*f*, 292
Amyotrophic lateral sclerosis, 223–227
  ALS1 (SOD1) gene in, 297
  animal model of, 89, 228
  cell death pathways in, 223–224, 232–234, 233*f*
    phases of, 234, 235*t*
    therapeutic strategies aimed at, 234, 235*t*
  clinical features of, 223–224
  defect in, molecular definition of, 297
  epidemiology of, 223–224
  familial, 224
    genetic defect in, 12
    genetic linkage analysis in, 227–228
    penetrance in, 227
  free radical pathology in, 229–230
  juvenile-onset, 224
  motor neuron degeneration in, 223–224, 232–234, 233*f*
  neurogenetics, recent advances in, 293–294
  pathogenesis of, hypotheses of, 226–227
  pathology of, 224

D

E